DESIGNERS' GUIDES TO THE EUROCODES

DESIGNERS' GUIDE TO EUROCODE 5: DESIGN OF TIMBER BUILDINGS

EN 1995-1-1

Eurocode Designers' Guide series

Designers' Guide to Eurocode: Basis of Structural Design. EN 1990. Second edition. H. Gulvanessian, J.-A. Calgaro and M. Holický. 978-0-7277-4171-4. Published 2012.

Designers' Guide to Eurocode 1: Actions on Buildings. EN 1991-1-1 and -1-3 to -1-7. H. Gulvanessian, P. Formichi and J.-A. Calgaro. 978-0-7277-3156-2. Published 2009.

Designers' Guide to Eurocode 1: Actions on Bridges. EN 1991-1-1, -1-3 to -1-7 and EN 1991-2. J.-A. Calgaro, M. Tschumi and H. Gulvanessian. 978-0-7277-3158-6. Published 2010.

Designers' Guide to EN 1991-1.4. Eurocode 1: Actions on Structures, General Actions. Part 1-4 Wind actions. N. Cook. 978-0-7277-3152-4. Published 2007.

Designers' Guide to Eurocode 2: Design of Concrete Structures. EN 1992-1-1 and EN 1992-1-2 General rules and rules for buildings and structural fire design. R.S. Narayanan and A.W. Beeby. 978-0-7277-3105-0. Published 2005.

Designers' Guide to EN 1992-2. Eurocode 2: Design of Concrete Structures. Part 2: Concrete bridges. C.R. Hendy and D.A. Smith. 978-0-7277-3159-3. Published 2007.

Designers' Guide to Eurocode 3: Design of Steel Buildings. EN 1993-1-1, -1-3 and -1-8. Second edition. L. Gardner and D. Nethercot. 978-0-7277-4172-1. Published 2011.

Designers' Guide to EN 1993-2. Eurocode 3: Design of Steel Structures. Part 2: Steel bridges. C.R. Hendy and C.J. Murphy. 978-0-7277-3160-9. Published 2007.

Designers' Guide to Eurocode 4: Design of Composite Steel and Concrete Structures. EN 1994-1-1. Second edition. R.P. Johnson. 978-0-7277-4173-8. Published 2012.

Designers' Guide to EN 1994-2. Eurocode 4: Design of Composite Steel and Concrete Structures. Part 2 General rules for bridges. C.R. Hendy and R.P. Johnson. 978-0-7277-3161-6. Published 2006.

Designers' Guide to Eurocode 5: Design of Timber Buildings. EN 1995-1-1. J. Porteous and P. Ross. 978-0-7277-3162-3. Published 2013.

Designers' Guide to Eurocode 6: Design of Masonry Structures. EN 1996-1-1. J. Morton. 978-0-7277-3155-5. Published 2012.

Designers' Guide to Eurocode 7: Geotechnical Design. EN 1997-1 General rules. R. Frank, C. Bauduin, R. Driscoll, M. Kavvadas, N. Krebs Ovesen, T. Orr and B. Schuppener. 978-0-7277-3154-8. Published 2004.

Designers' Guide to Eurocode 8: Design of Structures for Earthquake Resistance. EN 1998-1 and EN 1998-5. General rules, seismic actions, design rules for buildings, foundations and retaining structures. M. Fardis, E. Carvalho, A. Elnashai, E. Faccioli, P. Pinto and A. Plumier. 978-0-7277-3348-1. Published 2005.

Designers' Guide to Eurocode 8: Design of Bridges for Earthquake Resistance. EN 1998-2. B. Kolias, M.N. Fardis and A. Pecker. Published 2012.

Designers' Guide to Eurocode 9: Design of Aluminium Structures. EN 1999-1-1 and -1-4. P. Tindall and T. Höglund. Published 2012.

Designers' Guide to EN 1991-1-2, EN 1992-1-2, EN 1993-1-2 and EN 1994-1-2. T. Lennon, D.B. Moore, Y.C. Wang and C.G. Bailey. 978-0-7277-3157-9. Published 2007.

www.icevirtuallibrary.com
www.eurocodes.co.uk

DESIGNERS' GUIDES TO THE EUROCODES

DESIGNERS' GUIDE TO EUROCODE 5: DESIGN OF TIMBER BUILDINGS
EN 1995-1-1

JACK PORTEOUS
Consultant, UK

PETER ROSS
Consultant, UK

Series editor
Haig Gulvanessian CBE

Published by ICE Publishing, One Great George Street, Westminster, London SW1P 3AA

Full details of ICE Publishing sales representatives and distributors can be found at:
www.icevirtuallibrary.com/info/printbooksales

Eurocodes Expert

Structural Eurocodes offer the opportunity of harmonised design standards for the European construction market and the rest of the world. To achieve this, the construction industry needs to become acquainted with the Eurocodes so that the maximum advantage can be taken of these opportunities.

Eurocodes Expert is an ICE and Thomas Telford initiative set up to assist in creating a greater awareness of the impact and implementation of the Eurocodes within the UK construction industry.

Eurocodes Expert provides a range of products and services to aid and support the transition to Eurocodes. For comprehensive and useful information on the adoption of the Eurocodes and their implementation process please visit www.eurocodes.co.uk

www.icevirtuallibrary.com

A catalogue record for this book is available from the British Library

ISBN 978-0-7277-3162-3

ICE Publishing is a division of Thomas Telford Ltd, a wholly-owned subsidiary of the Institution of Civil Engineers (ICE).

Associate Commissioning Editor: Jennifer Saines
Production Editor: Imran Mirza
Market Specialist: Catherine de Gatacre

FSC
www.fsc.org
MIX
Paper from responsible sources
FSC® C013604

Typeset by Academic + Technical, Bristol
Printed and bound by CPI Group (UK) Ltd, Croydon, CR0 4YY

Preface

EN 1995 is the Eurocode for the design of timber and wood-based materials, and is subdivided into EN 1995-1 for general design and EN 1995-2 for bridge design. EN 1995-1 is also divided into two parts:

- EN 1995-1-1: 2004 + A1: 2008, 'Eurocode 5: Design of timber structures – Part 1-1: General – Common rules and rules for buildings'
- EN 1995-1-2, 'Eurocode 5: Design of timber structures – Part 1-2: General – Structural fire design'.

EN 1995-1-1: 2004 + A1: 2008, 'Eurocode 5: Design of timber structures – Part 1-1: General – Common rules and rules for buildings', is the code that describes the principles and design rules to be used for the design of timber and wood-based materials in building and civil engineering structures, and this guide primarily covers the content of this code. In this guide the document is referenced as EN 1995-1-1. Where a structural fire design is to be undertaken, guidance is given in Chapter 12 on the principles, requirements and rules of EN 1995-1-2 that will apply.

General

The content of this guide covers the main design requirements of EN 1995-1-1, and, where it is considered appropriate, background information to clarify the application and any limitation of use of certain design rules is given. A considerable number of the design rules in EN 1995-1-1 have been derived empirically, and the importance of using the units for specific functions as specified in the rules is stressed. Where matters are delegated for national choice, the requirement is given in the UK National Annex to EN 1995-1-1, and reference is included in the guide to the content of this document. Non-contradictory complementary information as well as guidance on subjects not covered by EN 1995-1-1 is published in PD 6693-1: 2012, Incorporating Corrigendum No. 1, Recommendations for the design of timber structures to Eurocode 5: Design of timber structures – Part 1: General – Common rules and rules for buildings'. This document has recently been published, and, where the content will be of significance in the application of the rules in EN 1995-1-1 covered in this guide, reference has been made to this in the guide. To explain the requirements of the code, examples of elements of design problems are given in the guide, and, where design loads are used, these will have been derived from application of the relevant rules in EN 1990 and EN 1991.

CEN/TC 250, the Technical Committee of the European committee for standardisation (CEN) responsible for technical matters associated with this code, has identified some errors in the code as well as matters where it is felt clarification of interpretation is required. The points have still to be fully discussed within the Committee, and when agreed are expected to be issued in a corrigendum statement prior to the next full revision of the code after 2015.

Some of the matters considered to be of significance to design and likely to be incorporated in the corrigendum are briefly referred to in Appendix A and, where appropriate, in chapters in this guide.

Layout of this guide

The headings used in this guide up to Chapter 11 follow the section headings in EN 1995-1-1 and, where cross-referencing, text reproduction, figures, tables or equations from the code have been incorporated in the guide text, the information is printed in *italics*. The author's expressions where numbered have numbers prefixed by D (for *Designers' Guide*): for example, Equation (D6.7) in Chapter 6.

Chapter 12 covers topics in EN 1995-1-2, and the headings used relate to specific topics in that code. Where relevant, the above approach has again been used.

Acknowledgements

The authors wish to thank the various individuals who were familiar with the background to the development of EN 1995-1-1 and have given advice on topics covered in this guide.

J. Porteous
P. Ross

Contents

Designers' Guide to Eurocode 5: Design of Timber Buildings
ISBN 978-0-7277-3162-3

http://dx.doi.org/10.1680/dtb.31623.001

Introduction

The material given in this introduction draws on the foreword to the European standard EN 1995-1-1: 2004 + A1: 2008, 'Eurocode 5: Design of timber structures – Part 1-1: General – Common rules and rules for buildings', using the same headings as given in the code.

Background to the Eurocode programme

In 1975, in order to eliminate technical obstacles to trade and to harmonise technical specifications between EU member states it was decided by the Commission of the European Communities that a set of harmonised technical rules for the design of building and civil engineering works would be produced. This resulted in the Structural Eurocode programme that produced design standards generally consisting of a number of parts, and comprising the following 10 documents:

EN 1990 'Eurocode: Basis of Structural Design'
EN 1991 'Eurocode 1: Actions on structures'
EN 1992 'Eurocode 2: Design of concrete structures'
EN 1993 'Eurocode 3: Design of steel structures'
EN 1994 'Eurocode 4: Design of composite steel and concrete structures'
EN 1995 'Eurocode 5: Design of timber structures'
EN 1996 'Eurocode 6: Design of masonry structures'
EN 1997 'Eurocode 7: Geotechnical design'
EN 1998 'Eurocode 8: Design of structures for earthquake resistance'
EN 1999 'Eurocode 9: Design of aluminium structures'

These documents are meant to work in harmony with national standards, and, in the case of timber design in the UK, the national design British standards have been withdrawn and replaced by the requirements of EN 1995.

Status and field of application of Eurocodes

The Eurocodes function as reference documents that cover several purposes. They provide structural design rules, information that is relevant for contract specification purposes and give a framework for drawing up harmonised technical specifications.

The structural design rules cover the design requirements for products, structural components and complete structures for commonly used forms of construction. Where unusual forms of construction are to be used or the design rules do not cover the particular design condition, specialist advice must be used.

National standards implementing Eurocodes

In the UK, national standards are published by the British Standards Institute (BSI), and the requirement for implementing the Eurocodes is that the national standard will comprise the complete Eurocode text as published by the European Committee for Standardisation (CEN), with no alterations, and all annexes included. The national standard may include a national title page and a national foreword preceding the content of the Eurocode, and may be followed by a National Annex.

Items left open in a Eurocode for national choice are referred to as Nationally Determined Parameters (NDPs) and it is only these items and any guidance on the use of the informative annexes in the Eurocode that can be referred to in the National Annex.

The National Annex may also include reference to non-contradictory complementary information (NCCI) that can be used to assist with the application of the Eurocode rules.

Links between Eurocodes and harmonised technical specifications (ENs and ETAs) for products

There must be consistency between the technical rules in the Eurocodes and the harmonised technical specifications for construction products. Also, where Eurocodes are referred to in the CE marking of a construction product, the accompanying information shall include a clear statement mentioning which NDPs have been taken into account.

Additional information specific to EN 1995-1-1

EN 1995 is a limit state concept code that, for the design of new structures, has to be used in conjunction with EN 1990: 2002 and the relevant parts of EN 1991.

It is used in conjunction with the partial factor method referred to in EN 1990, and acceptable levels of reliability will be achieved by using the numerical values recommended for the partial factors and other reliability parameters.

An outline of the main requirements of EN 1990 and EN 1991 that are relevant to the use of EN 1995-1-1 is given in appropriate chapters in this guide.

CEN/TC 250, the Technical Committee of the European Committee for Standardisation (CEN) responsible for technical matters associated with this code, has identified some errors in the code as well as matters where it is felt that clarification of interpretation is required. The points have still to be fully discussed within the committee, and when agreed are expected to be issued in a corrigendum statement prior to the next full revision of the code, which will take place beyond 2015.

Some of the matters considered to be of significance to design and likely to be incorporated in the amendment are outlined in Appendix A and, where appropriate, referred to in the guide.

National Annex for EN 1995-1-1

The clauses in EN 1995-1-1 where national choice is allowed are given in the National Annex to BS EN 1995-1-1: 2004 + A1: 2008 (incorporating National Amendment No. 2), the UK National Annex to 'Eurocode 5: Design of timber structures – Part 1-1: General – Common rules and rules for buildings', as follows:

Clauses 2.3.1.2(2)P, 2.3.1.3(1)P, 2.4.1(1)P, 6.1.7(2), 6.4.3(8), 7.2(2), 7.3.3(2), 8.3.1.2(4), 8.3.1.2(7), 9.2.4.1(7), 9.2.5.3(1), 10.9.2(3), 10.9.2(4)

Clause	Comment
2.3.1.2(2)P	Assignment of loads to load-duration classes
2.3.1.3(1)P	Assignment of timber constructions to service classes
2.4.1(1)P	Partial factors for material properties
6.1.7(2)	Modification factor for influence of cracks on shear resistance
6.4.3(8)	Tensile stresses perpendicular to grain in double tapered, curved and pitched cambered beams
7.2(2)	Limiting values for deflection of beams
7.3.3(2)	Vibrations in residential floors
8.3.1.2(4)	Lateral load-carrying capacity of nails in end grain
8.3.1.2(7)	Species sensitive to splitting in nailed joints
9.2.4.1(7)	Racking resistance of wall diaphragms
9.2.5.3(1)	Modification factors for bracing systems
10.9.2(3)	Erection tolerances for trusses: maximum bow
10.9.2(4)	Erection tolerances for trusses: maximum deviation from vertical alignment

Guidance is also given on the use of the Eurocode informative annexes, i.e. *Annex A*, *Annex B* and *Annex C*.

Non-contradictory complementary information

NCCI as well as guidance on subjects not covered by EN 1995-1-1 is published in PD 6693-1: 2012, Incorporating Corrigendum No. 1, 'Recommendations for the design of timber structures – Part 1: General – Common rules and rules for buildings'.

Where the content of this document is of significance in the application of the rules in EN 1995-1-1 covered in this guide, reference has been made to it in the guide.

Designers' Guide to Eurocode 5: Design of Timber Buildings
ISBN 978-0-7277-3162-3

http://dx.doi.org/10.1680/dtb.31623.005

Chapter 1
General

This chapter is concerned with the general aspects of EN 1995-1-1, 'Eurocode 5: Design of timber structures – Part 1-1: General – Common rules and rules for buildings', covered in *Section 1*. The following clauses are addressed:

- Scope *Clause 1.1*
- Normative references *Clause 1.2*
- Assumptions *Clause 1.3*
- Distinction between principles and application rules *Clause 1.4*
- Terms and definitions *Clause 1.5*
- Symbols *Clause 1.6*

Guidance is also given on the member axes convention adopted in the code under:

- Conventions for member axes

1.1. Scope

EN 1995 is the Eurocode for the design of buildings and civil engineering works in timber or wood-based panels joined together using mechanical fasteners or adhesives. In the context of the code, timber is interpreted to cover solid timber that is sawn, planed or in pole form, glued laminated timber or wood-based structural products (e.g. laminated veneer lumber (LVL)). The Eurocode comprises two parts:

EN 1995-1 'General'
EN 1995-2 'Bridges'

EN 1995-1 is subdivided into two parts:

EN 1995-1-1 'General – Common rules and rules for buildings'
EN 1995-1-2 'General – Structural fire design'

The basic rules for timber design are given in EN 1995-1-1, and, for bridge design to EN 1995-2, unless specifically stated otherwise in that code, the rules in Part 1-1 will also apply. The design rules in Part 1-1 only apply when the structure or its elements are not subjected to prolonged exposure to temperatures over 60°C, as the rules used for creep behaviour above this temperature will no longer apply. For building structures exposed to fire, the rules are different and are covered in EN 1995-1-2. In this guide the text primarily covers the rules in EN 1995-1-1, but in Chapter 12 an outline is given of the approach to the design of timber members in fire conditions.

Design of a timber structure will also require reference to EN 1990, particularly for loading combinations, to EN 1991 for loading, to code National Annexes, to non-contradictory complementary information and to normative references.

1.2. Normative references

EN 1995-1-1 makes reference to other standards (normative references), and these are listed in the code.

Where reference is made in this guide to any of these standards and it is dated in the text, the dated copy of the standard must be used. Otherwise the latest issue of the standard will apply.

1.3. Assumptions

The general assumptions in EN 1995-1-1 are the same as those given in EN 1990: 2002, and are summarised as follows:

1 Qualified and experienced designers must be used in the design of the structure
2 All stages of the construction process must be undertaken by skilled and experienced personnel
3 Adequate supervision and quality control must be provided through all stages of the project
4 Materials and products must comply fully with the specified requirements
5 The structure shall be used as designed and adequately maintained throughout its design life.

The owner/user of the building must be made fully aware of his/her responsibilities against item 5.

Additional requirements for timber and wood products are also given in *Section 10* of EN 1995-1-1.

1.4. Distinction between principles and application rules

The difference between principles and application rules is defined in EN 1990: 2002 clause 1.4, and applies to all codes. Clause numbers that are followed by the letter 'P' are principles, and where there is no letter 'P' the clause is an application rule. The designer has no choice but to follow the full requirement of clauses marked 'P', but can choose not to follow application rules. However, where application rules are not followed, the designer is responsible for demonstrating that their alternative proposal will result in a design that is fully compliant with the principle and will produce a design that is equivalent in terms of serviceability, structural integrity and durability. An important point to note is that in such a situation the design cannot be claimed to be fully compliant with the Eurocode requirements, and this may prove to be a problem if a CE marking is required for the design or substantiation of a product.

1.5. Terms and definitions

Clause 1.5.2

In addition to the terms listed in clause 1.5 in EN 1990: 2002, particular terms specific to timber design are listed under *clause 1.5.2*.

1.6. Symbols

Clause 1.6

The symbols used in EN 1995-1-1 are listed in *clause 1.6*.

In the UK, the full stop is traditionally used to define a decimal point in a number, but in the Eurocodes the procedure follows the requirements of ISO 3898 (ISO, 1997), which adopts a comma system. In this guide, the traditional UK procedure to define numbers is used (e.g. 'three-and-a-half mm' would be written as 3.5 mm not 3,5 mm).

1.7. Conventions for member axes

Design practice in the UK is to show the *z–z* axis as the longitudinal axis of a member, *x–x* as the major axis and *y–y* as the minor axis of the cross-section. The convention used in Eurocode 5 is for the *x–x* axis to follow the longitudinal axis, *y–y* to be the major axis and *z–z* the minor axis of the cross section. The Eurocode 5 convention has been followed in this guide.

Based on this convention the axes for some of the common cross sections used in timber design will be as shown in Figure 1.1, where $I_y \geq I_z$ unless otherwise noted.

Figure 1.1. Axes for some of the common timber sections. (Parts reproduced from EN 1995-1-1, with permission from British Standards Institution, 2004)

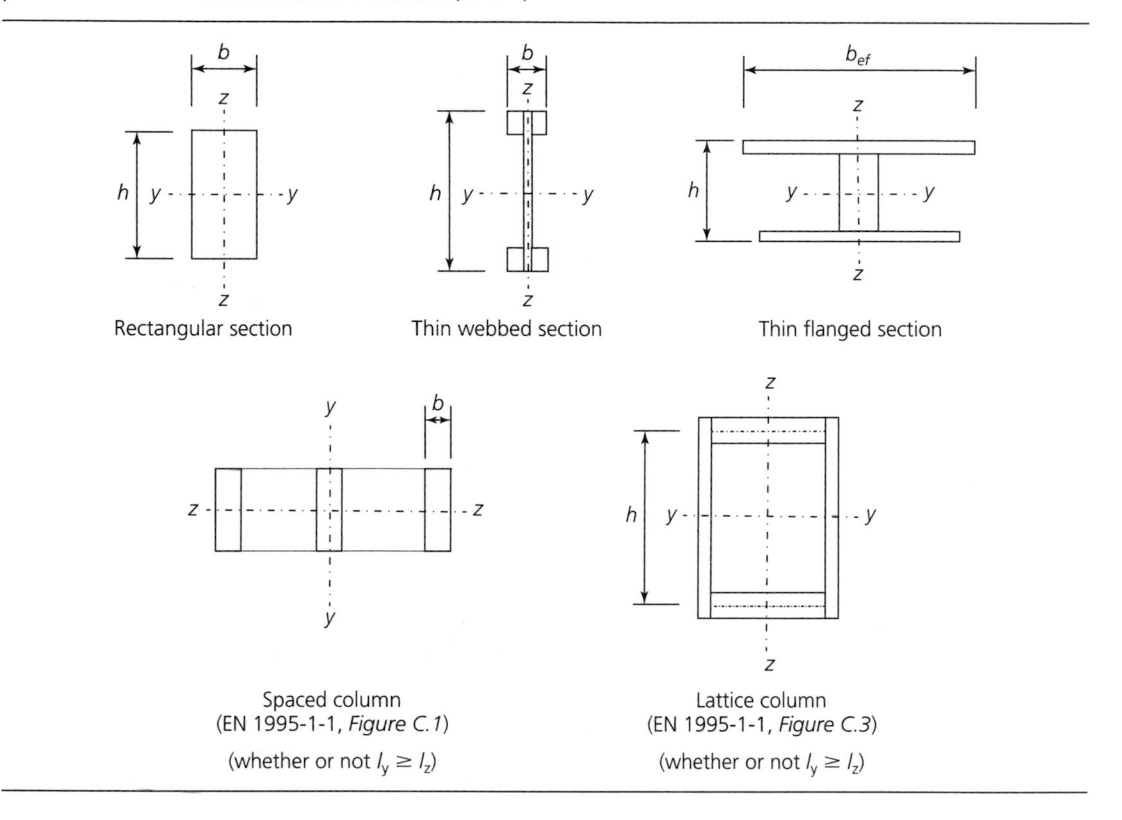

REFERENCE

ISO (1997) ISO 3898: 1997. Basis of design for structures – Notation – General symbols. International Organization for Standardization, Geneva.

Designers' Guide to Eurocode 5: Design of Timber Buildings
ISBN 978-0-7277-3162-3

http://dx.doi.org/10.1680/dtb.31623.009

Chapter 2
Basis of design

This chapter addresses the basis of design covered in *Section 2* of EN 1995-1-1, also drawing on the relevant content of Section 2 of EN 1990: 2002. The following clauses are addressed:

- Requirements *Clause 2.1*
- Principles of limit state design *Clause 2.2*
- Basic variables *Clause 2.3*
- Verification by the partial factor method *Clause 2.4*

It is important to stress that EN 1990 is the principle Eurocode that defines the framework within which design with any material must be undertaken and to be able to carry out a design using timber (or any other construction material) it is recommended that the content of EN 1990 is studied in addition to the requirements of *Section 2* of EN 1995-1-1.

2.1. Requirements

With any design, EN 1990 requires that the general assumptions summarised in Section 1.3 in this guide are fully complied with, that due skill and care appropriate to the circumstances are exercised, and any degradation in performance due to durability effects are fully taken into account in the design process. On this basis, to comply with the requirements of EN 1995-1-1 a structure shall be designed, constructed and maintained over its design working life in such a way that it will, with appropriate degrees of reliability and in an economical way:

- have adequate strength and stability
- meet all serviceability requirements
- have adequate durability
- have adequate structural resistance to the effects of fire over the agreed fire resistance period
- for any accidental action that has been agreed to be taken into account, exhibit adequate robustness.

For a detailed consideration of each of the above requirements, the reader is referred to the relevant comment given under Section 2.1 in the *Designers' Guide to Eurocode: Basis of structural Design: EN 1990* (Gulvanessian *et al.*, 2012). With a timber design, the requirements of EN 1990 have to be met, and the supplementary provisions given in *Section 2* of EN 1995-1-1 have also to be taken into account.

In EN 1995, the design methodology is reliability based and uses a limit states design format, which is fundamentally different to the permissible stress design approach that has traditionally been used for timber design in the UK.

With the limit states design concept, limit states are defined for the structure (e.g. strength states for bending, bearing and shear), and design checks are undertaken at each state to demonstrate that it will either be satisfactory or unsatisfactory. If any limit state is exceeded, the design is unsatisfactory and will fail. In this approach, contrary to the design procedure used in the permissible stress approach, stress states are checked for resistance and behaviour at the loading condition associated with the limit state being considered. It has been the UK's intention for some time to replace the permissible stress approach with a limit state design code, and this is now achieved by the adoption of EN 1995 as the UK timber design code.

Eurocode 1990 requires the structure and its elements to be checked at the following limit states:

Ultimate limit states: ■ these are associated with structural failure/collapse and equilibrium considerations (e.g. strength failure, buckling instability and overturning instability).

Serviceability limit states: ■ these are associated with behaviour of the structure under service conditions (e.g. deflection, vibration and appearance).

The basic requirement is to demonstrate for each limit state that when the structure is subjected to its design load condition the design resistance at that state will not be exceeded. For example, when a timber beam is subjected to bending at the ultimate limit bending state it must be shown that the maximum bending stress in the beam will not be greater than the design bending strength. This can be done by the use of probabilistic design methods, but for normal design the '*partial factor method*' referred to in *clause 2.4* is used, and is the method referred to in this guide. An explanation of the partial factor method is given in Section 2.4 in this guide, with design actions and design strengths covered in Sections 2.3 and 2.4.

Clause 2.4

Before design commences, the design working life of the structure must be agreed between the designer and the client, and indicative durations for different categories of structure are given in EN 1990 (clause 2.3, Table 2.1). Timber buildings will normally come under category 4 in the table, with an indicative design working life of 50 years.

2.2. Principles of limit state design

Clause 2.2

Clause 2.2 in EN 1995-1-1 basically relates to the significance of stiffness in timber analysis and the values to be used for stiffness properties at the ultimate and serviceability limit states.

Because the stiffness behaviour of timber is affected by creep and load duration, in addition to normal considerations the design models used for analysis at different limit states must also allow for the effect of these factors.

2.2.1 General

Clause 2.2.1

Clause 2.2.1 states the factors that have to be taken into account, as relevant, in design models at the limit states.

2.2.2 Ultimate limit states

At ultimate limit states the primary objective of the structural analysis is to determine the force distribution in the structural elements under a design loading condition.

Clause 2.2.2(1)P

Due to the risk of brittle failure with timber structures, as stated in *clause 2.2.2(1)P* a linear elastic model is to be used for the analysis at these states, and it can be a first- or a second-order type. A first-order analysis ignores instability and out of alignment effects, and a second-order analysis takes these and other effects into account. Where a first-order analysis is used, providing the internal force distribution is not affected by the stiffness distribution (e.g. all members have the same creep properties), the effect of creep will not be relevant, and mean values of stiffness properties shall be used. Where the internal force distribution is affected by the stiffness distribution (e.g. composite members made from elements with different creep properties), creep and load duration effects will have an effect, and are taken into account by using final mean values of stiffness adjusted to the load component causing the largest stress-to-strength ratio, and are referred to in *clause 2.3.2.2(2)*. If a second-order analysis is to be used, the design value of stiffness, referred to in *clause 2.4.1(2)P*, will apply. Consideration of the required loading and stiffness properties to be used in the analysis of timber structures at these states is given in Chapter 5 in this guide.

Clause 2.3.2.2(2)
Clause 2.4.1(2)P

2.2.3 Serviceability limit states

At serviceability limit states, the main objective of the analysis is to calculate the deformation of the structure and its elements and, as explained in Chapter 7 in this guide, there are two types of deformation that have to be determined. Firstly there is the instantaneous deformation, u_{inst}, which is the elastic deformation that occurs when subjected to the design loading condition, and for this analysis mean values of stiffness properties should be used. Then there is the final deformation, u_{fin}, which is the combined instantaneous, u_{inst}, and creep deformation, u_{creep}, and the stiffness used for this analysis is dependent on whether or not the structure consists of members or components having the same or different creep behaviour. If the creep behaviour is the same throughout the structure, u_{creep} is calculated using the same stiffness as for u_{inst} but the design loading condition will be derived from the quasi-permanent loading combination, referred to in Section 2.3.1 of this Guide, and for the instantaneous deformation the characteristic loading combination is used. When the creep behaviour is different throughout the structure the requirements of *clauses 2.2.3(3)* and *2.2.3(4)* are unclear and are to be revised as stated in Appendix A in this guide. Based on the revision, u_{creep} is to be calculated using the quasi-permanent loading combination with stiffness properties based on the final mean stiffness values as defined in *2.3.2.2(1)*, and u_{inst}, based on mean stiffness values, will be calculated using the difference between the characteristic and the quasi-permanent combinations of actions. An alternative simpler option suggested in this guide for this condition is to analyse the structure based on final mean stiffness properties but under the action of the characteristic loading combination and a value for the final deformation will be derived directly from this single analysis. It is a quicker method than the revised code procedure but will generally result in a larger value for the deformation.

Clause 2.2.3(3)
Clause 2.2.3(4)

Clause 2.3.2.2(1)

Consideration of the required loading and stiffness properties to be used in the analysis of timber structures at these states is given in Chapter 5 in this guide.

When dealing with vibration analyses, mean values of stiffness shall be used.

2.3. Basic variables

General
EN 1990 requires all relevant design situations to be considered, and they are classified in the code as follows:

- persistent design situations – conditions of normal use (i.e. self-weight and imposed loading, including wind and snow)
- transient design situations – temporary conditions (e.g. during construction or repair)
- accidental design situations – exceptional conditions (e.g. explosion or impact)
- seismic – conditions arising from a seismic event.

For normal timber structures in the UK there will be no requirement to design for seismic conditions, and this condition is not referred to in this guide.

Design checks have to be undertaken at the ultimate and the serviceability limit states for relevant design situations, and for the ultimate limit states EN 1990 requires that, where relevant, the following design conditions are addressed:

- Equilibrium (EQU) – confirm that stability is maintained.
- Strength (STR) – confirm that strength and/or buckling instability will not arise. The effects of displacement must be taken into account where they affect the strength or stability behaviour of the structure.
- Geotechnical (GEO) – confirm that the foundations will provide the strength and stiffness required by the structure.
- Fatigue (FAT) – confirm that there will be no failure due to fatigue effects.

For timber structures designed in accordance with EN 1995-1-1, fatigue effects are taken to be enveloped by ensuring that the design complies with the rules in the code.

2.3.1 Actions and environmental influences

2.3.1.1 General

Actions
Actions (i.e. loads) used in design are obtained from the relevant parts of EN 1991, and are classified on a time basis as follows:

- Permanent actions (G) – these do not vary with time (e.g. self-weight and indirect actions caused by shrinkage and/or settlement effects).
- Variable actions (Q) – these vary with time (e.g. imposed loads, wind loads and snow loads).
- Accidental actions (A) – usually of short duration and high magnitude (e.g. impact or explosion; also includes fire).

The values given in EN 1991 for these actions are characteristic values, defined with the suffix k (i.e. G_k, Q_k and A_k), and these are effectively equivalent to the load values given in British standards.

In addition to characteristic variable actions, Q_k, other representative values of variable actions are used in design, and are referred to as:

- Combination value ($\psi_0 Q_k$) – the probability of characteristic values of variable actions arising at the same time in any loading combination is unrealistic and the ψ_0 value, which is probability based, converts the characteristic value of an accompanying variable action to the maximum value that is considered will arise at the same time as the characteristic value of the leading variable. It is used for the characteristic combinations of irreversible serviceability limit states and for ultimate limit states.
- Frequent value ($\psi_1 Q_k$) – variable actions are cyclic in nature and the ψ_1 factor converts the characteristic value of an accompanying variable action to the maximum value that will occur over a period of 1% of the design life. It is used for the frequent combination of reversible serviceability limit states and for the verification of ultimate limit states involving accidental actions.
- Quasi-permanent value ($\psi_2 Q_k$) – the ψ_2 factor converts a variable action into the equivalent permanent action that is considered to be supported by the building over the design working life. It is used for the verification of frequent and long-term effects of serviceability limit states and for the representation of variable actions in accidental combinations at the ultimate limit states. For timber design, it is used to calculate the creep loading on structures.

Design values of actions

The design value of an action is the value used in a design and is obtained by multiplying the value of the action by a partial factor, γ. For permanent and variable actions these are defined as:

permanent design action: $G_d = \gamma_G G_k$

variable design actions: $Q_d = \gamma_Q Q_k$ and/or $\gamma_Q \psi_i Q_k$

where the suffix d defines a design action; γ_G is for permanent actions and γ_Q for variable actions; and, for the appropriate design condition, values for γ_G and γ_Q are obtained from Tables NA.A1.2 (A), NA.A1.2 (B), NA.A1.2 (C) and NA.A1.3 in the National Annex to EN 1990: 2002 + A1: 2005.

When a structure is subjected to a design action, internal stress resultants (e.g. moments, axial forces, stress or strain) and structural deformations (e.g. deflections and rotations), referred to as 'design effects', will be generated. Where the design effect arises from variable actions acting with permanent actions, to obtain the design value of the action the permanent action must be classified as favourable or unfavourable. Permanent actions should be classified as favourable when they act favourably opposite to the variable action or as unfavourable when they act unfavourably with the variable action and larger values of γ_G are used for the unfavourable classification.

Combinations of actions

To obtain the largest design effect, EN 1990 defines the combinations of the design values of permanent and variable actions that are to be used for designs at the ultimate and serviceability limit states. Those to be used for ultimate limit states are given in the National Annex to EN 1990, and those for serviceability limit states are referred to in this guide. The combinations given in the following sections exclude actions due to pre-stressing, as this is not generally relevant to timber design in the UK. For the same reason, the combination of actions to be used for seismic design are not referred to.

Combinations of actions for ultimate limit states

With ultimate limit states, for each design situation the combinations for persistent and transient design situations, referred to in EN 1990 as **fundamental combinations**, are as follows:

$$\sum_{j\geq 1} \gamma_{G,j} G_{k,j} \text{ '+' } \gamma_{Q,1} Q_{k,1} \text{ '+' } \sum_{i>1} \gamma_{Q,i} \psi_{0,i} Q_{k,i} \qquad \text{(EN 1990: 6.10)} \qquad \text{(D2.1)}$$

or, as an alternative, the less favourable of the following combination expressions may be considered for the STR and GEO limit states:

$$\sum_{j\geq 1} \gamma_{G,j} G_{k,j} \text{ '+' } \gamma_{Q,1} \psi_{0,1} Q_{k,1} \text{ '+' } \sum_{i>1} \gamma_{Q,i} \psi_{0,i} Q_{k,i} \qquad \text{(EN 1990: 6.10a)} \qquad \text{(D2.2a)}$$

$$\sum_{j\geq 1} \xi_j \gamma_{G,j} G_{k,j} \text{ '+' } \gamma_{Q,1} Q_{k,1} \text{ '+' } \sum_{i>1} \gamma_{Q,i} \psi_{0,i} Q_{k,i} \qquad \text{(EN 1990: 6.10b)} \qquad \text{(D2.2b)}$$

where the functions are as previously defined and

'+' means 'to be combined with'
$\sum$ means 'the combined effect of'
ξ is a reduction factor for unfavourable permanent actions (guidance on the value used is given in the National Annex to EN 1990)
$Q_{k,1}$ is the leading variable action
$Q_{k,i}$ is an accompanying variable action.

These alternatives are given in the National Annex to EN 1990 (Tables NA.A1.2 (A) and NA.A1.2 (B)) together with the associated partial factors, and, having selected the equation to be used for the fundamental combination, to determine the action configuration that will produce the largest design effect (i.e. the design loading condition) each variable action must be taken in turn to be the leading variable. Also, with timber designs, as duration of load and moisture content effects (which are accounted for by the k_{mod} factor referred to in Section 2.3.2 in this guide) affect the design strength, k_{mod} must be taken into account. This can be seen by considering the general case where a design action $F_{d,i}$ on a timber structure induces a design stress $f(g)F_{d,i}$ at the critical stress position, in which $f(g)$ will be a function of the loading configuration and the geometry of the cross section. If the modification factor for the loading condition is $k_{mod,i}$ and the design strength is $f_{x,d}$, the design verification relationship will be

$$f(g)F_{d,i} \leq k_{mod,i} f_{x,d} \qquad \text{(D2.3a)}$$

Clause 3.1.3

In accordance with the requirements of *clause 3.1.3*, the value of k_{mod} is a function of the action having the shortest duration in the loading combination as well as the moisture content of the timber, and to determine the design condition Equation D2.3a must be adjusted as follows:

$$f(g)F_{d,i}/k_{mod,i} \leq f_{x,d} \qquad \text{(D2.3b)}$$

The design loading condition will be the loading arrangement i that results in the largest value of F_d/k_{mod} and, because of the low values of k_{mod} associated with permanent actions, action configurations comprising only permanent actions have to be taken into account.

Where a static equilibrium analysis is required in addition to the strength analysis, alternative combinations are permitted depending on whether the verification does or does not involve the resistance of structural members. Where separate equilibrium and strength analyses are to be undertaken, the equilibrium analysis shall be based on the loading combination given against Equation D2.1 using the partial factors given in Table NA.A1.2 (A) and the strength analysis using the appropriate combination from Table NA.A1.2 (B). Where the verification of static equilibrium will also involve the resistance of structural members, as stated in the bottom section of Table NA.A1.2 (A), as an alternative to the use of two separate verifications a combined verification based on the design values of actions given in Table NA.A1.2 (A) using the partial factors listed in that section should be used.

When considering accidental actions (and excluding seismic actions), the combinations should either include for the design value of the accidental action, A_d (for fire or impact), or refer to the situation after an accidental event when $A_d = 0$. The combination to be used is given in EN 1990 and the National Annex to EN 1990, and is

$$\sum_{j\geq 1} G_{k,j} \text{ '+' } A_d \text{ '+' } (\psi_{1,1} \text{ or } \psi_{2,1})Q_{k,1} \text{ '+' } \sum_{i>1} \psi_{2,i}Q_{k,i} \qquad \text{(EN 1990: 6.11b)} \qquad \text{(D2.4)}$$

where functions are as previously defined, and A_d will be the leading action.

In this guide, where design values have been calculated in examples for strength states at ultimate limit states, the fundamental combination given in Equation 6.10 in EN 1990 has been used.

Combinations of actions for serviceability limit states

With these states the combination of actions are associated with the behaviour of the structure and its effect on the user under normal service conditions and the primary situations to be considered are vibration and deformation behaviour.

Clause 7.3

For vibration behaviour the design loading requirements are incorporated into the design verification procedure in *clause 7.3* in EN 1995-1-1 and the National Annex to EN 1995-1-1, and are referred to in Section 7.3 in this guide.

With deformation behaviour, three combinations of actions are given in EN 1990 that have to be considered:

- characteristic combination – used for irreversible limit states
- frequent combination – used for reversible limit states
- quasi-permanent combination – used for creep displacements.

An irreversible limit state is one in which some consequences of exceeding specified service requirements will remain after the actions are removed (e.g. cracking of finishes), and a reversible limit state is one in which there will be no consequences of exceeding such requirements after the actions are removed.

The characteristic combination is mainly intended to be used where, if the limit state criterion is exceeded, there will be significant distress to the structure/fabric or unacceptable irreversible displacements will arise. This is the condition used in EN 1995-1-1 for calculating the instantaneous deflection, and the design value of the combined actions is obtained from

$$\sum_{j\geq 1} G_{k,j} \text{ '+' } Q_{k,1} \text{ '+' } \sum_{i>1} \psi_{0,i}Q_{k,i} \qquad \text{(EN 1990: 6.14b)} \qquad \text{(D2.5)}$$

The frequent combination is mainly intended to be used where no consequences of actions exceeding the specified service requirements will remain when the actions are removed. This is not referred to in EN 1995-1-1 but is permitted in EN 1990, and can be used where it is agreed between the designer and the client. For this condition, the design value of the combined actions is obtained from

$$\sum_{j\geq 1} G_{k,j} \text{ '+' } \psi_{1,1}Q_{k,1} \text{ '+' } \sum_{i>1} \psi_{2,i}Q_{k,i} \qquad \text{(EN 1990: 6.15b)} \qquad \text{(D2.6)}$$

When long-term deflections are to be determined (i.e. creep deformation), the quasi-permanent loading combination applies, and the design value of the combined actions is obtained from

$$\sum_{j\geq 1} G_{k,j} \text{ '+' } \sum_{i>1} \psi_{2,i}Q_{k,i} \qquad \text{(EN1990: 6.16b)} \qquad \text{(D2.7)}$$

This relationship is applicable to both reversible and irreversible serviceability limit states for the creep element of the deflection. Where the creep behaviour is not constant across the structure, the loading requirement will be as stated in Section 2.2.3 and in Chapter 5 in this guide.

2.3.1.2 Load duration classes

The strengths of timber and wood products are dependent on the period of time over which they are loaded. The longer the load duration the greater will be the reduction in strength. To be able to take this effect into account in timber design, EN 1995-1-1 has defined five load duration classes, which are slightly different to the duration of load categorisation used in BS 5268 Part 2 (BSI, 2002), and are given in *Table NA.1* in the National Annex to EN 1995-1-1. Actions have to be assigned to one of the load duration classes referred to in the table, and the effect this has on the design strength is taken into account by the use of the modification factor, k_{mod}, as discussed later in this chapter.

Clause 2.3.1.2

For stiffness calculations, *Clause 2.3.1.2* also requires actions to be assigned to a load duration class. This relates to the derivation of the design loading to be used in the deformation calculation, as, excluding the stiffness property used for creep deformation, the stiffness property as given in EN 1995-1-1 is not a function of the load duration class structure.

2.3.1.3 Service classes

The strength of timber and of wood-based products reduces as the moisture content of the material increases, and, to include for this effect in design, these materials must be assigned a 'service class'. As they are hygroscopic, their moisture content is a function of the relative humidity of the surrounding air and of the air temperature, and, in an attempt to categorise the different environmental conditions that can be expected in design, three service conditions are used. They are given in EN 1995-1-1 (*clause 2.3.1.3*):

Clause 2.3.1.3

Service class 1 – is characterised by a moisture content in the materials corresponding to a temperature of 20°C and the relative humidity of the surrounding air only exceeding 65% for a few weeks per year.

NOTE: In service class 1 the average moisture content in most softwoods will not exceed 12%.

Service class 2 – is characterised by a moisture content in the materials corresponding to a temperature of 20°C and the relative humidity of the surrounding air only exceeding 85% for a few weeks per year.

NOTE: In service class 2 the average moisture content in most softwoods will not exceed 20%.

Service class 3 – is characterised by climatic conditions leading to higher moisture contents than in service class 2.

The service class to be used in UK designs for common design conditions is given in *Table NA.2* in the National Annex to EN 1995-1-1, and the effect of moisture content on the timber or wood-based product strength is taken into account in the design by the modification factor, k_{mod}, obtained from *Table 3.1*.

2.3.2 Materials and product properties

2.3.2.1 Load-duration and moisture influences on strength

As stated in Sections 2.3.1.2 and 2.3.1.3 in this guide, strength properties are affected by load duration and moisture content (service class) effects, and these are taken into account in design by the use of the modification factor, k_{mod}. This factor converts the values of strength properties derived under specified test conditions to those that will apply under different load duration and service classes in the design condition, as described in Section 2.4.1 in this guide.

The value to be used for the factor under different load duration and service class conditions is given in EN 1995-1-1 (*Table 3.1*).

A timber or wood-based product strength is defined by its characteristic value (which is generally taken to be its 5 percentile value), and is derived from tests undertaken at service class 1 conditions (sc1) over a relatively short period of time (e.g. of the order of seven minutes for strength tests in accordance with the requirements of EN 408 (BSI, 2010)). Although not stated in EN 1995-1-1, the value of k_{mod} for this condition will be 1. Where a design situation only involves permanent actions, the value of k_{mod} is that given in the column headed 'Permanent action' in the table, and if the load combination consists of actions belonging to different load duration classes the value will be that associated with the shortest duration action in the combination.

As will be seen from *Table 3.1*, the longer the load duration class the lower the strength, and as the moisture content increases there will generally also be a strength reduction. For instantaneous classes of load duration (e.g. wind and impact) there will be a strength increase.

The value of k_{mod} also depends on the type of material being used, and where a connection is formed from two timber-based elements having different k_{mod} values (e.g. k_{mod1} and k_{mod2}) the value used in design should be

$$k_{mod} = \sqrt{k_{mod1} k_{mod2}}$$

2.3.2.2 Load duration and moisture influences on deformations

When creep occurs in a structure, stiffness properties will be affected, and, depending on the type of structure being considered, and in particular where members or components have different creep behaviour, final mean stiffness values rather than mean stiffness values should be used in final deformation and stress distribution analyses. Values for these properties are given in *clauses 2.3.2.2(1)* and *2.3.2.2(2)*.

Clause 2.3.2.2(1)
Clause 2.3.2.2(2)

Creep occurs after instantaneous deformation has taken place, and is the deformation that arises when the structure is subjected to its loading over a period of time. In timber or wood-based products, creep occurs due to the combined effects of temperature, load duration, moisture content and stress level. As the temperature and/or the stress level in a structure increases, the rate of creep also increases, but at temperatures of less than 60°C and for stresses up to serviceability limit states levels, for design purposes, it can be assumed that the creep rate will reduce to zero within the design working life and that all creep deformation will have taken place. Material, moisture content and load duration effects influence the value of the creep deformation, and in EN 1995-1-1 these are taken into account through the use of the deformation factor, k_{def}, referred to in *clause 3.1.4* and in Section 3.1 in this guide. Values for k_{def} have been derived from tests on timber and wood-based products subjected to long-term actions under service class 1, 2 and 3 conditions where the instantaneous deflection, u_{inst}, and creep deflection, u_{creep} are measured, and k_{def} is obtained from

Clause 3.1.4

$$k_{def} = \frac{u_{creep}}{u_{inst}} \tag{D2.8}$$

As stated in Section 3.1 in this guide, the value of k_{def} is obtained from *Table 3.2* for timber and wood-based materials, and when timber is installed at or near its fibre saturation point and is likely to dry out under load it should be modified as required by *clause 3.2(4)*. When dealing with connections, the value must be doubled in accordance with the requirements of *clauses 2.3.2.2(3)* and *2.3.2.2(4)*.

Clause 3.2(4)
Clause 2.3.2.2(3)
Clause 2.3.2.2(4)

The types of structure referred to in *clause 2.3.2.2(1)* consist of members or components having different creep values, and the calculation of u_{creep} under these conditions is complex. The requirement in EN 1995-1-1 for this condition is unclear, and to better define it, as stated in Section 2.2.3 in this guide, the statements in *clauses 2.2.3(3)* and *2.2.3(4)* and *2.3.2.2(1)* are to be revised as detailed in Appendix A, requiring that u_{creep} is calculated under the quasi-permanent combination of actions with the stiffness of each element and connection based on the final mean value obtained from *Equations 2.7, 2.8* and *2.9*, which are functions of k_{def}. The instantaneous deformation, u_{inst}, for these types of structure will still be based on mean stiffness values, but is calculated using the difference between the characteristic and the quasi-permanent

Clause 2.3.2.2(1)
Clause 2.2.3(3)
Clause 2.2.3(4)
Clause 2.3.2.2(1)

combination of actions. As stated in Section 2.2.3, an alternative simpler option suggested in this guide for calculating the final deformation for this type of structure is to analyse the structure based on final mean stiffness properties but under the action of the characteristic loading combination, and a value for the final deformation will be derived directly from the analysis. It is a quicker method, but will generally result in a larger value than will be obtained using the revised procedure.

For ultimate limit states, where the distribution of member forces and moments is affected by the stiffness in the structure (e.g. composite members made from materials having different creep properties), creep behaviour must also be taken into account, and this is achieved by using the final mean values as defined in *Equations 2.10, 2.11* and *2.12*. These values are again functions of the k_{def} factor, but are also modified by the ψ_2 factor for the quasi-permanent value of action causing the largest stress in relation to strength. The ψ_2 factor is referred to in Section 2.3.1.1 in this guide and in the National Annex to EN 1990.

The reader is also referred to Chapter 5 in this guide for the application of the content of *clause 2.3.2.2* in a structural analysis.

Clause 2.3.2.2

2.4. Verification by the partial factor method

The design must be verified at the serviceability and ultimate limit states, and EN 1990 permits this to be undertaken using a deterministic or a probabilistic approach. In this guide, the deterministic approach is followed, using the partial factor method referred to in Section 6.0 in EN 1990.

In the partial factor method, design values of actions, E_{fd}, are derived at each of the limit states as described in subclauses in *clause 2.4* in EN 1995-1-1 and covered in the related subsections given in this guide. For ultimate limit states, design resistances, R_d, are calculated by dividing material strengths by partial factors, and for serviceability limit states R_d will normally be a deformation or vibration limit given in EN 1995-1-1 or the National Annex to EN 1995-1-1. Verification is then undertaken at the relevant states to demonstrate that E_{fd}, will not exceed R_d, i.e.

Clause 2.4

$$Ef_d \leq R_d \tag{D2.9}$$

2.4.1 Design value of material property

Strength property

The design value, X_d, of the strength property of timber or of a wood-based product is obtained from *Equation 2.14*:

$$X_d = k_{mod} \frac{X_k}{\gamma_M} \tag{2.14}$$

where k_{mod} is the modification factor, X_k is the characteristic value of the strength property X and γ_M is the partial factor for a material property.

The function of k_{mod} is to take into account the effects of load duration and moisture content as discussed in Section 2.3.1 and related subsections in this guide.

The characteristic strength, X_k, is obtained from product standards (e.g. EN 338 (BSI, 2009)), as this type of information is not given in EN 1995-1-1. A requirement of product standards is that strengths are derived under **service class 1** conditions and that the duration of the load test should be a relatively short period, as stated in Section 2.3.2.1 in this guide.

The purpose of the material property factor, γ_M, is to cover for unfavourable deviations in strength, and values for this factor are given in *Table NA.3* in the National Annex to EN 1995-1-1.

There are also other factors not referred to in *Equation 2.14* that affect the design value (instability factors, size factors, system factor, etc.), and the rules in EN 1995-1-1 and the guidance given in the relevant sections is this guide show how these should be taken into account.

Stiffness property

The design member stiffness property, E_d or G_d, of timber or wood-based products is obtained from *Equations 2.15* and *2.16*:

$$E_d = \frac{E_{mean}}{\gamma_M} \qquad (2.15)$$

$$G_d = \frac{G_{mean}}{\gamma_M} \qquad (2.16)$$

where E_{mean} is the mean value of the modulus of elasticity, G_{mean} is the mean value of the shear modulus and γ_M is the partial factor for the material property.

As with strength properties, values for E_{mean} and G_{mean} are obtained from European standards and/or product standards, and are not given in EN 1995-1-1.

For analyses at the serviceability limit states, the stiffness property is derived with $\gamma_M = 1$, and, when undertaking a second-order analysis of a structure, γ_M will be the value given in *Table NA.3* in the National Annex to EN 1995-1-1 for the material being used.

2.4.2 Design value of geometrical data

Design values of geometrical data, a_d, are the sizes to be used in the analysis and design of the timber or wood-product structures, and are defined as

$$a_d = a_{nom} \qquad \text{(D2.10)}$$

where a_d is the nominal reference dimension and a_{nom} is the nominal value given in product standards, drawings or specifications.

Design values of deviation from straightness of column and beam members must also be limited to the maximum allowances permitted in *Section 10* of EN 1995-1-1, and the effects of these deviations have been incorporated into the relevant design equations given in EN 1995-1-1.

2.4.3 Design resistances

The design value, R_d, of a resistance is obtained from *Equation 2.17*:

$$R_d = k_{mod} \frac{R_k}{\gamma_M} \qquad (2.17)$$

where k_{mod} is the modification factor, R_k is the characteristic value of the load-carrying capacity and γ_M is the partial factor for a material property.

Clause 2.4.1

The function of k_{mod} is the same as referred to in *clause 2.4.1*, and is discussed in Section 2.3.1 and related subsections in this guide.

When dealing with resistances, the characteristic resistance, R_k, will generally be calculated from design rules given in EN 1995-1-1. For example, the characteristic splitting capacity ($F_{90,Rk}$) for the arrangement shown for softwood timbers in *Figure 8.1* in EN 1995-1-1 will be calculated from *Equation 8.4* as follows:

$$F_{90,Rk} = 14bw\sqrt{\frac{h_e}{1 - h_e/h}} \qquad (8.4)$$

Clause 8.1.4(3)

and the functions are as described in *clause 8.1.4(3)*.

Clause 2.4.1

The γ_M factor fulfils the same function as referred to in *clause 2.4.1*, and is obtained from the values given in *Table NA.3* in the National Annex to EN 1995-1-1.

REFERENCES

BSI (2002) BS 5268-2: 2002. Structural use of timber – Part 2: Code of practice for permissible stress design, materials and workmanship. BSI, London.
BSI (2009) BS EN 338: 2009. Structural timber. Strength classes. BSI, London.
BSI (2010) BS EN 408: 2010. Timber structures – Structural timber and glued laminated timber – Determination of some physical and mechanical properties. BSI, London.
Gulvanessian H, Calgaro J-A and Holicky M (2012) *Designers' Guide to Eurocode: Basis of Structural Design: EN 1990*, 2nd edn. ICE Publishing, London.

Designers' Guide to Eurocode 5: Design of Timber Buildings
ISBN 978-0-7277-3162-3

http://dx.doi.org/10.1680/dtb.31623.021

Chapter 3
Material properties

This chapter references the properties of the various timber and wood-based materials that can be designed in accordance with the rules in EN 1995-1-1. Due to the range of timber species and grades, and the variety of wood-based materials and fasteners, the properties are given in a series of European standards, noted below in the following clauses:

- General — *Clause 3.1*
- Solid timber — *Clause 3.2*
- Glued laminated timber — *Clause 3.3*
- Laminated veneer lumber — *Clause 3.4*
- Wood-based panels — *Clause 3.5*
- Adhesives — *Clause 3.6*
- Metal fasteners — *Clause 3.7*

3.1. General

- Strength and stiffness parameters: strength calculations should be based on a linear stress–strain relationship, although a non-linear relationship may be used for parts of members subject to compression.
- The strength modification factor: as explained in Section 2.3.1.2, the strength of all timber-based materials depends upon the duration of the load and the moisture content of the material. *Table 3.1* gives the various values of the strength modification factor k_{mod} for the materials commonly used in design.
- The deformation factor: the deformation of timber elements under sustained load increases with time, but at a decreasing rate. This effect, known as creep, is allowed for by the deformation factor k_{def}, given in *Table 3.2* for the commonly used materials in the various strength classes. Note that timber in service class 3, or which is installed 'green' (i.e. fully saturated) and then dries out under load (*clause 3.2(4)*), will undergo increased creep deflection.

Clause 3.2(4)

3.2. Solid timber

EN 14081 (BSI, 2006a) covers the two methods of strength grading: visual grading, and machine grading. For **visual grading**, it might be thought that new European-wide rules would be introduced as part of a unified system, but this proved to be a step too far. Instead, the standard lays down the general form of grading rules, and lists the various national grading standards. EN 1912 (BSI, 2004a) then allocates the national grades for various species to strength classes.

The strength class system was introduced primarily for the softwoods, which have broadly similar strength profiles. For this reason, it was possible to develop the system shown in Table 3.1, based on EN 338, which defines strength classes C14 to C50 (C denotes coniferous (i.e. softwood), and 14 is the characteristic bending strength in N/mm^2). Although softwood classes are listed to C50, it would be difficult to find softwoods strong enough to enter classes above C35.

Much of the current commercial supply is now **machine graded**: a process in which the timber (up to about 75 mm thickness) is bent by a known force about the weaker axis, and the deflection measured. The timber strength is determined by known relationships between species strength and stiffness. The system allows timber to be graded directly to a strength class (with some visual overrides).

Table 3.1. Strength and stiffness properties and density values for structural timber strength classes (C14 to C35 and D30 to D70)

		Softwood species									Hardwood species					
		C14	C16	C18	C20	C22	C24	C27	C30	C35	D30	D35	D40	D50	D60	D70
Strength properties: N/mm²																
Bending	$f_{m,k}$	14	16	18	20	22	24	27	30	35	30	35	40	50	60	70
Tension parallel	$f_{t,0,k}$	8	10	11	12	13	14	16	18	21	18	21	24	30	36	42
Tension perpendicular	$f_{t,90,k}$	0.4	0.4	0.4	0.4	0.4	0.4	0.4	0.4	0.4	0.6	0.6	0.6	0.6	0.6	0.6
Compression parallel	$f_{c,0,k}$	16	17	18	19	20	21	22	23	25	23	25	26	29	32	34
Compression perpendicular	$f_{c,90,k}$	2.0	2.2	2.2	2.3	2.4	2.5	2.6	2.7	2.8	8.0	8.1	8.3	9.3	10.5	13.5
Shear	$f_{v,k}$	3.0	3.2	3.4	3.6	3.8	4.0	4.0	4.0	4.0	4.0	4.0	4.0	4.0	4.5	5.0
Stiffness properties: kN/mm²																
Mean modulus of elastic parallel	$E_{0,mean}$	7	8	9	9.5	10	11	11.5	12	13	11	12	13	14	17	20
5% modulus of elasticity parallel	$E_{0,5}$	4.7	5.4	6.0	6.4	6.7	7.4	7.7	8.0	8.7	9.2	10.1	10.9	11.8	14.3	16.8
Mean modulus of elasticity perpendicular	$E_{90,mean}$	0.23	0.27	0.30	0.32	0.33	0.37	0.38	0.40	0.43	0.73	0.80	0.86	0.93	1.13	1.33
Mean shear modulus	G_{mean}	0.44	0.5	0.56	0.59	0.63	0.69	0.72	0.75	0.81	0.69	0.75	0.81	0.88	1.06	1.25
Density: kg/m³																
Density	ρ_k	290	310	320	330	340	350	370	380	400	530	540	550	620	700	900
Mean density	ρ_{mean}	350	370	380	390	410	420	450	460	480	640	650	660	750	840	1080

Data from EN 338:2009, Table 1 (BSI, 2009)

Table 3.2. Commonly available sizes of softwood timber

Sawn thickness (to tolerance class 1): mm	Machined thickness (to tolerance class 2): mm	Sawn width (to tolerance class 1): mm									
		75	100	125	150	175	200	225	250	275	300
		Machined width (to tolerance class 2): mm									
		72	97	120	145	170	195	220	245	270	295
22	19		■	■	■	■	■	■			
25	22	■	■	■	■	■	■	■			
38	35	■	■	■	■	■	■	■			
47	44	■	■	■	■	■	■	■	■		×
63	60		■	■	■	■	■	■			■
75	72		■	■	■	■	■	■	■	■	■
100	97		■		■		■	■	■		■
150	145				■		■				

X applies to sections with a sawn width or a sawn thickness
Data from EN 336 (BSI, 2003) and Porteous and Kermani (2007)

The strength class system avoids the necessity of a designer choosing a species when this is best left to the supplier: for example, internal framing work covered by finishes may be specified simply as 'structural softwood, graded C16 in accordance with EN 14081'.

Strength classes (D18 to D70) for (deciduous) hardwoods are also included, but the hardwoods are a more diverse grouping, and a few species are effectively down-graded to some degree so that all the properties comply with the strength class. Hardwoods are generally chosen on other properties (such as appearance or durability) besides their strength, and so it is usual to name a particular species in the material specification, as well as the strength class. No species are listed in EN 1912 for D18 or D24. For the strength properties of European oak and sweet chestnut, reference should be made to PD 6993-1.

The commonly available sizes of softwood are given in Table 3.2, based on EN 336 (BSI, 2003). These are 'target' sizes; tolerances on these are given as

- (sawn timber, T1): dimensions <100 mm: −1/+3 mm, >100 mm: −2 mm/+4 mm
- (planed timber, T2): dimensions <100 mm: −1/+1 mm, >100 mm: −1.5 mm/+1.5 mm.

There is of course a loss of around 3–5 mm for planing. The target sizes may generally be used in design calculations without reference to the minus tolerance.

Lengths of up to 5.5–6.0 m are generally available: above this length, pieces would have to be end finger-jointed, or enquiries made for a particular species such as Douglas fir, usually available in large baulks.

The grade stresses for depth in bending and width in tension for solid timber apply to a reference size of 150 mm. For dimensions less than this, the characteristic values of $f_{m,k}$ and $f_{t,0,k}$ may be increased by the depth factor k_h, given in *clause 3.2(3)*. Note that the factor for glulam has a different reference depth (see below).

Clause 3.2(3)

3.3. Glued laminated timber

The range of available timber supply sizes can be considerably increased by the use of glulam (glued laminated) timber. Glulam members are essentially made up of solid timber laminates, glued together on their long faces. The laminates are typically 4–5 m long and 35–50 mm thick, and joined in their length by finger joints to EN 385 (BSI, 2001). Fabricators are capable of supplying single components that are limited only by transport and erection considerations. The fabrication of glulams is a specialist process, where the issues of production control relate mainly to the adhesive, as referred to in Section 3.6 of this guide.

Table 3.3. The properties of homogeneous and combined glulam strength classes

Glulam strength class	Homogeneous glulam			Combined glulam		
	GL 24h	GL 28h	GL 32h	GL 24c	GL 28c	GL 32c
Bending strength						
$f_{m,g,k}$: N/mm^2	24	28	32	24	28	32
Tension strength						
$f_{t,0,g,k}$: N/mm^2	16.5	19.5	22.5	14	16.5	19.5
$f_{t,90,g,k}$: N/mm^2	0.4	0.45	0.5	0.35	0.4	0.45
Compression strength						
$f_{c,0,g,k}$: N/mm^2	24	26.5	29	21	24	26.5
$f_{c,90,g,k}$: N/mm^2	2.7	3.0	3.3	2.4	2.7	3.0
Shear strength						
$f_{v,g,k}$: N/mm^2	2.7	3.2	3.8	2.2	2.7	3.2
Modulus of elasticity						
$E_{0,g,mean}$: kN/mm^2	11.6	12.6	13.7	11.6	12.6	13.7
$E_{0,g,0.5}$: kN/mm^2	9.4	10.2	11.1	9.4	10.2	11.1
$E_{90,g,mean}$: kN/mm^2	0.39	0.42	0.46	0.32	0.39	0.42
Shear modulus						
$G_{g,mean}$: kN/mm^2	0.72	0.78	0.85	0.59	0.72	0.78
Density						
$\rho_{g,k}$: kg/m^3	380	410	430	350	380	410

Data from EN 1194 (BSI, 1999) and Porteous and Kermani (2007)
Note: For strength-related calculations, take $G_{g,0.5} = E_{0,g,0.5}/16$
Mean density is taken to be the average of the mean density of the inner and outer laminates, based on BS EN 338 (BSI, 2009)

Clause 6.4.3

Clause 3.5.3.1

Glulams can be made of a single laminate grade throughout (homogeneous), or with lower-grade laminates near the centre of the member (combined), which may give an economic advantage for larger beams. A range of standard sizes is generally held as stock, but most glulam members are made to order. It is also possible to fabricate single- or double-tapered beams (see *clause 6.4.3*) or curved beams, as arches or portals. In order to bend the individual laminates without fracturing them, BS 5268 recommends (*clause 3.5.3.1*) a limit to beam curvature of around 150–180t (depending on the grade), where t is the laminate thickness. A revision to EN 14080, shortly to be published, recommends a limit to beam curvature of around 200t based on the declared characteristic strength of the finger joints. Thin laminates, of course, produce a more than proportional increase in cost.

The design information for glulam strengths and stiffnesses is given in EN 1194 (BSI, 1999) for standard classes GL24, GL28 and GL32, and is reproduced here in Table 3.3.

Clause 3.3(3)

As for solid timber, the size of the member has to be taken into account by the factor k_h, given in *clause 3.3(3)* for bending and tension calculations, but here the reference depth is 600 mm.

3.4. Laminated veneer lumber (LVL)

Laminated veneer lumber is made from softwood veneers around 3 mm in thickness, which are peeled from the log, end jointed with a scarf, and then glued together to form long sheets of various thickness between 25 mm and 90 mm. Supply in Europe is dominated by Kerto-LVL, manufactured in Finland by Finnforest.

There are two basic panel types: Kerto-S, which has the grain of all the veneers running in the long direction, and Kerto-Q, which contains a few cross-grain veneers. Kerto-S has a higher in-plane bending strength than Kerto-Q, since all the veneers are contributing, but Kerto-Q

Table 3.4. Kerto strength and stiffness properties and density values

	Symbol	Units	Kerto-S	Kerto-Q
Characteristic values				
Bending				
Edgewise	$f_{m,0,edge,k}$	N/mm^2	44.0	32.0
Size effect parameter*	s	N/mm^2	0.12	0.12
Flatwise	$f_{m,0,flat,k}$		50.0	36.0
Tension				
Parallel to grain	$f_{t,0,k}$	N/mm^2	35.0	26.0
Perpendicular to grain	$f_{t,90,k}$	N/mm^2	0.8	6.0
Compression				
Parallel to grain	$f_{c,90,k}$	N/mm^2	35.0	26.0
Perpendicular to grain edgewise	$f_{c,90,edge,k}$	N/mm^2	6.0	9.0
Perpendicular to grain flatwise	$f_{c,90,flat,k}$	N/mm^2	1.8	1.8
Shear				
Edgewise	$f_{v,90,edge,k}$	N/mm^2	4.1	4.5
Flatwise	$f_{v,90,flat,k}$	N/mm^2	2.3	1.3
Modulus of elasticity				
Parallel to grain	$E_{0,k}$	N/mm^2	11 600	8 800
Shear modulus				
Edgewise	$G_{0,k}$	N/mm^2	400	400
Density	ρ_k	kg/m^3	480	480
Mean values				
Modulus of elasticity				
Parallel to grain	$E_{0,mean}$	N/mm^2	13 800	10 500
Shear modulus				
Edgewise	$G_{0,mean}$	N/mm^2	600	600
Density	ρ_{mean}	kg/m^3	510	510

Adapted from Porteous and Kermani (2007)
* s is the size effect exponent referred to in *clause 3.4*

Clause 3.4

has better cross-grain stability. The material is produced in accordance with EN 14374 (BSI, 2004b), and the strength and stiffness properties are given in Table 3.4.

Clause 3.4(3)
Clause 3.4(4)

For bending and tension, the size of the member is to be taken into account (*clauses 3.4(3)* and *3.4(4)*) with a reference depth in bending of 300 mm, and a reference length in tension of 3 m. The size effect factor s is to be declared by the manufacturers.

Although standard sizes are quoted, the material is actually made in sheets 2.4 m wide by 26 m long, and may be ordered up to this size.

3.5. Wood-based panels

Of the various panels that fall under this heading, OSBs (oriented strand boards) are the most used for structural purposes. The boards are made up of outer layers of strands oriented in the long direction, with an inner layer of randomly oriented strands, all held together with adhesive. In the proprietary form of Sterling Board, the material is available in various thickness between 8 mm and 25 mm, and in standard 1.2 × 2.4 m panels, although it can be supplied in panels up to 2.4 × 4.8 m.

There are three grades defined in EN 300 (BSI, 2006b):

- OSB1 – general-purpose boards for use in dry conditions
- OSB2 – load-bearing boards for use in dry conditions
- OSB3 – load-bearing boards for use in humid conditions.

Table 3.5. Properties of OSB2 and OSB3

Thickness, t: mm	Density, ρ: kg/m^3	Bending: N/mm^2		Tension: N/mm^2		Compression: N/mm^2		Shear: N/mm^2	
		$f_{m,0,k}$	$f_{m,90,k}$	$f_{t,0,k}$	$f_{t,90,k}$	$f_{c,0,k}$	$f_{c,90,k}$	$f_{v,k}$	$f_{v,rol,k}$
>6–10	550	18.0	9.0	9.9	7.2	15.9	12.9	6.8	1.0
>10–18	550	16.4	8.2	9.4	7.0	15.4	12.7	6.8	1.0
>18–25	550	14.8	7.4	9.0	6.8	14.8	12.4	6.8	1.0

Data from EN 300 (BSI, 2006b)

For timber frame construction (see Chapter 13 of this guide) where the panels are used for racking stability, and installed close to the external wall cavity, OSB3 is usually specified. Table 3.5 lists the properties of grades OSB2 and OSB3.

3.6. Adhesives

Although wood components are easily glued together, the adhesives used in the past for furniture were vulnerable in damp conditions. The invention in the 1930s of adhesives that were water-resistant enabled the fabrication of composite timber building elements, such as glulam, for use in the various service classes.

The adhesives most commonly used are from the phenolic and aminoplastic range, and classified for use in accordance with EN 301 (BSI, 2006c). Type 1 is specified for high-hazard conditions (service class 3) such as exterior exposure, or interior warm and damp conditions. Type 11 may only be used in low-hazard conditions (service classes 1 and 2) when protected from direct sun and rain or in controlled interior environments. Resorcinol formaldehyde, the most durable of the type 1 adhesives, produces a dark glue line, and is relatively expensive. As an alternative, phenol formaldehyde is also a type 1 adhesive, but is light in colour. Cheaper adhesives, such as urea formaldehyde, can be used in low-hazard conditions.

The fabrication of glued components requires a high degree of control throughout the entire process, including:

- the moisture content and surface preparation of the timber
- the pot life and distribution of the adhesive
- the temperature and humidity of the environment
- the clamping pressures and times.

The work should therefore be carried out by specialists under controlled factory conditions. Since the formaldehyde adhesives are thermosetting, thin joints are necessary to prevent excessive heat build-up reducing the final strength, and while they have a very high adhesion to timber, they do not adhere to metal. However, the more recently introduced ranges of resin adhesives will bond to most materials, including steel, and are much more tolerant of thicker beds. They can thus be used, for example, for bonding threaded rods into holes drilled into the timber, which is a useful method of joining larger components.

All the above adhesives require the timber to be dry (i.e. a moisture content of 20% or less, depending on the adhesive), but the most recently developed range of polyurethane adhesives are moisture-tolerant. However, the effect of the potential movement of the drying timber on the integrity of the glue line would have to be considered.

3.7. Metal fasteners

EN 14592 (BSI, 2008) specifies the requirements (and test method where relevant) for materials, geometry, strength, stiffness and durability (i.e. corrosion resistance) of 'dowel-type' fasteners (i.e. fasteners which join two pieces of wood as a dowel, and are primarily loaded in shear). The requirements cover the range of fasteners below.

3.7.1 Nails

Nails are the easiest and quickest fixing to install, and (in softwoods) are generally fired pneumatically from nail guns fed by 'cartridge belts'. They are made from steel wire with a minimum ultimate tensile strength of 600 N/mm^2, and with diameters commonly in the range 2.5–6 mm. The main types are round (smooth shank) or annular ring-shank, the latter giving higher withdrawal strengths (to be declared by the manufacturers, based on tests).

In softwoods (with density generally $< 500\ kg/m^3$), nails up to 6 mm in diameter need not be predrilled. When nails are installed in predrilled holes, the code gives an increased lateral loading capacity, and allows closer spacing, but this may not be an economic option. Hardwoods have a density generally $> 500\ kg/m^3$, and so will require pre-drilling.

When corrosion resistance is required, nails may be galvanised, or supplied from a (more limited) range in austenitic stainless steel.

3.7.2 Screws

While screws are a little more labour-intensive to install than nails, they have a greater strength against axial withdrawal. They are manufactured, as for nails, in steel or stainless steel, with diameters in the range 3–6 mm, and lengths up to 150 mm. Coach screws will extend this range up to 18 mm in diameter and 250 mm in length. They can also be unscrewed without damage, which may be a design aim (e.g. for the eventual replacement of external decking boards).

When calculating the strength of rolled thread screws, the effective diameter, d_{ef}, should be used, defined in *clause 8.7.1(3)* as 1.1 times the inner (root) diameter of the thread (d_1). EN 14592 allows the inner diameter to be between 60% and 90% of the outer thread diameter d. Thus, detailed calculations of screw capacity either have to be based on information from a potential manufacturer or the conservative assumption that $d_{ef} = 2/3d$. In addition, the length of the smooth shank is unspecified in the standard, but the lengths l and l_g are to be declared by the manufacturer. This would be important, for instance in the design of a screw that fixed a thin steel plate to timber.

Clause 8.7.1(3)

3.7.3 Dowels

In appearance, dowels (compared with bolt heads) are relatively unobtrusive. Round steel dowels may be smooth or fluted (with a minimum fluted diameter of $0.95d$). The flutes (together with a possible slight end chamfer) make installation easier, since the allowable tolerance in drilling the dowel holes (see *clause 10.4.4*) is virtually nil. The problem of accuracy becomes more critical in a large group, where it may be possible to arrange a temporary lay-up of the members one over the other, and drill the holes in one pass, or to use modern CNC machines that can drill the holes with the necessary accuracy.

Clause 10.4.4

It is also possible to obtain dowels with a cutting end that will drill through relatively thin steel flitch plates, put in position as blanks.

3.7.4 Bolts

For larger assemblies, bolts have a strength advantage over dowels, in terms of the rope effect, and for triangulated structures, such as roof trusses, the increased bolt slip contribution to the deflection (which would arise if the permission in *clause 10.4.3(1)* to drill bolt holes 1 mm oversize was used) may be acceptable.

Clause 10.4.3(1)

The size of the specified washers ($3d$, where d is the bolt diameter) is sometimes objected to on visual grounds. A check on the criticality of the rope effect in the relevant fastener equations could be made, or the bolts treated as dowels, which might allow the washer area required to be reduced.

For corrosion resistance, bolts may be galvanised, or manufactured in austenitic stainless steel. (In the latter case, it may be more economical to fabricate the shank from rod, and then thread each end to receive a washer and nut.)

3.7.5 Punched metal plate fasteners

The punched metal plate fastener (or 'Gangnail' – the original proprietary market name) was invented in the USA and introduced to the UK in the 1960s: trusses made with them rapidly replaced the traditional cut roofs of rafters and purlins. It is estimated that today some 80 million or so trussed rafters stand in the UK alone. The principle is simple: steel plates around 1 mm thick are set under presses that punch the teeth out. To make a truss, for example, the plates are set in position at the nodes on a table, pre-cut truss members are placed on top of them, and then a hydraulic press moves down the table, pressing the plates home. There are now several proprietary systems of a broadly similar form.

The design of the plates is generally carried out by the fabricator, using software supplied by the system owner.

REFERENCES

BSI (1999) BS EN 1194: 1999. Timber structures. Glued laminated timber. Strength classes and determination of characteristic values. BSI, London.

BSI (2001) BS EN 385: 2001. Finger jointed structural timber. Performance requirements and minimum production requirements. BSI, London.

BSI (2003) BS EN 336: 2003. Structural timber. Sizes, permitted deviations. BSI, London.

BSI (2004a) BS EN 1912: 2004 + A4: 2010. Structural timber. Strength classes. Assignment of visual grades and species. BSI, London.

BSI (2004b) BS EN 14374: 2004. Timber structures – Structural laminated veneer lumber – Requirements. BSI, London.

BSI (2006a) EN 14081-1: 2005 + A1: 2011. Timber structures. Strength graded structural timber with rectangular cross section. General requirements. BSI, London.

BSI (2006b) BS EN 300: 2006. Oriented Strand Boards (OSB). Definitions, classification and specifications. BSI, London.

BSI (2006c) BS EN 301: 2006. Adhesives, phenolic and aminoplastic, for load-bearing timber structures. Classification and performance requirements. BSI, London.

BSI (2008) BS EN 14592: 2008 + A1: 2012. Timber structures. Dowel-type fasteners. Requirements. BSI, London.

BSI (2009) BS EN 338: 2009. Structural timber. Strength classes. BSI, London.

Porteous J and Kermani A (2007) *Structural Timber Design to Eurocode 5*. Blackwell, Oxford.

Designers' Guide to Eurocode 5: Design of Timber Buildings
ISBN 978-0-7277-3162-3

http://dx.doi.org/10.1680/dtb.31623.029

Chapter 4
Durability

This short chapter relates to the issue of durability of timber and metal under various environmental conditions. The two clauses are:

- Resistance (of timber) to biological organisms — *Clause 4.1*
- Resistance (of metal) to corrosion — *Clause 4.2*

4.1. Resistance to biological organisms

The basic principle given in *Clause 4.1(1)* is that the timber shall either:

Clause 4.1(1)

- have adequate natural durability for the particular hazard class or
- be given a preservative treatment.

4.1.1 The hazard (or use) classes

Hazard classes for timber (the hazard being the development of fungal or insect attack) are defined in EN 335-1 (which in the 2006 edition now refers to them as use classes; BSI, 2006), and are summarised in Table 4.1 (column 1). They can be compared with the service classes, given in *clause 2.3.1.3* and summarised in Table 4.1 (column 2). Service classes for timber are defined in terms of the ambient temperature and relative humidity of the surrounding air. For the lower two classes, limits are then given for the average moisture content that the timber will attain. Typical construction forms for the service classes are given in *clause NA.2.2* in the National Annex to EN 1995-1-1 and illustrated in Figure 4.1. From this table it can be seen that timber in service classes 1 and 2 (i.e. internal to the building, such as the roof, walls and floor construction) will generally have a moisture content of less than 20%. Timber external to the building or fully exposed to the weather (service/use class 3) will routinely attain a moisture content in excess of 20%, while timber submerged in water (use classes 4 and 5) will obviously attain a very high permanent moisture content. For structural design purposes, use classes 4 and 5 are covered by the requirements of service class 3.

Clause 2.3.1.3

Clause NA.2.2

4.1.2 Fungal and insect attack of wood

There are a small number of fungi and insects that can attack timber, using it as a food source. The minute spores of the various rots are carried in the air, and, alighting on wood may germinate, if the surface moisture content is greater than 20%. The wood-boring beetles, by and large, prefer damper timber, although the house longhorn beetle (*Hylotrupes bajulus*) active in areas in and around Surrey, is an exception. Not all wood species are equally vulnerable to attack, due to the degree of natural resistance to fungal growth provided by extractives in the heartwood, the core of the trunk (Figure 4.2). The natural durability of important species is given in EN 350-2 (Table 3) (BSI, 1994), which allocates each species to one of five classes. A small extract from the table for common species in relation to fungal decay is given in Table 4.2, together with the species treatability, again allocated as classes. Treatment to improve durability (if required) is generally by means of chemical solutions applied under pressure.

4.1.3 Timber in use/service classes 1 and 2

From Table 4.1 it can be seen that these classes include timber in internal situations. If, therefore, the weather envelope is soundly constructed to resist water penetration from the roof, walls and floor, and is well enough insulated or ventilated to avoid areas of significant condensation, then the internal timber will dry down to well below 20%, and there will be no significant risk of rot or beetle attack (except in the geographical area noted above). The species may be chosen without

Table 4.1. Comparison between the service classes and the use classes in EN 335-1

Clause 2.3.1.3
Clause NA.2.2

Extract from EN 335-1: 2006 Definitions of use classes	Extract from Eurocode 5 *Clause 2.3.1.3*: service classes	UK National Annex, *clause NA.2.2*: Timber constructions in service classes
1. Situation in which the wood is under cover, not exposed to weather and wetting	1. Characterised by a moisture content in the materials corresponding to a temperature of 20°C and the relative humidity of the surrounding air only exceeding 65% for a few weeks per year	Warm roofs Intermediate floors Timber frame walls (external and party)
2. Situation in which the wood is under cover and not exposed to weather but where high environmental humidity can lead to occasional, but not persistent, wetting	2. Characterised by a moisture content in the materials corresponding to a temperature of 20°C and the relative humidity of the surrounding air only exceeding 85% for a few weeks per year	Cold roofs Ground floors Timber frame external walls External uses protected from direct wetting
3. Situation in which the wood is not under cover or in contact with the ground. It is either continually exposed to the weather or is protected from the weather but subject to wetting	3/4/5. Characterised by climatic conditions leading to higher moisture contents than in service class 2	External uses, fully exposed, but not in continuous ground contact
4. Situation in which the wood is in contact with ground or freshwater and thus permanently exposed to wetting		River structures Jetty supports Fenceposts in the ground (BS 8417, Table 1)
5. Situation in which the wood is permanently exposed to saltwater		Marine structures, piers, jetties, etc. (BS 8417, Table 1)

regard to its durability classification or whether some pieces include areas of sapwood, and no preservative treatment will generally be necessary. Most building-related timber structures fall into this category.

Although the distinction between these two service classes does not affect the durability of the timber, it is significant in relation to the stiffness of a member under creep conditions (see Table 3.2).

4.1.4 Timber in use/service class 3

Timber in this class, such as cladding or fully external structures, will get wet frequently enough to make conditions favourable for fungal and/or beetle attack. Following the basic principle in Section 4.1, it is necessary to consider whether a particular species has adequate natural durability, or whether durability could be achieved by the application of preservative.

When considering natural durability, it must be remembered that the durability of the sapwood in any species is never better than class 4 or 5 (see Table 4.2), and commercial suppliers of small-section softwood do not guarantee to exclude it. Guidance on the design life of heartwood timber in various use classes is given in BS 8417 (Table 3) (BSI, 2011). This must be regarded as 'broad-brush' advice – much depends on the standard of detailing (to avoid water traps), and cladding will generally last longer than decking because (apart from the question of wear) the water will shed more quickly from a vertical surface. Consideration should also be given to a replacement strategy – cladding and decking may be more easily replaced than primary structure.

Figure 4.1. Typical construction forms in service classes

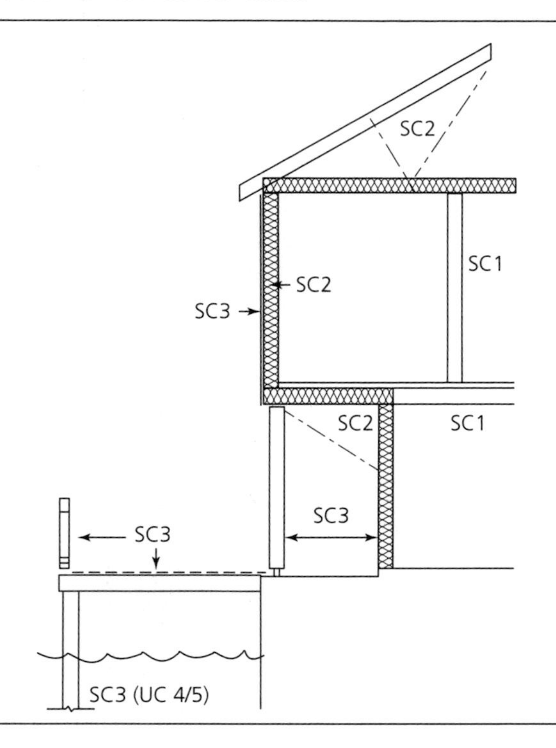

The softwoods

It can be seen from Table 4.2 that the most common species (spruce, redwood) are only **slightly durable**. In addition, spruce can be *difficult* to treat. Larger pieces of, for example, imported Douglas fir (**moderately durable**) can be purchased sap-free. The only softwood with a **durable** rating is western red cedar. It is therefore often used for cladding, and for light frameworks, since it is not a particularly strong timber.

The tropical hardwoods

There are a great variety of species, and the commercial range contains some of the most durable. They are often imported as planks, intended for joinery work (e.g. iroko), but a few (e.g. greenheart) are available in large sections.

Figure 4.2. Heartwood and sapwood in a trunk

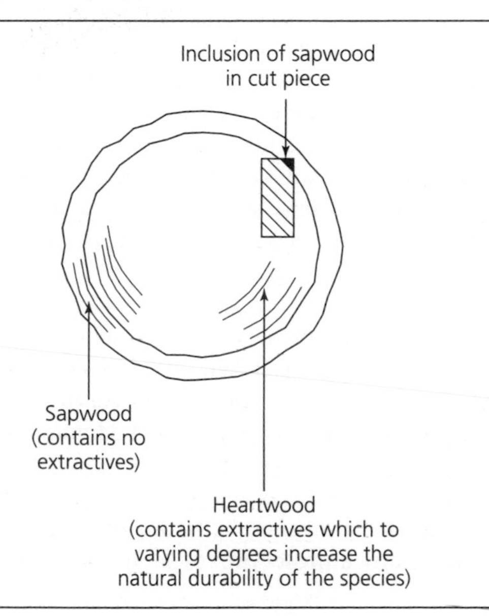

Table 4.2. Durability and treatability classes of common species

	Extracts from EN 350-2								
	Durability class					Treatability class			
	1 VD	2 D	3 MD	4 SD	5 ND	1 E	2 ME	3 MD	4 D
Softwoods									
Sitka spruce				H	H		S	SH	
Redwood			H	H		S		H	H
Douglas fir (imported)			H					S	H
Larch (European)			H	H			S		H
Western red cedar (North America)		H						SH	H
Western red cedar (UK)			H					SH	H
(All species)				S	S*				
Tropical hardwoods									
Iroko	H	H							
Balau (yellow)		H							
Teak	H								
Greenheart	H								
Temperate hardwoods									
Oak		H			S				
Sweet chestnut		H			S				
(Ash/beech/birch)					H				

* Spruce sapwood is visually indistinguishable from heartwood
Key: VD, very durable; D, durable; MD, moderately durable; SD, slightly durable; ND, not durable; E, easily treated; ME, moderately easy; MD, moderately difficult; D, difficult; H, heartwood; S, sapwood

The temperate hardwoods

Although there are many species in the commercial range of the European forest trees, most are durability class 4 or 5. The exceptions are oak and sweet chestnut, both **durable**. Most larger sections are sold 'green' (see Section 3.1 of this guide), but can be ordered to a specific cutting schedule, omitting sapwood if required.

4.1.5 Timber in use classes 4 and 5 service class 3

Timber installed below the water level becomes saturated, and is too wet to support the growth of the 'above-ground' rots. Thus, many nineteenth-century structures stand successfully on timber piles installed below the ground water level. In open water, however, timber is vulnerable to attack by marine borers, including (class 4, fresh water) gribble and (class 5, sea-water) *Teredo navalis*. If a post emerges from the water, the moisture content will drop with height to a much lower equilibrium value in the free air, at some level achieving the optimum value for rot growth. Thus, the most likely level of failure will be just above the water level, or, in the equivalent case of fence posts, the ground level. If the post is supported above the water, or the ground level, avoiding permanent contact with water, the use class drops from 4 to 3.

4.1.6 Preservatives

The selection and specification of wood preservatives for wood is outside the scope of this guide. Recommendations will be found in BS 8417 (BSI, 2011), which takes into account the use class, the likelihood and consequences of failure, the desired service life, and the treatability of the wood being used. Reference should be made to the *WPA Manual* (Wood Protection Association, 2012).

The chemicals, for obvious reasons, are generally toxic, and recent regulations have increasingly restricted the use of some well-known preservatives such as creosote and copper chrome arsenic (CCA).

4.2. Resistance to corrosion

This clause gives examples of the minimum specification for the corrosion protection of metal fastener components, varying from no treatment to zinc coatings to the use of stainless steel. When specified, the majority of fasteners such as nails and screws are supplied in austenitic stainless steel, for its combination of corrosion resistance and ductility, in grades A2 ('A' for austenitic) and A4 (formerly grades 302 and 316). The latter is more expensive, but is genuinely 'stainless'. A2 steel will stain to some degree in moist conditions, but will not lose profile.

Table 4.1 gives examples of minimum specifications for material protection against corrosion protection for fasteners, and as stated in Appendix A the proposal is to replace note a in the table with the revised statement given in the appendix.

REFERENCES

BSI (1994) BS EN 350-2: 1994, Durability of wood and wood-based products. Natural durability of solid wood. Guide to natural durability and treatability of selected wood species of importance in Europe.

BSI (2006) BS EN 335-1: 2006, Durabiltity of wood and wood-based products. Definitions of use classes. General. BSI, London.

BSI (2011) BS 8417: 2011, Preservation of wood. Code of practice. BSI, London.

WPA (2012) *WPA Manual: Industrial Wood Preservation – Specification and Practice*, 2nd edn. WPA, Castleford.

Designers' Guide to Eurocode 5: Design of Timber Buildings
ISBN 978-0-7277-3162-3

http://dx.doi.org/10.1680/dtb.31623.035

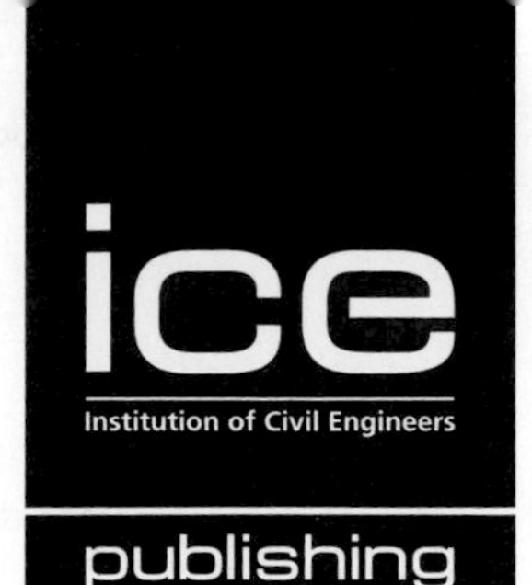

Chapter 5
Basis of structural design

This chapter concerns the subject of structural analysis, and includes guidance on how the behaviour of connections shall be taken into account when deriving the internal forces in the structure. It covers the material in *Section 5* of EN 1995-1-1, and addresses the following clauses:

- General *Clause 5.1*
- Members *Clause 5.2*
- Connections *Clause 5.3*
- Assemblies *Clause 5.4*

5.1. General

Timber structures should be analysed using appropriate design models based on linear elastic methods, generally involving the application of statics, to determine member forces and moments. Member validation is achieved by compliance with the relevant strength rules in EN 1995-1-1 for the types of element being designed (e.g. the rules for columns, beams and racking), as appropriate.

The guidance given for structural analysis in *Section 5* is related to the requirements of timber structures that function as frames that are either laterally rigid or able to sway, and for arches. In timber, the common types of laterally rigid framed structures are generally associated with roof trusses (including trussed rafters) and floor trusses. Framed structures with the potential for sway under lateral or instability loading are portal frames that are free to sway laterally, pitched portal frames and arches. These structures are commonly manufactured from glued laminated timber, laminated veneered lumber or timber and wood panel composites, and, in order to simplify detailing and construction and to accommodate moment transfer around the structure, are normally two or three pinned. Some examples of the above types of structures are shown in Figure 5.1.

Much use is made of platform-framed structures for the construction of timber-framed housing and residential accommodation. In these structures, timber studs provide the vertical support for the floor and roof structures, and lateral resistance is achieved by fixing sheathing material to the studs to form racking walls. The lateral loading is generally transferred to the walls through diaphragm action in the floor and the roof structures. This type of structure is also shown in Figure 5.1, and design models are normally based on the assumption that plastic theory can be applied.

The design models used for all types of timber structure must take into account the effects of the variables that influence the behaviour of timber and wood product structures, and, where it is considered necessary, test results can also be used. Apart from the effect of actions, the main variables to be taken into account in design models are the effects of moisture, creep and load duration, and by following the design rules in EN 1995-1-1 these effects will be covered.

For framed and arch-type structures, the internal forces within the timber structure must be derived from a global analysis, and EN 1991-1-1 requires a linear elastic analysis model to be used. Normally, a first-order linear elastic analysis will be acceptable for deriving member forces, and second-order effects due to deviation from straightness in members will implicitly be taken into account by compliance with the relevant member design rules given in *Section 6* of the Eurocode. For some types of structure (e.g. sway-type structures) this will not adequately

Figure 5.1. (a–d) Framed and (e) platform structures

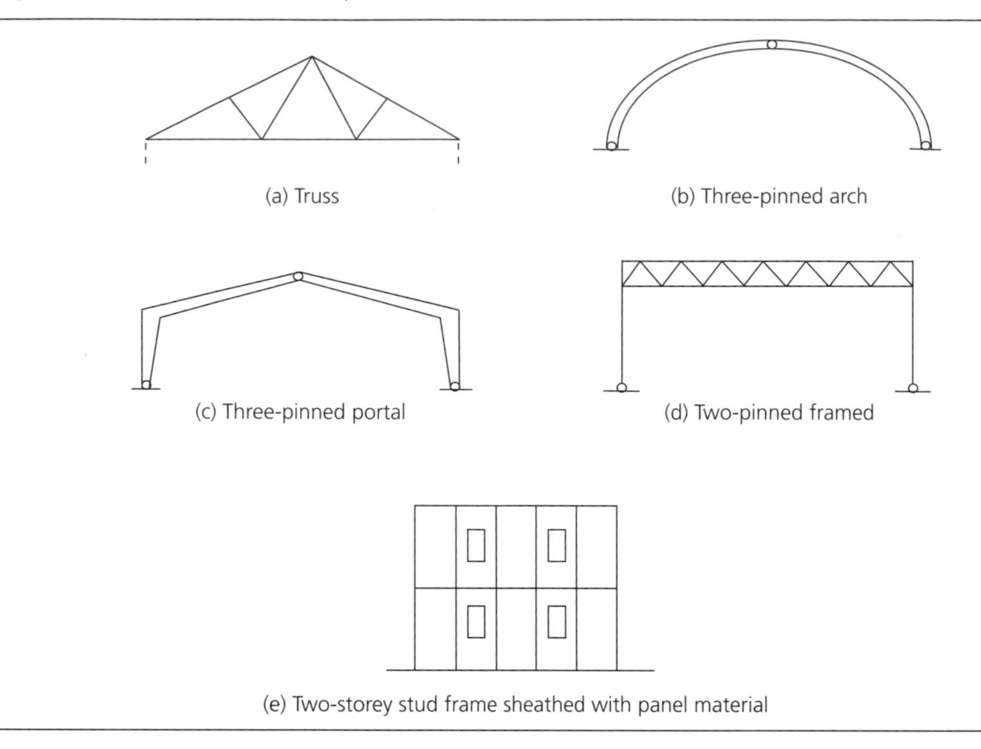

account for the effects of changes in the structure geometry, and in these cases a second-order linear elastic analysis should be used. This is still based on the assumption that linear elastic material behaviour applies, but the analysis also takes into account the effects of changes in the structure geometry under the design loading as well as instability effects. Some guidance on initial sway imperfections are given in *clause 5.4.4*; however, rules for the effect of out-of-plane geometric imperfections on behaviour are not yet given in EN 1995-1-1, and this has still to be fully addressed in the code. Typical structures likely to require the inclusion of second-order effects are pitch portal frames and arches, and the initial out-of-alignment criteria to be used in the analysis of these types are given in *clause 5.4.4*.

Clause 5.4.4

Clause 5.4.4

In any analysis the stiffness properties to be used for members and their connections must be appropriate for the type of structure being analysed, and the EN 1995-1-1 requirements are given in *clauses 2.2.2(1)P* and *2.3.2.2* and discussed in Section 5.4 in this guide.

Clause 2.2.2(1)P
Clause 2.3.2.2

Where connections in a structure exhibit adequate ductility, the internal forces can be derived from elastic–plastic methods. No guidance is given to the designer on what is meant by 'adequate ductility'; however, where metal dowel-type fasteners as referred to in *Section 8* are used and their design condition is based on ductile rather than brittle failure, such connections should be considered to be able to satisfy this requirement.

EN 1991-1-1 requires the effects of deformations at connections (i.e. slip and rotation) to be taken into account when determining the internal force effects (stresses) in the structure. For the types of connection referred to in EN 1995-1-1 the effect of connection behaviour will be achieved by modelling the translational and/or rotational stiffness of the connection in the global analysis as mentioned in Section 5.4.2. The stiffness properties of the connections are derived from the appropriate slip properties in *clause 7.1*, referred to in Chapter 7 in this guide, and the designer must also take into account the requirements of *clauses 2.2.2(1)P* and *2.3.2.2*, as stated above.

Clause 7.1
Clause 2.2.2(1)P
Clause 2.3.2.2

5.2. Members

There is no requirement to include for the effects of inhomogeneities of the timber or wood product in the global analysis. The grading requirements of the materials limit the extent and nature of the defects that will be acceptable for structural use, and, by ensuring member design is carried out in accordance with the relevant design rules in EN 1995-1-1, the effects of

Figure 5.2. Effective cross-section in a joint with multiple bolts

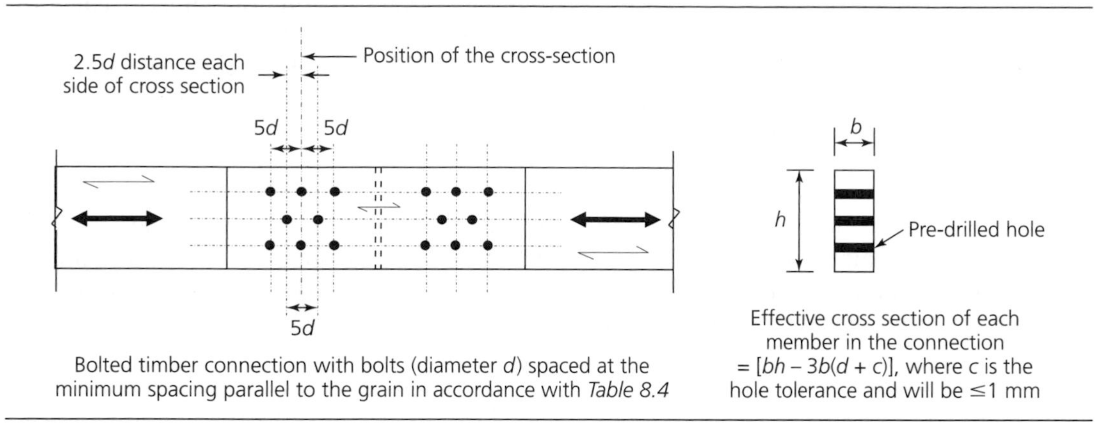

these defects are implicitly covered. The only situation where additional requirements are placed on the designer relates to connection design, where *clause 10.4.1* requires that, in the region of a connection, wane, splits, knots and other defects are limited. It is left to the designer to decide what will be acceptable, and this will generally depend on the type of connection being used.

Clause 10.4.1

There are limits on the maximum deviation from straightness allowed in members, and these are given in *clause 10.2* and referred to in Section 10.2 in this guide. When the global analysis is undertaken using a first-order linear elastic analysis, as stated in Section 5.1, the members are verified using the forces and moments from the analysis, and the effects of these deviations on strength and stability behaviour are implicitly incorporated within the design rules in the code. The designer must ensure that these limits are clearly specified and complied with, as the design rules are only valid if they are not exceeded.

Clause 10.2

Where the cross-section of a member is reduced in section, this has to be taken into account in the member design. If it is a local effect (e.g. loss of area at a connection), the member can be considered to have the full cross-section when modelling the global structural analysis, but the reduced section has to be used in the member strength verification.

With connections, if pre-drilling is used the loss of area caused by the pre-drilling has to be taken into account. The exception to this is where the connection is subjected to a compression force, and the pre-drilled holes are filled with fasteners having a greater stiffness than the timber or wood product. This will clearly be the case for connections using nails or screws or metal dowels, but will not apply with bolts, as the pre-drilled hole will not be filled. Also, when using nails or screws of 6 mm diameter or less, providing there is no pre-drilling, the full cross-section of the member can be used.

If connections are formed with multiple fasteners, and the minimum fastener spacing parallel to the grain, as defined in *Section 8*, is being used, all holes within a distance of half of the minimum spacing along the grain direction from the position being checked must be assumed to occur at the check position. An example of this is shown in Figure 5.2.

5.3. Connections

Connections have to be designed to take the forces and moments arising from the global structural analysis, and the designer must decide whether they are to be modelled as pinned or rotationally rigid. Guidance on which should be used is given in *Clause 5.4.2*. If the connections are modelled to be rotationally rigid, the design verification must include for the effects of the moments. Design verification of a connection will be obtained by following the design rules in *Section 8*.

Clause 5.4.2

5.4. Assemblies

5.4.1 General

For timber structures subjected to normal loading conditions, including wind loading, only static methods of analysis need be used. When the structure is framed such that sway effects are prevented, the global analysis is undertaken using a first-order linear elastic method. If the connections in the structure are taken to be rotationally rigid, the analysis must include for moment effects, and if they are pin-jointed, a linear elastic pin-jointed analysis is appropriate. The pin-jointed linear elastic form of analysis is also able to be used for the analysis of fully triangulated trusses formed with punched-metal plate fasteners, commonly referred to in the UK as trussed rafters. This is referred to in EN 1995-1-1 as a '*simplified analysis*', and the requirements of this method are described in *clause 5.4.3*.

Clause 5.4.3

Where vibrating loads are to be supported by the structure (e.g. vibrations arising from machinery), a dynamic analysis is likely to be required. For human-induced vibrations in residential buildings, validation requirements are covered by complying with the requirements of *clause 7.3.3* and the relevant clauses in the National Annex to EN 1995-1-1.

Clause 7.3.3

5.4.2 Frame structures

This clause gives the modelling procedure to be used for timber-framed structures, and follows the methodology normally applied when modelling for computer-based analysis. System lines are set up to form the line diagram to be used for the model, and the points of intersection of these lines at connections are commonly referred to as nodes. The lines must be positioned to coincide with the centre-lines of the main members of the structure, and, for the others (e.g. internal members in trusses), they must lie within the member profile. Where in a member the system line and the centre-line do not coincide, the effect of any eccentricity must be taken into account in the member strength verification. When eccentric connections occur in the modelling, as is commonly the case at truss supports, this can be taken into account by the use of fictitious elements. It is important that these elements are positioned to realistically mirror the joint behaviour and to prevent instability problems arising in the analysis. Guidance on modelling joints is well publicised, and background information is given by Nielsen (2003). Examples of the consequences of eccentricity between system and centre lines for typical truss details and joint models are shown in Figure 5.3.

Where a first-order linear elastic analysis is used, as stated in Section 5.2 there is no requirement to include for initial deviations from straightness, as the member strength validation rules implicitly include for these effects. For a second-order analysis, deviations in geometry are taken into account, and specific requirements for plane frames and arches are given in *clause 5.4.4*.

Clause 5.4.4

Clause 5.4.2(6) draws attention to the need to make sure that appropriate values of stiffness are used in the analysis for members and connections, and this information is covered in *clauses 2.2.2(1)P*, *2.2.3* and *2.3.2.2*. The values used for stiffness properties depend on whether the analysis is undertaken to determine deformations at the serviceability limit states or force

Clause 5.4.2(6)
Clause 2.2.2(1)P
Clause 2.2.3
Clause 2.3.2.2

Figure 5.3. Modelling effects and details

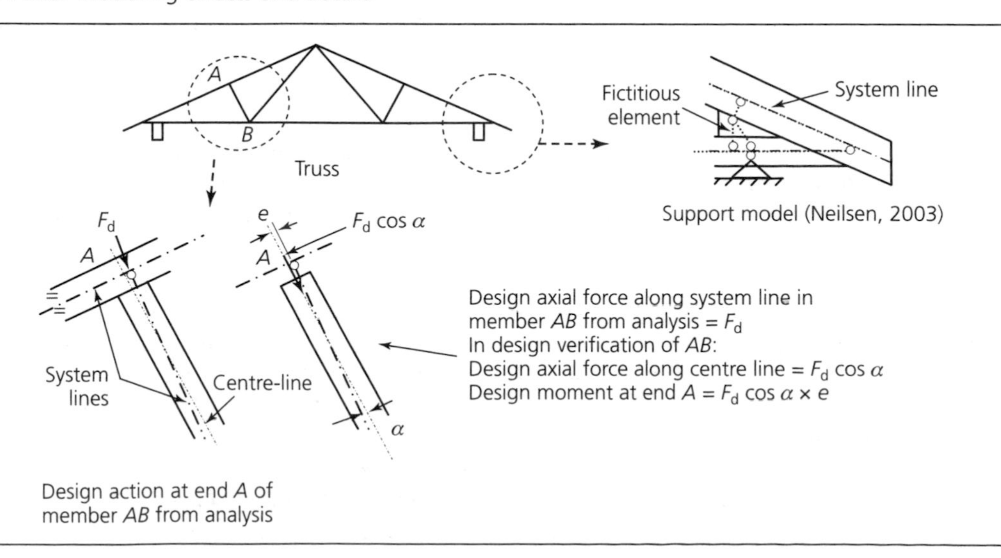

distributions at the ultimate limit states, and guidance on the stiffness properties for these different requirements is summarised as follows:

- **Deformation analysis at the serviceability limit states.** In an analysis to calculate the instantaneous deformation, mean values of the appropriate properties (i.e. the mean modulus of elasticity ($E_{0,mean}$), the mean shear modulus (G_{mean}) and the mean slip modulus (K_{ser})) should be used.
  In an analysis to calculate the final deformation (i.e. the summation of the instantaneous and the creep deformations), the values will depend on whether the members or components of the structure have the same or different values of creep behaviour (i.e. their respective deformation factor, k_{def}, is the same or different). If they all have the same k_{def} value, mean values as defined above are used, and if they are different, final mean values as defined in *clause 2.3.2.2(1)* must be adopted. The final mean values are: modulus of elasticity ($E_{0,mean}/(1 + k_{def})$), shear modulus ($G_{mean}/(1 + k_{def})$) and slip modulus ($K_{ser}/(1 + k_{def})$). Because *clauses 2.3.2.2(3)* and *2.3.2.2(4)* require the value of k_{def} to be doubled for timber as well as wood-based connections, excluding the case where a structure is formed from steel gusset plate connections (where it can be argued that k_{def} should be able to be used), the connection stiffness will always differ from the member stiffness, and will require such structures to be analysed using final mean values.
- **Global force analysis at the ultimate limit states.** For a first-order linear elastic analysis, as stated in *clause 2.2.2(1)P*, the values used for the stiffness properties will depend on whether the members or components of the structure have the same or different time-dependent properties (i.e. value of k_{def}). When k_{def} is the same throughout, mean stiffness values are used, but when it is not, final mean values adjusted to the load component causing the largest stress-to-strength ratio, as defined in *clause 2.3.2.2(2)*, must be used. The final mean values are: modulus of elasticity ($E_{0,mean}/(1 + \psi_2 k_{def})$), shear modulus ($G_{mean}/(1 + \psi_2 k_{def})$) and slip modulus ($K_{ser}/(1 + \psi_2 k_{def})$). In these cases, the k_{def} value will be derived on the same basis as described in the above deformation analysis, and the value for ψ_2 will be the value associated with the action causing the largest stress-to-strength ratio. Values for ψ_2 are given in the National Annex to EN 1990, and where the largest ratio is caused by a permanent force, ψ_2 is replaced by 1. The resources required to determine the action causing the largest stress-to-strength ratio can be considerable, and adopting a value of 1 for ψ_2 for all design conditions within this category will normally alter stress values by a relatively small percentage, and should produce an acceptable result.
  For a second-order linear elastic analysis, *clause 2.2.2(1)P* requires the stiffness properties to be based on design values that are not adjusted for the duration of load. The rules in *clause 2.4.1(2)P* apply, and the values used are the elasticity modulus ($E_{0,mean}/\gamma_M$), the shear modulus (G_{mean}/γ_M) and the slip modulus (K_{ser}/γ_M). The value of the partial material factor, γ_M, is obtained from *Table NA.3* in the National Annex to EN 1995-1-1.
  Examples of the stiffness and loading conditions to be used for deformation and global force analyses for different design conditions are given in Table 5.1.
  Stiffness properties ($E_{0,mean}$ and G_{mean}) are obtained from relevant European standards (e.g. EN 338 and EN 300) and from manufacturers' literature.

Clause 2.3.2.2(1)
Clause 2.3.2.2(3)
Clause 2.3.2.2(4)
Clause 2.2.2(1)P
Clause 2.3.2.2(2)
Clause 2.2.2(1)P
Clause 2.4.1(2)P

Whether connections are rotationally stiff (rigid) or rotationally pinned, the axial force distribution around framed structures will not normally be significantly affected. What is affected is the deflection behaviour, with a rotationally pin-jointed structure deforming more than the same structure formed with rotationally rigid connections. The code allows the designer to decide whether connections in a structure can be designed as rotationally rigid or pinned. If relative rotation of the members within a connection will have no significant effect on the distribution of the member moments and force, it can be assumed to be rotationally rigid. Providing connections have been designed to resist their design moments and lateral forces in accordance with the rules in *Section 8*, normal types of framed structures should be able to be modelled as rotationally rigid. Semi-rigid behaviour can also be considered, and guidance on how to obtain the stiffness of semi-rigid connections for this type of analysis can be obtained from Porteous and Kermani (2007).

The code permits splice connections in truss (lattice) structures to be designed as rotationally rigid, providing the connections satisfy the design rules given in *clause 5.4.2(9)*.

Clause 5.4.2(9)

Table 5.1. Stiffness and loading requirements for a first-order and a second-order linear elastic analysis of timber and wood product structures[a]

Purpose of the analysis	Stiffness values for the elements of the structure (see *Table 3.2* and Chapters 2 and 3 in this guide for the value to be used for k_{def})	Design loading used in the analysis (EN 1990, National Annex to EN 1990 and Chapter 2 in this guide)
First-order linear elastic analysis		
(a) Calculate the instantaneous deformation at the SLS	(a) Mean values of the modulus of elasticity ($E_{0,mean}$), shear modulus (G_{mean}), slip modulus (K_{ser})	(a) Design loading derived from the characteristic loading combination:[b] $\sum_{j\geq 1} G_{k,j}$ '$\pm$' $Q_{k,1}$ '$\pm$' $\sum_{i>1} \psi_{0,i} Q_{k,i}$
(b) Calculate the final deformation at the SLS – all members and components have the same creep values. (For structures with connections, this will only apply when steel gusset plates or equivalent are being used.) (*Clauses 2.3.2.2(3)* and *2.3.2.2(4)*) and Appendix A (*clause 2.2.3(3)*)	(b) The stiffness properties for this analysis should be those given in (a) divided by k_{def}. However, to simplify the approach, the stiffness values used are those given in (a), and the effect of the k_{def} function is taken into account in the design loading	(b) As (a) plus the design loading combination derived from the quasi-permanent loading combination times k_{def}:[d] $k_{def}\left(\sum_{j\geq 1} G_{k,j} \text{ '}\pm\text{' } \sum_{i>1} \psi_2 Q_{k,i}\right)$
(c) (i) Calculate the final deformation at the SLS – members and components have different creep values. Add the instantaneous deformation and the creep deformation to obtain the final deformation. (Appendix A (*clauses 2.2.3(4)* and *2.3.2.2(1)*))	(c) (i,a) Mean values of the modulus of elasticity ($E_{0,mean}$), shear modulus (G_{mean}), slip modulus (K_{ser}) for the instantaneous deformation	(c) Design loading derived from the characteristic loading combination[b] minus the quasi-permanent loading combination: $\left(\sum_{j\geq 1} G_{k,j} \text{ '}\pm\text{' } Q_{k,1} \text{ '}\pm\text{' } \sum_{i>1} \psi_{0,i} Q_{k,i}\right) - \left(\sum_{j\geq 1} G_{k,j} \text{ '}\pm\text{' } \sum_{i>1} \psi_2 Q_{k,i}\right)$
	(i,b) Final mean values of the modulus of elasticity ($E_{0,mean}/(1 + k_{def})$), shear modulus ($G_{mean}/(1 + k_{def})$), slip modulus ($K_{ser}/(1 + k_{def})$) for the creep deformation. (The requirements of *clauses 2.3.2.2(3)* and *2.3.2.2(4)* will apply when calculating the deformation factor to be used for connections) (That is, two analyses required)	(i,b) Design loading derived from the quasi-permanent loading combination: $\sum_{j\geq 1} G_{k,j}$ '$\pm$' $\sum_{i>1} \psi_2 Q_{k,i}$
Or, as an alternative (conservative option suggested in this guide): (ii) Calculate the final deformation at the SLS – members and components have different creep values	(ii) Final mean values of the modulus of elasticity ($E_{0,mean}/(1 + k_{def})$), shear modulus ($G_{mean}/(1 + k_{def})$), slip modulus ($K_{ser}/(1 + k_{def})$) to obtain the final deformation. (The requirements of *clauses 2.3.2.2(3)* and *2.3.2.2(4)* will apply when calculating the deformation factor to be used for connections) (That is, one analysis required)	(ii) As (a)

(d) Calculate the global force distribution at the ULS – all members have the same creep values. For structures with connections, this will only apply when steel gusset plates or an equivalent are being used. (*Clause 2.2.2(1)P*)	(d) As (a)	(d) Design loading combination derived from the **fundamental** loading combination to be used, i.e. Equations 6.10, or 6.10a or 6.10b in EN 1990
(e) Calculate the global force distribution at the ULS – members have different creep values. (*Clauses 2.2.2(1)P* and *2.3.2.2(2)*)	(e) Final mean values adjusted to the load component causing the largest ratio of stress to strength: modulus of elasticity ($E_{0,mean}/(1+\psi_2 k_{def})$), shear modulus ($G_{0,mean}/(1+\psi_2 k_{def})$), slip modulus ($K_{ser}/(1+\psi_2 k_{def})$). If the largest ratio of stress to strength is due to permanent action, ψ_2 should be replaced by 1.[c] (The requirements of *clauses 2.3.2.2(3)* and *2.3.2.2(4)* will apply when calculating the deformation factor to be used for connections)	(e) As (d)
Second-order linear elastic analysis		
(a1) Calculate the global force distribution at the ULS. (*Clause 2.2.2(1)P*)	(a1) Design values of the modulus of elasticity modulus ($E_{0,mean}/\gamma_M$), shear modulus (G_{mean}/γ_M), slip modulus (K_{ser}/γ_M). The value of the partial material factor, γ_M, is obtained from *Table NA.3* in the National Annex to EN 1995-1-1	(a1) As (d)

SLS, serviceability limit state; ULS, ultimate limit state

[a]The term 'structures' when referred to in this table means any type of structure, including simply supported and continuous beams

[b]In this guide, the characteristic loading combination is used, in accordance with the requirements of *clause 2.2.3(2)*. Where an agreement can be reached between the designer and the client that a reversible limit state condition is applicable, the frequent combination given in Equation 6.15b in EN 1990 can be used instead of the characteristic combination

[c]Where variable loading is likely to result in the largest stress-to-strength ratio, a value $\psi_2 = 1$ can be used in the analysis, and the stresses obtained will not significantly differ from those obtained using the ψ_2 value associated with the highest stress-to-strength ratio

[d]The stiffness properties used in the creep deformation analysis are mean values divided by k_{def}, and to be able to combine the instantaneous and creep deformations in a single analysis, they are retained as mean values, and the quasi-permanent loading (used for the creep deformation) is multiplied by the k_{def} factor

Figure 5.4. Requirements for a simplified analysis

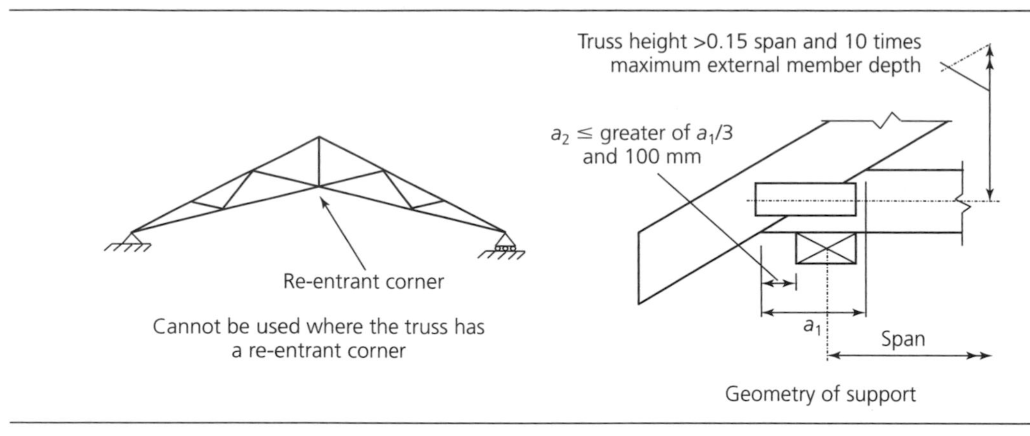

5.4.3 Simplified analysis of trusses with punched-metal plate fasteners

Clause 5.4.3

Clause 5.4.3 states that the simplified method of analysis can only be used for trusses formed with punched-metal plate fasteners, and gives conditions that must be complied with:

- there are no re-entrant angles in the external profile
- the bearing width is situated within the length a_1, and the distance a_2 in Figure 5.4 is not larger than $a_1/3$ or 100 mm, whichever is the greater
- the truss height is greater than 0.15 times the span and 10 times the maximum external member depth.

When using this method of analysis, the truss is modelled assuming all joints are pin jointed, and a first-order linear elastic analysis is carried out with the design loading applied **at the joint node positions**. The axial forces in the members are obtained from the analysis. Where members are subjected to moment and shear forces from lateral loading (e.g. the effect of permanent and applied variable loading on members), they are analysed as beams that are supported at their node positions and subjected to the lateral design load. If a member is only a single-bay length, it is designed as a simply supported beam, and, if continuous, as a continuous beam resting on simple supports at the node positions. To take into account the effect of deflection of nodes at support positions and partial fixity at connections, the moments in the members at positions of continuity over supports are reduced by 10%. These members are then validated against the combined moment and shear forces plus the axial force from the pin jointed analysis.

5.4.4 Plane frames and arches

Where structures are able to deform and by so doing generate additional internal force and instability effects, commonly referred to as second-order effects, these must be taken into account in the analysis. Framed structures that can sway laterally and arches come into this category.

In such situations it will be acceptable to analyse these types of structures using a first-order linear elastic analysis, taking second-order effects into account by subsequent additional analysis. Member strength validation can be carried out using the results from these analyses in conjunction with the relevant member design rules in EN 1995-1-1. An example of how second-order effects can be taken into account on this basis is given by Mortensen (1995). Alternatively, computer software based on elastic theory incorporating second-order effects can be used.

Clause 5.4.4
Clause 2.2.2(1)P
Clause 2.4.1(2)P

Where a second-order elastic analysis is to be undertaken, *clause 5.4.4* gives guidance on the initial out-of-alignment displacements and rotations that must be incorporated in the model, and, as required by *clause 2.2.2(1)P*, the stiffness properties have to be derived using the rules in *clause 2.4.1(2)P*. A summary of the design stiffness values and the loading conditions required for a second-order linear elastic analysis is given in Table 5.1.

Clause 5.4.4

The deviations from straightness to be used in the model are defined in *clause 5.4.4*, and referred to below. They are shown for a pitched portal frame and an arch in Figure 5.5, where the initial deviations in geometry are as follows:

Figure 5.5. Examples of assumed initial deviations in geometry for a second order analysis. Based on EN 1995-1-1 (*Figure 5.3*). Reproduced with permission from BS EN 1995-1-1, © British Standards Institution 2004

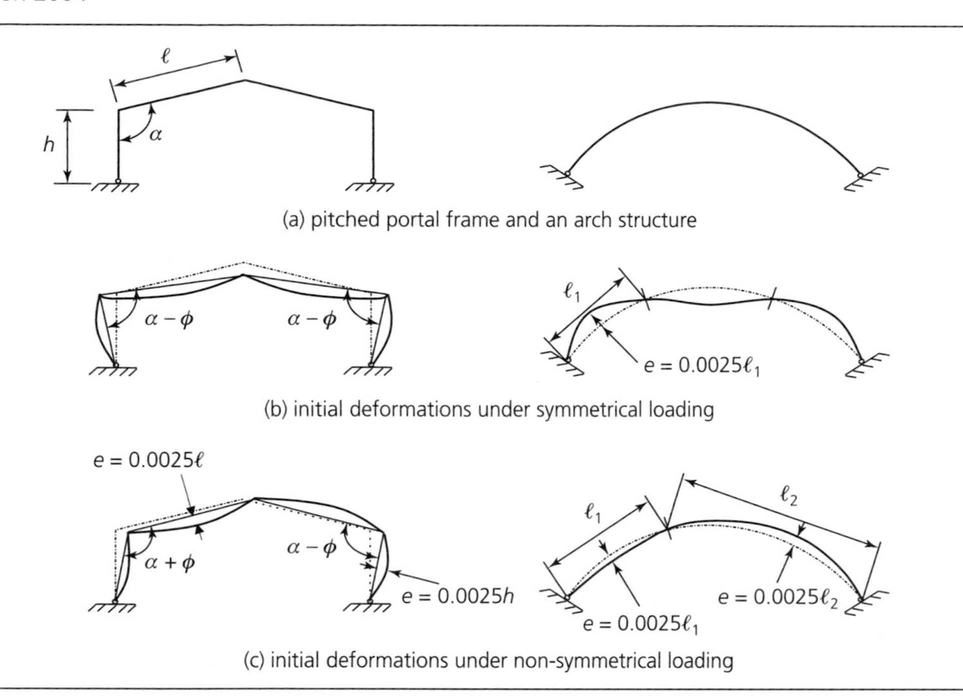

- An angle of inclination ϕ is applied to the structure, or relevant parts. In addition, a sinusoidal curvature is to be applied between the positions of zero bending moment in the structure, and the maximum eccentricity within the curved length will be e, where the value of e is given below.
- The value of ϕ should not be less than:

  ϕ 0.005 radians for $h \leq 5$ m
  ϕ $0.005 \times (5/h)^{0.5}$ radians for $h > 5$ m

  where h is the length of the member or the height of the structure in metres.
- The value of e should not be less than:

  $e \times 0.0025 \times \ell$

  where ℓ is the member length in metres, and is shown in Figure 5.5.

Where portal frames of one or more storeys are being used and are free to sway laterally, no guidance is currently given on the allowances to be made; however, it is suggested in this guide that incorporating the initial out-of-alignment displacements and rotations in the model as shown in Figure 5.6 should produce an acceptable result. The values for ϕ are as defined for Figure 5.5, and the initial out-of-alignment limits are based on the requirements of *clause 10.2*.

Clause 10.2

Figure 5.6. Initial imperfections assumed for a framed structure

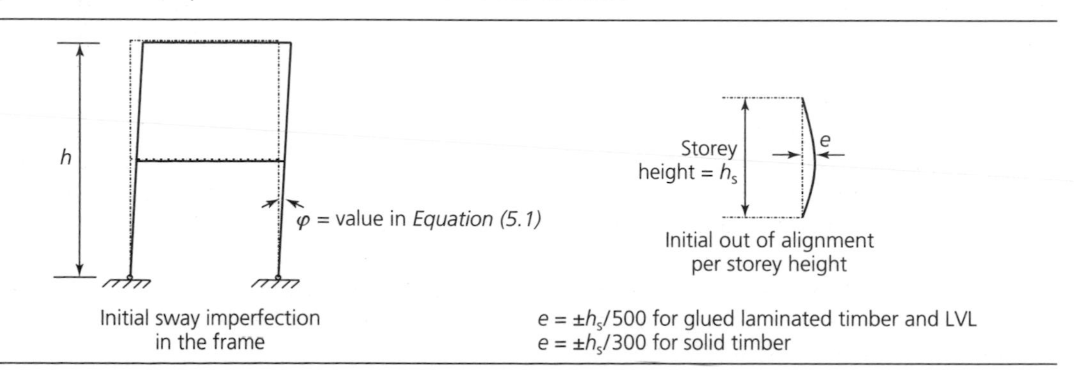

REFERENCES

Mortensen L (1995) Plane frames and arches. In *Timber Engineering: STEP 2* (Blass HJ, Aune P, Choo BS *et al.* (eds)). Centrum Hout, Almere.

Nielsen J (2003) Trusses and joints with punched metal plate fasteners. In *Timber Engineering* (Thelandersson S and Larsen HJ (eds)). Wiley, Chichester, ch. 19.

Porteous J and Kermani A (2007) *Structural Timber Design to Eurocode 5*. Blackwell, Oxford.

Designers' Guide to Eurocode 5: Design of Timber Buildings
ISBN 978-0-7277-3162-3

http://dx.doi.org/10.1680/dtb.31623.045

Chapter 6
Ultimate limit states

This chapter is concerned with the requirements for member design at ultimate limit states. It covers the material in *Section 6* of EN 1995-1-1, and addresses the following clauses:

- Design of cross-sections subjected to stress in one principal direction — *Clause 6.1*
- Design of cross-sections subjected to combined stresses — *Clause 6.2*
- Stability of members — *Clause 6.3*
- Design of cross-sections in members with varying cross-section or curved shape — *Clause 6.4*
- Notched members — *Clause 6.5*
- System strength — *Clause 6.6*

6.1. Design of cross-sections subjected to stress in one principal direction

6.1.1 General

The sign convention used in EN 1995-1-1 to define the principal axes is different to that used for timber design in accordance with British standards, and is used in this guide. In EN 1995-1-1, the *x–x* axis is the longitudinal axis, the *y–y* axis is the major principal axis (the strong axis) and the *z–z* axis is the minor principal axis (the weak axis), as shown in Figure 6.1. However, the UK practice of using a decimal point to display a fraction rather than the European practice of using a comma is followed in this guide. Although the stress convention used in the Eurocode shows all stresses as being positive, **in applying the rules, normal design practice will apply (e.g. tensile and compressive stresses will have opposite signs)**.

Material properties are obtained from the relevant normative references or other approved documents referred to in Chapter 3 of this guide, and design strengths will be calculated by following the rules in *Section 2*.

The theory used in the sub-clauses of *clause 6.1* only applies to members that have constant cross-section, whose grain runs essentially parallel to the length of the member and will only be subjected to stresses of one type under the effect of an applied action. For example, tension actions can only be applied along the centroidal axis of a member to ensure that a uniform tensile stress will occur. If the same action is applied eccentric to the axis, combined axial tension and bending stresses will arise, and the verification requirements for combined stress conditions will be as given in *clause 6.2*.

Clause 6.1

Clause 6.2

6.1.2 Tension parallel to the grain

The design tensile stress, $\sigma_{t,0,d}$, at a cross-section is obtained by dividing the design force at the stress position, $N_{t,0,d}$, acting along the centroidal axis of the member by the net cross-sectional area at that position, A:

$$\sigma_{t,0,d} = N_{t,0,d}/A \qquad \text{(D6.1)}$$

The net cross-sectional area is the cross-sectional area of the member less any cut-outs at the section as well as material removed to fit fasteners or connectors. The net area must be symmetrical about the *y–y* and *z–z* axes to prevent bending stresses, and the requirements for area loss associated with fasteners are given in *clause 5.2* and discussed in Section 5.2 of this guide.

Clause 5.2

Figure 6.1. Member axes

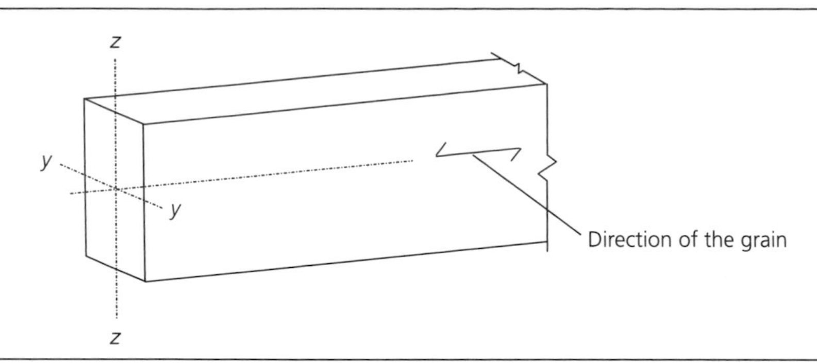

The mode of failure associated with the characteristic tension strength, $f_{t,0,k}$, will invariably be a brittle failure condition, and the design tensile strength, $f_{t,0,d}$, is derived from

$$f_{t,0,d} = k_{mod}k_{sys}k_h f_{t,0,k}/\gamma_M \qquad \text{(D6.2)}$$

where k_{mod}, k_h and γ_M are as defined in Chapters 2 and 3 in this guide, and k_{sys}, the system strength factor (which will only apply if the member forms part of a system), is discussed in *clause 6.6* and in Section 6.6 in this guide.

Clause 6.6

6.1.3 Tension perpendicular to the grain

When timber is stressed in this manner, it is generally as a consequence of other types of stress condition, and some examples of where tension stresses perpendicular to the grain will arise, together with the relevant clauses in this guide which discuss the design condition, are:

- Radial tensile stresses arising from bending moments in the apex of double tapered, curved and pitched cambered beams. The requirements for these types of situation are covered in Section 6.4 of this guide.
- Tension fracture caused by shear forces at notch positions, which is referred to in Section 6.5.2 of this guide.
- Tension forces leading to tension fracture of timber in connections, which is covered in Section 8.1.4 of this guide.

Where required, the design tensile strength perpendicular to the grain, $f_{t,90,d}$, is derived from the characteristic tension strength perpendicular to the grain, $f_{t,90,k}$, as follows:

$$f_{t,90,d} = k_{mod}k_{sys}f_{t,90,k}/\gamma_M \qquad \text{(D6.2a)}$$

where k_{mod}, k_{sys}, and γ_M are as referred to in Section 6.1.2.

6.1.4 Compression parallel to the grain

Clause 6.1.4
Clause 6.3.2(1)
Clause 6.3.2

The rules for compression parallel to the grain given in *clause 6.1.4* only apply to members that will not be affected by lateral instability. This will be the case when the relative slenderness ratio, λ_{rel}, about both axes, as defined in *clause 6.3.2(1)*, is ≤ 0.3. If $\lambda_{rel} > 0.3$ lateral instability can arise and the verification procedure given in *clause 6.3.2* must be followed.

Design compressive stress parallel to the grain, $\sigma_{c,0,d}$, is obtained by dividing the design force, $N_{c,0,d}$, acting along the centroidal axis of the member by the net cross-sectional area at the stress position, A:

$$\sigma_{c,0,d} = N_{c,0,d}/A \qquad \text{(D6.3)}$$

Clause 5.2

The net cross-sectional area will be the full cross-sectional area less any cut-outs as well as any material removed for fasteners and connectors, and to prevent bending stresses it has to be symmetrical about the y–y and z–z axes. Requirements for fasteners are given in *clause 5.2*, and discussed in Section 5.2 of this guide.

Figure 6.2. Member on (a) continuous and (b) discrete supports. (Reproduced with permission from EN 1995-1-1, © British Standards Institute, 2004)

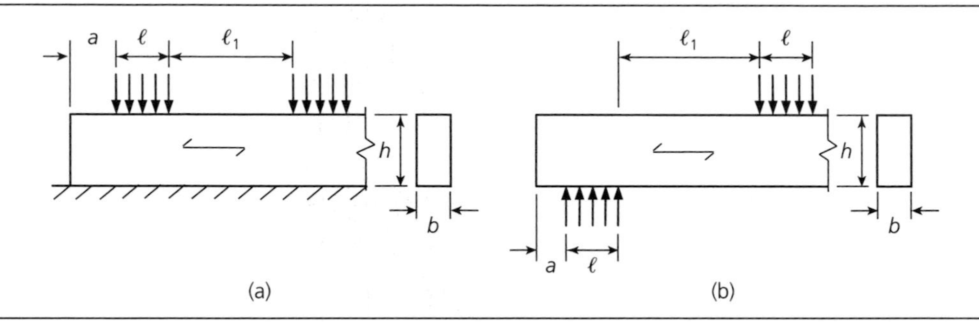

The characteristic compression strength parallel to the grain, $f_{c,0,k}$, will normally involve yielding of the material, and the design strength, $f_{c,0,d}$ is derived from

$$f_{c,0,d} = k_{mod}k_{sys}f_{c,0,k}/\gamma_M \tag{D6.4}$$

where k_{mod}, k_{sys} and γ_M are as referred to in Section 6.1.2.

6.1.5 Compression perpendicular to the grain

Clause 6.1.5

For this stress condition, *clause 6.1.5* requires that the design compression stress perpendicular to the grain, $\sigma_{c,90,d}$, does not exceed $k_{c,90}$ times the design compression strength perpendicular to the grain, $f_{c,90,d}$:

$$\sigma_{c,90,d} \leq k_{c,90}f_{c,90,d} \tag{6.3}$$

To obtain the design compression stress, $\sigma_{c,90,d}$, the design compressive force perpendicular to the grain, $F_{c,90,d}$, is divided by the effective contact area perpendicular to the grain, A_{ef}. The effective contact area is the product of the member width, b, and the effective contact length parallel to the grain, ℓ_{ef}, which can be considered to be the maximum length over which the compression stress is taken to act after dispersal within the member. It is dependent on the loading and support configuration, and is derived by increasing the actual contact length, ℓ (in mm), by 30 mm at each end, but by no more than a, ℓ or $\ell_1/2$, where the symbols are as shown in Figure 6.2. On this basis, the design bearing stress is independent of the type of material being stressed.

The design compressive strength perpendicular to the grain, $f_{c,90,d}$, is dictated by strain limitations at the serviceability limit state rather than strength, and is obtained from

$$f_{c,90,d} = k_{mod}k_{sys}f_{c,90,k}/\gamma_M \tag{D6.5}$$

where k_{mod}, k_{sys}, and γ_M are as referred to in Section 6.1.2.

Clause 6.1.5(3)
Clause 6.1.5(4)

The characteristic compressive strength of solid softwood timber and glued laminated timber loaded perpendicular to the grain, $f_{c,90,k}$, is derived from tests on short samples in accordance with EN 408 (BSI, 2010), where the samples are loaded uniformly across their bearing area and the strength is calculated at a compressive strain of approximately 3%. At this condition the factor $k_{c,90}$ is set equal to 1.0. For loading and support conditions that permit stress to disperse across the depth, the strain will be lower than that from the short sample uniform strain test set-up, and by allowing it to increase to the EN 408 requirement, the bearing strength will also increase. EN 1995-1-1 has adopted this approach, with strength being increased through the use of the factor $k_{c,90}$. Because the influence of the annual ring orientation when loaded perpendicular to the grain is less favourable for solid timber than glued laminated softwood timber, $k_{c,90}$ is smaller for solid timber (Blass and Gorlacher, 2004). For the conditions defined in *clauses 6.1.5(3)* or *6.1.5(4)* of EN 1995-1-1 as well as the relevant clauses included in PD 6693-1, the value of $k_{c,90}$ will be as given in Table 6.1. For all other conditions, $k_{c,90}$ should be taken to be 1.0. Reference should also be made to the content of Appendix A (*clause 6.1.5(4)*) in this guide for clarification of the code requirement for members on discrete supports.

Table 6.1. Values for $k_{c,90}$ for different support conditions

Loading condition (see Figure 6.2; h is the member depth)	Material	Value of factor $k_{c,90}$	Reference
Members on continuous supports, provided that $\ell_1 \geq 2h$ (Figure 6.2(a))	(a) Solid softwood timber (b) Glued laminated softwood timber	(a) 1.25 (b) 1.5	*Clause 6.1.5(3)* *Clause 6.1.5(3)*
Members on discrete supports, provided that $\ell_1 \geq 2h$ (Figure 6.2(b))	(c) Solid softwood timber (d) Glued laminated softwood timber and $\ell \leq 400$ mm	(c) 1.50 (d) 1.75	*Clause 6.1.5(4)* *Clause 6.1.5(4)*
Members on discrete supports supporting a uniformly distributed load	(e) Solid softwood timber (f) Glued laminated softwood timber	(e) 1.50 (f) 1.75	PD 6693-1 PD 6693-1
The bottom chord members of trusses reinforced with punched-metal plate fasteners over a support	(g) Solid softwood timber	(g) 1.5	PD 6693-1

Clause 6.1.5(3)
Clause 6.1.5(3)
Clause 6.1.5(4)
Clause 6.1.5(4)

Example 6.1: Calculation of the bearing stress and strength of a beam

A 50 mm (b) wide by 300 mm (h) deep softwood timber beam supports at mid-span another beam which applies a characteristic permanent load of 2.25 kN and a characteristic variable medium duration load of 3.35 kN. The 300 mm-deep beam has a clear span of 3.80 m and a bearing length of 100 mm (l) at each end, is of strength class C20 and functions in service class 2 conditions. Check the bearing strength of this beam is satisfactory. Properties are obtained from BS EN 338: 2009.

Effective beam span:

$$L = 3.8 + 0.1 = 3.9 \text{ m}$$

Self-weight of the beam:

$$W_k = bhg\rho_m = 50 \times 300 \times 9.81 \times 390 \times 10^{-9} = 0.06 \text{ kN/m}$$

Design load at each end reaction:

$$F_d = \gamma_G(G_k + W_k L)/2 + \gamma_Q Q_k/2 = 1.35 \times (2.25 + 0.22)/2 + 1.5 \times 3.35/2 = 4.18 \text{ kN}$$

Effective contact length at the support:

$$l_{ef} = l + 30 \text{ mm} = 100 + 30 = 130 \text{ mm}$$

Effective contact area perpendicular to the grain:

$$A_{ef} = bl_{ef} = 50 \times 130 = 6.5 \times 10^3 \text{ mm}^2$$

Design compression stress:

$$\sigma_{c,90,d} = F_d/A_{ef} = 4.18/6.5 = 0.64 \text{ N/mm}^2$$

The characteristic compression strength, $f_{c,90,k} = 2.3$ N/mm^2; $k_{c,90} = 1.50$ (see Table 6.1); from *Table 3.1*, $k_{mod} = 0.8$; and from the National Annex to EN 1995-1-1, $\gamma_M = 1.3$.

Design compression strength:

$$f_{c,90,d} = k_{mod} f_{c,90,k}/\gamma_M = 0.8 \times 2.3/1.3 = 1.42\ \text{N/mm}^2$$

Stress ratio:

$$\sigma_{c,90,d}/k_{c,90} f_{c,90,d} = 0.64/1.50 \times 1.42 = 0.30 < 1 \qquad \textit{(Equation 6.3)}$$

and thus OK.

6.1.6 Bending

Clause 6.1.6
Clause 6.3.3(2)
Clause 6.3.3

The rules for bending in *clause 6.1.6* only apply to members not affected by lateral torsional instability, which will be the case when the relative slenderness ratio for bending about the strong axis, $\lambda_{rel,m}$, as defined in *clause 6.3.3(2)*, is ≤ 0.75. Where $\lambda_{rel,m} > 0.75$, lateral torsional instability can arise, and the verification procedure given in *clause 6.3.3* must be followed.

When a member is subjected to a design moment, M_d, about its strong or weak axis the design bending stress, $\sigma_{m,d}$, will be

$$\sigma_{m,d} = M_d/W \qquad \text{(D6.6)}$$

where W is the section modulus about the axis of bending.

For this stress condition the verification requirement is $\sigma_{m,d} \leq f_{m,d}$, where $f_{m,d}$ is the design bending strength of the material:

$$f_{m,d} = k_{mod} k_{sys} k_h f_{m,k}/\gamma_M \qquad \text{(D6.7)}$$

where k_{mod}, k_{sys} and γ_M are as referred to in Section 6.1.2 and k_h is defined in Chapter 3. Under this stress state, failure will normally be a brittle tensile bending condition caused by defects in the tension zone of the member.

Where there is combined bending about the *y–y* and the *z–z* axes, *Equations 6.11* and *6.12* will apply. The verification rules in the Eurocode are based on elastic theory; however, depending on the materials and the cross-sectional profile being stressed, an element of yielding and stress redistribution is permitted at the maximum stress position. This is achieved through the use of the modification factor k_m. If the material is solid timber, glued laminated timber or laminated veneer lumber (LVL) and the cross-section is rectangular, $k_m = 0.7$. For all other cross-sections or for wood-based structural products (e.g. plywood or OSB) of any cross-section, no yielding is permitted, and $k_m = 1$.

6.1.7 Shear

The shear stress across the depth of any cross-section is taken to comply with elastic theory, and for a rectangular section subjected to a vertical shear force it will vary parabolically across the depth, with the greatest value occurring at the neutral axis position. For composite sections the shear stress is still calculated in accordance with elastic theory but the stress distribution will differ, and this condition is referred to in Sections 9.1.1 and 9.1.2 of this guide.

Because of changes in moisture content, cracks can develop along the grain direction, and to take account of the effect of this on the shear strength of the member, EN 1995-1-1 proposes that the member width is reduced. This is done by multiplying the member width, b, by a crack factor, k_{cr}, to obtain an effective width, b_{ef}. For members of solid rectangular section, width b and depth h, subjected to a design shear force, V_d, the design shear stress at the neutral axis, τ_d, when loaded as shown in Figure 6.3, will be

$$\tau_d = 1.5 V_d/k_{cr} bh \qquad \text{(D6.8)}$$

where V_d, is the design shear force on the member. To take advantage of the effect of the bearing stress at the support position, any loading within a bearing cone zone of 45° (as shown in

Figure 6.3. Conditions at a support for which concentrated loads may or may not be disregarded when calculating the end shear force

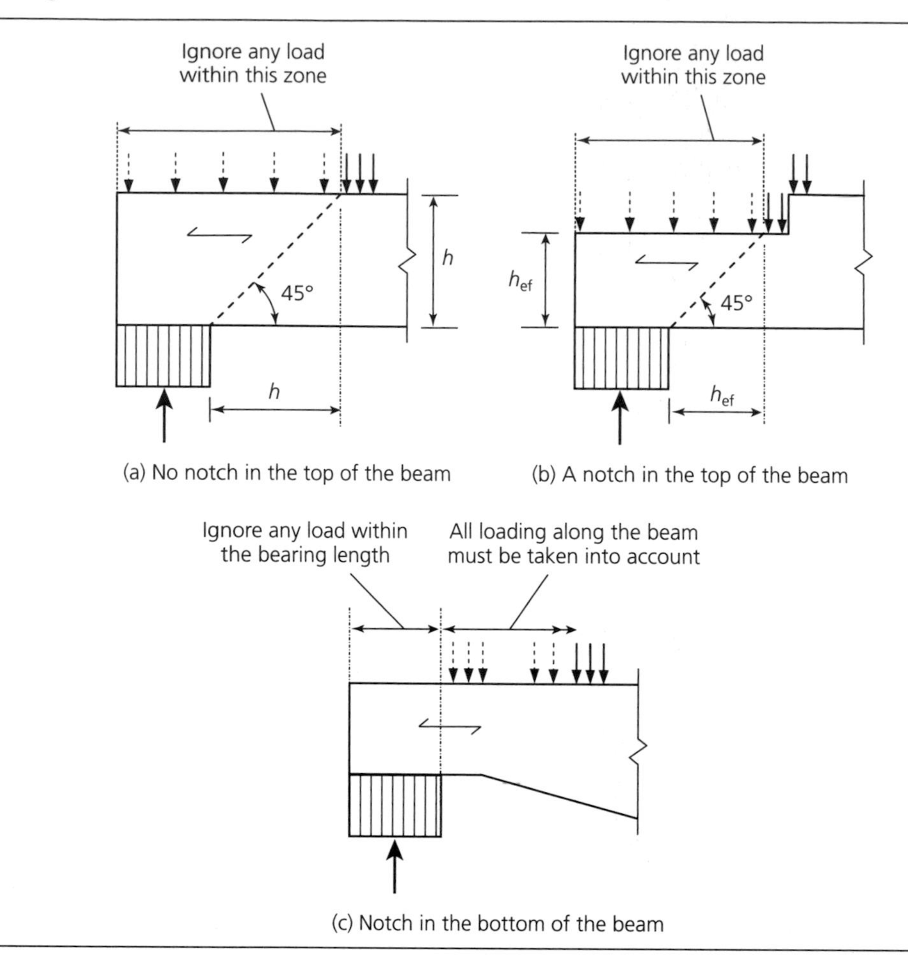

Figures 6.3(a) and 6.3(b)), which includes any load over the bearing area, can be ignored. Where there is a notch in the member and it is on **the same side as the support**, as shown in Figure 6.3(c), brittle failure due to tension fracture at the corner of the notch can occur, and the load reduction effect outside the bearing length is not allowed.

k_{cr} is dependent on the type of material being used, and for solid timber and glued laminated timber, $k_{cr} = 0.67$. For LVL sections compliant with EN 14374 and for wood-based products (e.g. plywood, OSB and particle board) compliant with EN 13986 (BSI, 2004), the splitting effect can be ignored, and $k_{cr} = 1.0$.

It is to be noted the above is a national choice item that has been accepted for use in the UK but not by all EU countries.

The verification condition to be satisfied is $\tau_d \leq f_{v,d}$, where $f_{v,d}$ is the design shear strength of the material and

$$f_{v,d} = k_{mod}k_{sys}f_{v,k}/\gamma_M \qquad \text{(D6.9)}$$

where $f_{v,k}$ is the characteristic shear strength and k_{mod}, k_{sys} and γ_M are as referred to in Section 6.1.2.

Clause 6.1.7(1)P

Where the shear stress is perpendicular to the grain as shown in *Figure 6.5(b)*, the member will be sheared in rolling shear, and for timber the characteristic shear strength in this direction is stated in *clause 6.1.7(1)P* to be '*approximately equal to twice the tensile strength perpendicular to grain*'.

6.1.8 Torsion

Clause 6.1.8

Torsional strength relationships for solid and rectangular sections subjected solely to torsion stresses are given in *clause 6.1.8*. With these sections the effect of warping can be ignored, and, when subjected to a design torsional moment, T_d, from the application of Saint-Venant torsion theory the design torsional stress, $\tau_{tor,d}$, and the angle of twist per unit length of the member, θ/ℓ, will be

solid circular section: $\tau_{tor,d} = 2T_d/\pi r^3 \quad \theta/\ell = 2T_d/\pi r^4 G_{mean}$

solid rectangular section: $\tau_{tor,d} = T_d/k_2 hb^2 \quad \theta/\ell = T_d/k_1 hb^3 G_{mean}$

where r is the radius of the circular cross-section; h and b are the cross-sectional sizes of the rectangular section (with h being the larger of the two dimensions); G_{mean} is the material shear modulus; and k_1 and k_2 are constants as determined by Timoshenko and Goodier (1951). Values for k_1 and k_2 for different ratios of h/b are given in Table 6.2, and approximate solutions can be obtained from

$k_1 = (1 - 0.63b/h)/3$ (when $h/b \geq 1.5$)

$k_1 = k_2$ (when $h/b \geq 2.5$)

With the rectangular section the maximum torsional stress will occur halfway along each side h.

The design condition to be satisfied is $\tau_{tor,d} \leq k_{shape} f_{v,d}$, where $f_{v,d}$ is the design shear strength of the material and

$$f_{v,d} = k_{mod} k_{sys} f_{v,k}/\gamma_M \quad \text{(D6.10)}$$

where k_{mod}, k_{sys} and γ_M are as referred to in Section 6.1.2.

Depending on the cross-sectional shape of the member, the torsional shear strength will be greater than the design shear strength, and is obtained by multiplying the design shear strength by a factor k_{shape}, defined in *Equation 6.15*. The values given in this equation for rectangular sections have been found to be incorrect, and the equation is to be revised as defined in Appendix A.

Torsional stress interacts with lateral shear stress, but no design verification rules are given in the Eurocode for a combined shear and torsion condition. In STEP lecture B4 (Aune, 1995), Equation D6.11 is proposed; however, the more conservative elastic stress combination given in Equation D6.12 in this guide should be able to be used:

$$\tau_{tor,d}/k_{shape} f_{v,d} + (\tau_{v,d}/f_{v,d})^2 \leq 1 \quad \text{(D6.11)}$$

$$\tau_{tor,d}/k_{shape} f_{v,d} + \tau_{v,d}/f_{v,d} \leq 1 \quad \text{(D6.12)}$$

where the functions are as previously defined, and the design shear stress will be the value of the shear stress occurring at the position of the design torsional stress.

Table 6.2. The deformation function (k_1) and stress function (k_2) for a rectangular section

h/b	k_1	k_2	h/b	k_1	k_2
1.0	0.1406	0.208	3.0	0.263	0.267
1.2	0.166	0.219	4.0	0.281	0.282
1.3	0.177	0.223	5.0	0.291	0.291
1.5	0.196	0.231	6.0	0.298	0.298
1.7	0.211	0.237	8.0	0.307	0.307
2.0	0.229	0.246	10.0	0.312	0.312
2.5	0.249	0.258	∞	0.333	0.333

6.2. Design of cross-sections subjected to combined stresses

6.2.1 General

Clause 6.2

The theory covered by the sub-clauses of *Clause 6.2* applies to members of constant cross-section that are subjected to stresses from combined actions or stresses acting in two or three of its principal axes.

Strength rules for combined stress conditions are based on linear elastic theory, making the assumption that linear interaction relationships will apply. However, where stress redistribution is permitted, a strength reduction factor or power function is incorporated in the combined strength relationship, enhancing the withstand capability. For combined axial compression and bending, yielding under compression stress is allowed, and the interaction relationship is based on a combination of plastic and elastic behaviour.

6.2.2 Compression stresses at an angle to the grain

Clause 6.2.2

The design of cross-sections subjected to compression stresses in two or more directions is referred to in *clause 6.2.2*. When a member of width b and depth h is loaded in compression by a design action F_{d} at an angle α to the grain, the design compression stress, $\sigma_{\mathrm{c},\alpha,\mathrm{d}}$, due to this load, as shown in Figure 6.4, will be

$$\sigma_{\mathrm{c},\alpha\mathrm{d}} = F_{\mathrm{d}} \cos(\alpha)/bh \tag{D6.13}$$

For this condition, the verification requirement is that $\sigma_{\mathrm{c},\alpha\mathrm{d}} \leq f_{\mathrm{c},\alpha,\mathrm{d}}$, where $f_{\mathrm{c},\alpha,\mathrm{d}}$ is the design compression strength based on the compression strengths referred to in Sections 6.1.4 and 6.1.5 in this guide and obtained from the strength criterion derived by Hankinson (1921) as defined in *Equation 6.16*, and is reproduced in Equation D6.14:

$$f_{\mathrm{c},\alpha,\mathrm{d}} \leq \frac{f_{\mathrm{c},0,\mathrm{d}}}{\dfrac{f_{\mathrm{c},0,\mathrm{d}}}{k_{\mathrm{c},90} f_{\mathrm{c},90,\mathrm{d}}} \sin^2 \alpha + \cos^2 \alpha} \tag{D6.14}$$

In Equation D6.14 it is suggested in this guide that $k_{\mathrm{c},90}$ should be taken as equal to 1.

6.2.3 Combined bending and axial tension

Clause 6.1.6

Clause 6.2.3

As the mode of failure under bending or axial tension will generally be a brittle failure condition, the verification requirement for combined bending and axial tension is based on elastic theory. However, with solid timber, glued laminated timber and LVL members of rectangular cross-section, an element of stress redistribution in bending is permitted, and is achieved by the use of the k_{m} factor referred to in *clause 6.1.6*. It takes into account the effect of inhomogeneities, and allows for stress redistribution and is applied to the secondary bending stress relationship, as shown in *Equations 6.17* and *6.18* in *clause 6.2.3*. For other cross-sections and for panel material, no stress redistribution is permitted.

Equations 6.17 and *6.18* check the tensile strength condition of the member. However, if there is a condition where lateral torsional instability can arise about the strong axis (i.e. $\lambda_{\mathrm{rel,m}} > 0.75$) the **compression bending strength condition** will also have to be checked. For this condition, in line with the stress convention used in EN 1995-1-1, it is suggested in this guide that, for a case where the tension and bending stresses arise from a common action, the following relationships can be used (remembering that the tensile and compression stresses will have opposite signs):

$$\sigma_{\mathrm{t},0,\mathrm{d}}/f_{\mathrm{t},0\mathrm{d}} + \sigma_{\mathrm{m,y,d}}/k_{\mathrm{crit}} f_{\mathrm{m,y,d}} + k_{\mathrm{m}} \sigma_{\mathrm{m,z,d}}/f_{\mathrm{m,z,d}} \leq 1 \tag{D6.15}$$

Figure 6.4. Compressive stresses at an angle to the grain

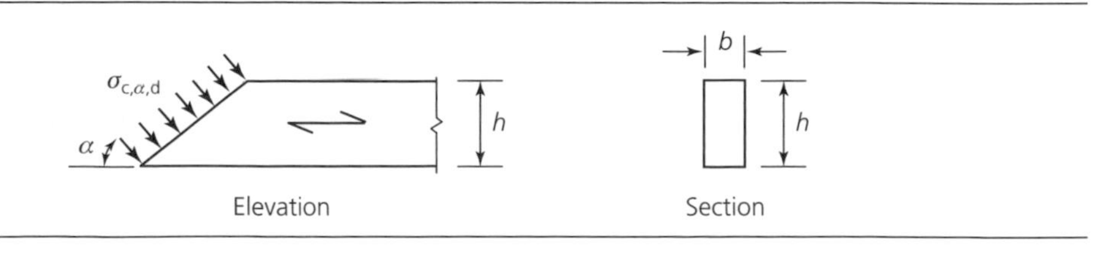

$$\sigma_{t,0,d}/f_{t,0d} + k_m\sigma_{m,y,d}/k_{crit}f_{m,y,d} + \sigma_{m,z,d}/f_{m,z,d} \leq 1 \quad \text{(D6.16)}$$

where k_{crit} is derived as described in Section 6.3.3, and bending and the axial tension stresses are determined using the methods given in Sections 6.1.6 and 6.1.2, respectively, in this guide.

Also, as stated in Appendix A, the tensile stress should be ignored and the member checked under bending using the design rules in Section 6.3, which would apply when the stresses arise from independent actions.

6.2.4 Combined bending and axial compression

This clause only applies to members that will not fail by instability under axial loading or by lateral torsional instability under bending, requiring the relative slenderness ratios for axial compression, $\lambda_{rel,y}$ and $\lambda_{rel,z}$, to be ≤ 0.3 and the relative slenderness ratio for bending, $\lambda_{rel,m}$, to be ≤ 0.75. Where these slenderness ratios exceed their limit, the verification requirements in *clause 6.3.2* and/or *6.3.3* must be used.

Clause 6.3.2
Clause 6.3.3

Because instability is prevented, advantage is taken of the ability to yield and redistribute stress under axial compression, enabling the combined axial and bending stress condition to be increased above that permitted by the application of elastic theory, as shown in *Equations 6.19* and *6.20*. For solid timber, glulam and LVL members of rectangular cross-section, an element of stress redistribution is also allowed in bending by the use of the k_m factor applied to the secondary bending stress relationship. This will not apply for other cross-sectional shapes or with panel material.

The methods for determining the bending and the axial compression stresses are given in Sections 6.1.6 and 6.1.4, respectively, in this guide.

6.3. Stability of members

6.3.1 General

The content of *clause 6.3* relates to the design of members that are affected by buckling instability. It applies where the design buckling strength under axial loading is less than the design compression strength referred to in Section 6.1.4 in this guide and/or the design lateral torsional strength under bending is less than the design bending strength referred to in Section 6.1.6.

Clause 6.3

Maximum deviations from straightness that are permitted for columns and beams in timber structures are given in *clause 10.2*, and, as stated in Section 5.2.1, there is no requirement for the designer to make allowance for these in the analysis process as their effect on strength and stability behaviour are implicitly incorporated in the design rules in EN 1995-1-1. A brief explanation of how this is achieved for axially loaded members and for members subjected to bending is given in Sections 6.3.2 and 6.3.3, respectively, in this guide.

Clause 10.2

The theory in *clauses 6.2.2* and *6.2.3* assumes that prismatic members are being analysed, covering the majority of the design situations that will arise in practice. If non-prismatic members are used, either conservative approximations have to be made by the designer to enable the rules to be applied or a second-order linear analysis can be undertaken.

Clause 6.2.2
Clause 6.2.3

Axial stability and lateral torsional stability strengths are functions of the member stiffness, and to calculate these the 5 percentile value of stiffness (e.g. $E_{0,05}$) rather than the mean stiffness value ($E_{0,mean}$) must be used. The 5 percentile value represents the statistical minimum value, while the mean stiffness is the average value of the stiffness property along the member length.

In deriving the stability strengths, *clause 6.3.1* states that the rules in *clause 6.3.2* apply to columns and those in *clause 6.3.3* apply to beams. There could, however, be cases where this generalised categorisation may not result in the most critical strength condition, and in such situations it is advisable for the designer to check the design stress against the requirements of both clauses.

Clause 6.3.1
Clause 6.3.2
Clause 6.3.3

Where lateral stability is provided by bracing members, the design procedure for the bracing members is covered in Chapter 9 of this guide.

6.3.2 Columns subjected to either compression or combined compression and bending

When a column is subjected to axial compression, its design compression strength will be the lesser of its axial compression strength and its axial stability strength (buckling strength), and the limit in EN 1995-1-1 above which the buckling strength will be less than the compression strength is when the relative slenderness ratio, λ_{rel}, about the buckling axis (i.e. the *y–y* and/or *z–z* axis) is greater than 0.3. For this condition, a column subjected to either compression or combined compression and bending must comply with the requirements of *Equations 6.23* and *6.24*, and where no moment exists they reduce to

$$\sigma_{c,0,d} \leq k_{c,y} f_{c,0,d} \qquad \text{(D6.17)}$$

$$\sigma_{c,0,d} \leq k_{c,z} f_{c,0,d} \qquad \text{(D6.18)}$$

where $\sigma_{c,0,d}$ is the design compression stress and $f_{c,0,d}$ is the design compression strength derived as described in Section 6.1.4.

Clause 10.2

The factors $k_{c,y}$ and $k_{c,z}$ are instability factors that convert the design compression strength of the member to its buckling strength, and are derived from *Equations 6.25–6.28*. These factors are dependent on the relative slenderness ratio of the member, λ_{rel}, and also take into account the effect of member defects; deviation from straightness (based on the limits stated in *clause 10.2*); and yielding across the section. This is achieved by the use of the β function, which is defined in *Equation 6.29*. For solid timber, $\beta = 0.2$, and for glued laminated timber and LVL, $\beta = 0.1$.

A typical plot of the instability factor k_c against λ_{rel} for C24 solid timber and GL 24h homogeneous glulam is shown in Figure 6.5, including for $\lambda_{rel} \leq 0.3$, over which the value of the instability factor will be 1.

The relative slenderness ratio for each axis is the ratio of the Euler buckling load (the elastic critical buckling load) of the member to its characteristic compression strength parallel to the grain, simplifying to the non-dimensional relationships given in *Equations 6.21* and *6.22*:

$$\lambda_{rel,y} \frac{\lambda_y}{\pi} \sqrt{\frac{f_{c,0,k}}{E_{0.05}}} \qquad \textit{(6.21)}$$

$$\lambda_{rel,x} \frac{\lambda_z}{\pi} \sqrt{\frac{f_{c,0,k}}{E_{0.05}}} \qquad \textit{(6.22)}$$

where λ_y is the slenderness ratio and $\lambda_{rel,y}$ is the relative slenderness ratio, both about the *y–y* axis (i.e. the member will deflect in the *z* direction); and λ_z is the slenderness ratio and $\lambda_{rel,z}$ is the relative slenderness ratio, both about the *z–z* axis (i.e. the member will deflect in the *y* direction).

Figure 6.5. Instability factor, k_c, for C24 timber ($\beta = 0.2$) and GL 24h glulam ($\beta = 0.1$)

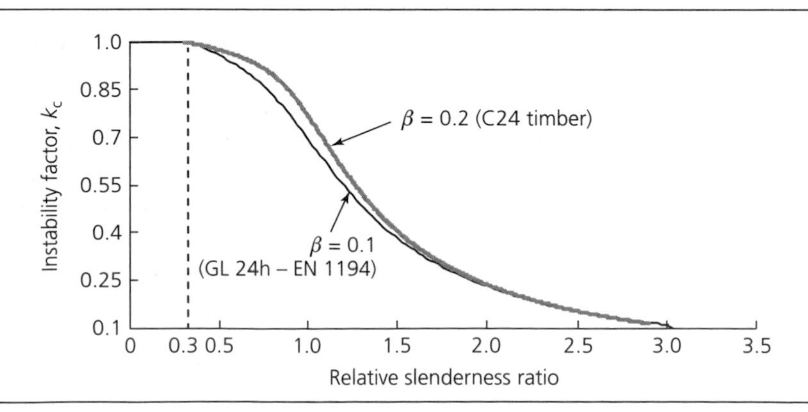

Figure 6.6. Effective length of compression members

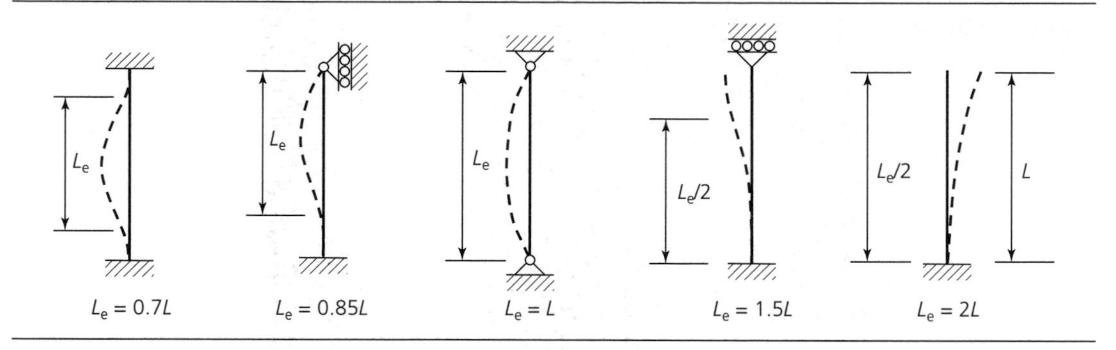

The slenderness ratio, λ, is the ratio of the effective length (buckling length), L_e, of the member divided by its radius of gyration, i, about the axis of bending. The effective length is the member length between adjacent points of zero bending moment, and, with the exception of compression members in trusses, EN 1995-1-1 gives no guidance on how this should be calculated. In the UK, recommended values for L_e have been given in BS 5268 for different support conditions, and this, with additional guidance, has been included in PD 6693-1. An extract from this document giving the effective lengths of compression members in frames subjected to the constraint conditions shown in Figure 6.6 is given in Table 6.3. It should be noted there will be some relaxation at rigid connections, and the values given in the table take this into account.

Values to be used for the effective length of members in trussed rafters can be obtained from PD 6693-1.

It may be more convenient for designers to derive the value of the instability factors using tabular data rather than by solving the Eurocode equations, and tables for timber, glulam and LVL members with slenderness ratios up to 240 from which k_c can be derived in this manner are given in Tables 6.8, 6.9 and 6.10 at the end of this chapter.

When a compression member is also subjected to bending, *Equations 6.23* and *6.24* will apply, but these assume the column will not fail by lateral torsion instability. This will only be the case when the relative slenderness ratio for bending, $\lambda_{rel,m}$, is ≤ 0.75, and for such a condition the compression stress and strength functions will be derived as described above, and the bending stress and strength functions will be obtained as defined in Section 6.1.6.

If $\lambda_{rel,m}$ is >0.75, the column has to be considered to function as a beam, and the requirements of Section 6.3.3 have to be followed.

When validating the strength at the ends of a column, buckling effects will not be relevant, and the requirements in Section 6.1.4 will apply.

Table 6.3. Effective length (L_e) of members of length (L) for the fixity conditions shown in Figure 6.6

Support condition at the ends of the member (see Figure 6.6)	Ratio L_e/L
Held effectively in position and direction at both ends	0.7
Held effectively in position at both ends and in direction at one end	0.85
Held effectively in position at both ends but not in direction	1.0
Held effectively in position and direction at one end and in direction but not position at the other end	1.5
Held effectively in position and direction at one end and completely free at the other end	2.0

Data from PD 6693-1-1

Example 6.2: calculation of a column with axial compression and bending about the *y–y* axis

A 150 mm (b) by 200 mm (h) column of strength class C20 is pin-jointed at each end and subjected to the design loading given below. The column is 4.50 m high (L), has its ends restrained in position but not in direction and functions in service class 2 conditions. Show that the column complies with the requirements of EN 1995-1-1 for the combined permanent and variable loading condition at the ultimate limit state. Timber properties are obtained from BS EN 338: 2009.

Design loading – combined

Permanent axial action, $N_d = 20.25$ kN; design variable axial action, $Q_d = 42$ kN; medium-term variable moment along the column about the y–y axis, $M_{y,d} = 2.49$ kN m; medium-term variable moment along the column about the z–z axis, $M_{z,d} = 1.56$ kN m.

Permanent loading condition

Axial stress:

$$\sigma_{c,0,d} = N_d/bh = 20.25 \times 10^3/150 \times 200 = 0.68 \text{ N/mm}^2$$

With $k_{mod,perm} = 0.6$, $\gamma_M = 1.3$, $f_{c,0,k} = 19$ N/mm^2, the design axial strength (*Equation D6.4*) will be

$$f_{c,0,d} = k_{mod,perm} f_{c,0,k}/\gamma_M = 0.6 \times 19/1.3 = 8.77 \text{ N/mm}^2$$

Including for instability effects: the effective length of column about each axis, $L_e = 1.0L = 4.5$ m (see Table 6.3), and the slenderness ratios will be

$$i_y = (I_y/bh)^{0.5} = ((bh^3/12)/bh)^{0.5} = (200^2/12)^{0.5} = 57.74 \text{ mm}$$

$$i_z = (I_z/bh)^{0.5} = ((hb^3/12)/bh)^{0.5} = (150^2/12)^{0.5} = 43.3 \text{ mm}$$

$$\lambda_y = L_e/i_y = 4500/57.74 = 77.9$$

$$\lambda_z = L_e/i_z = 4500/43.3 = 103.9$$

and

$$\lambda_{rel,y} = (\lambda_y/\pi)(f_{c,0,k}/E_{0,05})^{0.5} = (77.9/\pi)(19/6.4 \times 10^3)^{0.5} = 1.35$$

$$\lambda_{rel.z} = (\lambda_z/\pi)(f_{c,0,k}/E_{0,05})^{0.5} = (103.9/\pi)(19/6.4 \times 10^3)^{0.5} = 1.8$$

The instability factors, $k_{c,y}$ and $k_{c,z}$ about the y–y and z–z axes are obtained from *Equations 6.25–6.28* with $\beta = 0.2$:

$$k_y = 0.5(1 + \beta\,(\lambda_{rel,y} - 0.3) + \lambda^2_{rel,y}) = 0.5(1 + 0.2(1.35 - 0.3) + 1.35^2) = 1.52$$

$$k_z = 0.5(1 + \beta\,(\lambda_{rel,z} - 0.3) + \lambda^2_{rel,z}) = 0.5(1 + 0.2(1.8 - 0.3) + 1.8^2) = 2.27$$

$$k_{c,y} = 1/(k_y + (k_y^2 - \lambda^2_{rel,y})^{0.5}) = 1/[1.52 + (1.52^2 - 1.35^2)^{0.5}] = 0.45$$

$$k_{c,z} = 1/(k_z + (k_z^2 - \lambda^2_{rel,z})^{0.5}) = 1/[2.27 + (2.27^2 - 1.8^2)^{0.5}] = 0.27$$

The critical design condition from *Equations 6.23* and *6.24* will be

$$\sigma_{c,0,d}/k_{c,z} f_{c,0,d} = 0.68/0.27 \times 8.77 = 0.29 < 1$$

which is OK.

Combined permanent and variable loading condition
Axial stress:

$$\sigma_{c,0,d} = (N_d + Q_d)/bh = (20.25 + 42) \times 10^3/150 \times 200 = 2.08 \text{ N/mm}^2$$

With $k_{mod} = 0.8$, the design axial strength (*Equation D6.4*) will be

$$f_{c,0,d} = k_{mod} f_{c,0,k}/\gamma_M = 0.8 \times 19/1.3 = 11.69 \text{ N/mm}^2$$

Bending stresses about the y–y and z–z axes:

$$\sigma_{m,y,d} = M_{y,d}/W_y = 2.49 \times 10^6/(bh^2/6) = 2.49 \times 10^6/(150 \times 200^2/6) = 2.49 \text{ N/mm}^2$$

$$\sigma_{m,z,d} = M_{z,d}/W_z = 1.56 \times 10^6/(hb^2/6) = 1.56 \times 10^6/(200 \times 150^2/6) = 2.08 \text{ N/mm}^2$$

With $k_{mod} = 0.8$, $k_h = 1$ (both axes), $k_{sys} = 1$ and $f_{m,y,k} = 20$ N/mm^2 the design bending strength (*Equation 6.7*) will be

$$f_{m,y/z,d} = k_{mod} k_h k_{sys} f_{m,y,k}/\gamma_M = 0.8 \times 1 \times 1 \times 20/1.3 = 12.31 \text{ N/mm}^2$$

Check for lateral torsional instability about the y–y axis: the effective length of the column about this axis, $L_{ef} = 1.0L = 4.5$ m (see Table 6.4); $E_{0.05} = 6.4$ kN/mm^2; and the critical bending stress, $\sigma_{m,crit}$, (*Equation 6.32*) will be

$$\sigma_{m,crit} = 0.78b^2 E_{0.05}/(hL_e) = 0.78 \times 150^2 \times 6.4 \times 10^3/(200 \times 4500) = 124.8 \text{ N/mm}^2$$

$$\lambda_{rel.m} = (f_{m,k}/\sigma_{m,crit})^{0.5} = (20/124.8)^{0.5} = 0.4 < 0.75$$

so $k_{crit} = 1$.

Validation requirements for *Equations 6.23* and *6.24* and $k_m = 0.7$ (*clause 6.1.6*):

Clause 6.1.6

$$\sigma_{c,0,d}/k_{c,y} f_{c,0,d} + \sigma_{m,y,d}/f_{m,y,d} + k_m \sigma_{m,z,d}/f_{m,z,d} \leq 1 \qquad (\textit{Equation 6.23})$$

$$2.08/0.45 \times 11.69 + 2.49/12.31 + 0.7 \times 2.08/12.31 = 0.72 \leq 1$$

which is OK.

$$\sigma_{c,0,d}/k_{c,z} f_{c,0,d} + k_m \sigma_{m,y,d}/f_{m,y,d} + \sigma_{m,z,d}/f_{m,z,d} \leq 1 \qquad (\textit{Equation 6.24})$$

$$2.08/0.27 \times 11.69 + 0.7 \times 2.49/12.31 + 2.08/12.31 = 0.97 \leq 1$$

which is OK.

6.3.3 Beams subjected to either bending or combined bending and compression

When a beam is subjected to bending about its weak axis, the design strength will be the design bending strength as defined in Section 6.1.6. For a beam subjected to bending about the strong axis, the strength will be the lesser of its design bending strength and its critical bending strength (i.e. its lateral torsional buckling strength). The lateral torsional buckling strength is lower than the design bending strength and is a function of the relative slenderness ratio for bending of the beam, $\lambda_{rel,m}$. When $\lambda_{rel,m}$ is less than or equal to 0.75, the rules in EN 1995-1-1 state that lateral torsional instability will not occur, and for this condition the design bending strength in Section 6.1.6 will apply. When $\lambda_{rel,m}$ exceeds 0.75, the torsional instability strength is lower than the design bending strength, and the requirements of *clause 6.3.3* (*Equation 6.33*) have to be followed. For this condition, the validation requirement is

Clause 6.3.3

$$\sigma_{m,d} \leq k_{crit} f_{m,d} \qquad (6.33)$$

where $\sigma_{m,d}$ is the design bending stress and $f_{m,d}$ is the design bending strength derived as described in Section 6.1.6 in this guide.

The factor k_{crit} is an instability factor that converts the design bending strength of the member to its lateral torsional buckling strength, and is derived from the relationships given in *Equation 6.34*. It is dependent on $\lambda_{rel,m}$, and, unlike the instability factors for axial buckling, no function is included in the factor to take account of the effect of member defects including the effect of deviation from straightness. When $\lambda_{rel,m} < 0.75$, lateral torsional instability will not occur, and when $\lambda_{rel,m} > 1.4$, the beam is assumed to function elastically, and the critical buckling strength is derived in accordance with elastic buckling theory. For the condition $0.75 < \lambda_{rel,m} < 1.4$, the beam will fail in an inelastic manner, and the code adopts the approximation of a straight-line relationship between the above limits. A typical plot of the instability factor k_{crit} against λ_{rel} is shown in Figure 6.7, where it can be seen that when $\lambda_{rel} \leq 0.75$, k_{crit} will always be 1 and the design bending strength will apply. This will also be the case when lateral displacement of the compression face of the beam is prevented from lateral movement along its full length.

The relative slenderness ratio for bending, $\lambda_{rel,m}$, is defined in *Equation 6.30*, and is obtained from the characteristic bending strength, $f_{m,k}$, and the critical bending stress, $\sigma_{m,crit}$, as follows:

$$\lambda_{rel,m} \sqrt{\frac{f_{m,k}}{\sigma_{m,crit}}} \qquad (6.30)$$

$\sigma_{m,crit}$ is obtained by dividing the elastic critical moment, $M_{y,crit}$, by the section modulus of the beam, W_y, about the strong axis (y–y), and is defined in *Equation 6.31*. For a solid rectangular softwood timber beam, this expression has been simplified, and is given in *Equation 6.32*.

Figure 6.7. Instability factor k_{crit} plotted against the relative slenderness ratio for bending

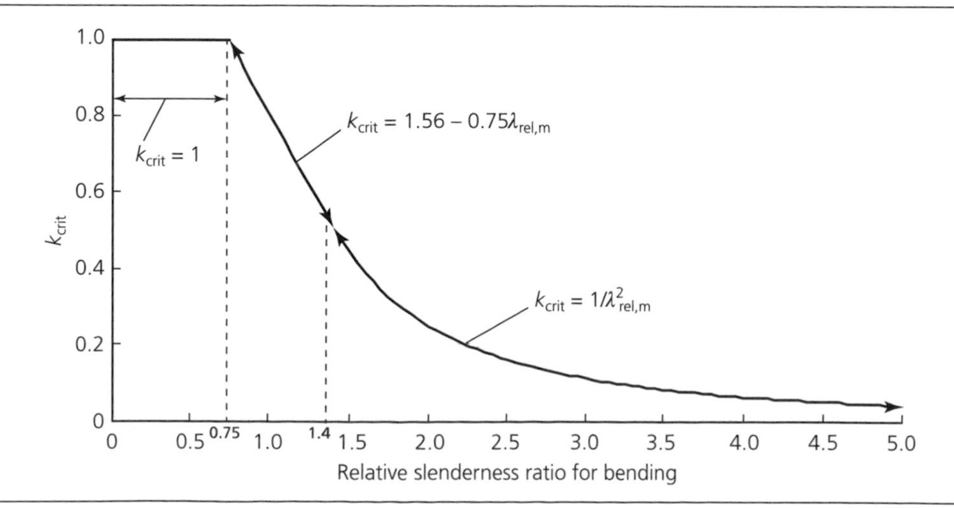

Table 6.4. Effective length, ℓ_{ef}, as a ratio of the design span of a beam

Beam end condition (restrained rotationally and laterally (out of plane movement) but free to rotate in plan – unless otherwise noted below)	Applied loading on a beam of depth h	ℓ_{ef}/ℓ (where ℓ is the design span unless noted below)[a]
Simply supported	Constant moment along the beam length	1.0
	Uniformly distributed load along beam	0.9
	Point load at mid-span of the beam	0.8
	Point loads at quarter and at three quarter positions on the beam	0.96
	Moment (M) at one end and ($-M/2$) at the other end of the beam	0.76
	Moment (M) at one end and no moment at the other end of the beam	0.53
Fully fixed at both ends	Uniformly distributed load along beam	0.78
	Point load at mid-span of the beam	0.64
Simply supported and torsionally and laterally restrained (out of plane) at mid-span	Point load at mid-span of the beam	0.56 (where the value used for denominator = $\ell/2$)
Cantilever (restrained as noted above and prevented from rotating in plan at the fixed end and free at the other end)	Uniformly distributed load along beam	0.5
	Point load at the free end of the beam	0.8

[a]The values for ℓ_{ef}/ℓ are valid when the beam load is applied at the level of the centroidal axis. If it is at the compression face, ℓ_{ef} must be increased by $2h$ and if it is at the tension face ℓ_{ef} may be reduced by $h/2$

When rectangular hardwood, glulam or LVL beams are used, *Equation 6.31* can be approximated to

$$\sigma_{m,crit} \frac{\pi b^2}{h\ell_{ef}} \sqrt{E_{0.05}G_{0.05}(1 - 0.63b/h)} \tag{D6.19}$$

where b and h are the beam width and depth, respectively; ℓ_{ef} is the effective length and $E_{0.05}$ and $G_{0.05}$ are the fifth percentile values of the modulus of elasticity parallel to the grain and the shear modulus, respectively. The effective length is the length between adjacent points of lateral support, and, from elastic buckling theory, for different loading and support conditions it can be expressed as a fraction of the design span, ℓ. Ratios of ℓ_{ef}/ℓ are given for a number of conditions in *Table 6.1*, and Table 6.4 gives this information together with ratios for other loading and support conditions.

Torsional instability will not occur if the beam length between adjacent lateral supports, l, is less than or equal to the effective length, ℓ_{ef}, when $\lambda_{rel,m} = 0.75$, and for solid rectangular timber beams or homogeneous glulam beams subjected to a uniform moment between adjacent positions of lateral support along their lengths, an approximate value for l for such a condition can be obtained from Figure 6.8. The figure is based on 25 mm-wide C22 solid timber to EN 338, and 90 mm-wide homogeneous glulam beams of strength class GL 24h to EN 1194, and a ratio of $E/G = 16$. For beams having a different width b, l will be obtained by multiplying the length from

Figure 6.8. Maximum beam length between lateral supports at which $k_{crit} = 1$ (for rectangular sections subjected to a uniform moment along the full length)

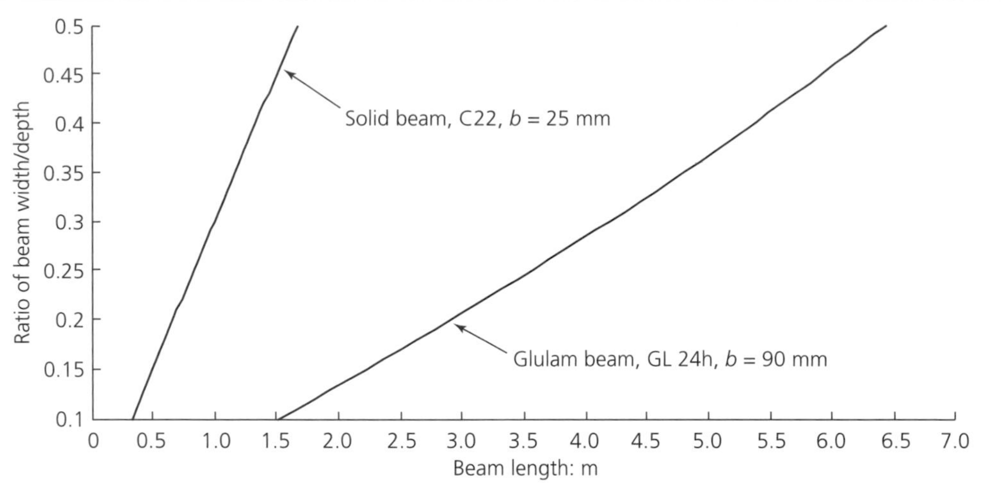

Table 6.5. Modification factor for rectangular timber and homogeneous glued laminated beams

Strength class to EN 338	C14	C16	C18	C20	C24	C27	C30
Factor	1.102	1.108	1.095	1.051	1.012	0.936	0.876
Homogeneous glued laminated beams to EN 1194	GL 28h	GL 32h	GL 36h				
Factor	0.93	0.886	0.844				

the graph by the ratio $b/25$ for a timber beam, and $b/90$ for a homogeneous glulam beam, where b is in mm. For other strength classes the length must be multiplied by the relevant factor given in Table 6.5, and, for alternative loading conditions, by the ℓ_{ef}/ℓ factor given in Table 6.4 and adjusted as noted at the foot of the table.

Example 6.3: use of Figure 6.8

A 100 mm (b) by 500 mm (h) deep GL 32h glued laminated beam functions in a floor structure. What is the maximum spacing for lateral bracing to permit the full bending strength of the beams to be used?

$b/h = 100/500 = 0.2$; from Figure 6.8, the GL 24h beam length ≈ 2.91 m, and applying the factor 100/90, length $= (10/9) \times 2.91 = 3.23$ m. For GL 32h, the factor in Table 6.5 is 0.886, so the spacing between bracing members $l \approx 3.23 \times 0.886$, say 2.86 m.

When beams are subjected to combined bending and axial compression, *Equation 6.35* will apply, but only for beams subjected to bending about the strong axis. Under this condition the compression stress and strength functions will be calculated as described in Section 6.3.2, and the bending stress and strength functions as defined above. If the relative slenderness ratio has a value approximating 0.75, the condition can also be considered to equate to that associated with *Equation 6.24* when there is no bending moment about the z–z axis, and, as the EN 1995-1-1 strength criterion is different for these conditions, it is suggested in this guide that the requirements of *Equations 6.24* and *6.35* should be satisfied.

No guidance is given in EN 1995-1-1 for the case where a beam is subjected to axial loading with bending about the strong and the weak axes.

Example 6.4: calculation of the lateral torsional buckling strength of a beam

A 100 mm wide (b) by 550 mm deep (h) GL 28c glued laminated beam, AB, supports another beam CD at mid-span as shown in Figure 6.9. Beam CD provides lateral restraint to beam AB at C (mid span of beam AB) and applies a vertical design load of 34.8 kN at the compression surface of the beam. Beam AB has an effective span of 6.85 m (L), and is restrained torsionally and laterally against out-of-plane movement at the end supports. The design load is combined permanent and medium-term variable loading, and the beam functions in service class 1 conditions. Confirm that the bending strength of beam AB will be acceptable.

Figure 6.9. Configuration of the beams in Example 6.4

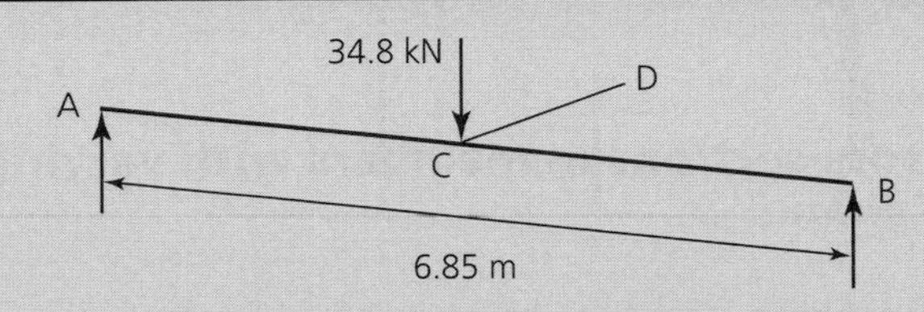

Effective beam length for bending (Table 6.4), ℓ_{ef}:

$$\ell_{ef} = 0.56(L/2) + 2h = 0.28 \times 6.85 + 2 \times 0.55 = 3.018 \text{ m}$$

From BS EN 1194:

$$E_{0,g,0.05} = 10.2 \text{ kN/mm}^2$$

$$G_{g,0.05} = (5/6) \times G_{g,mean} = (5/6) \times 0.72 = 0.6 \text{ kN/mm}^2 \text{ and } f_{m,g,k} = 28 \text{ N/mm}^2$$

So

$$\sigma_{m,crit} = \frac{\pi b^2}{h\ell_{ef}} \sqrt{E_{0.05}G_{0.05}(1 - 0.63b/h)} \qquad \text{(Equation D6.19)}$$

$$= (\pi \times 100^2/550 \times 3018) \times [10200 \times 600(1 - 0.63(100/550))]^{0.5} = 44.06 \text{ N/mm}^2$$

$$\lambda_{rel,m} = (f_{m,g,k}/\sigma_{m,crit})^{0.5} = (28/44.06)^{0.5} = 0.8 \qquad (\textit{Equation 6.30})$$

From *Equation 6.34*:

$$k_{crit} = 1.56 - 0.75 \times \lambda_{rel,m}$$

$$\lambda_{rel,m} = 1.56 - 0.75 \times 0.8 = 0.96$$

Design bending stress: adopting a mean density of beam $\rho_{g,m} = 410 \text{ kg/m}^3$, the design bending moment is

$$M_d = 34.8L/4 + \gamma_G \rho_{g,m} bhgL^2/8$$

$$= 34.8 \times 6.85/4 + 1.35 \times 0.410 \times 0.1 \times 0.55 \times 9.81 \times 6.85^2/8 = 61.35 \text{ kN m}$$

$$\sigma_{m,g,d} = M_d/(bh^2/6) = 61.35 \times 10^6/(100 \times 550^2/6) = 12.17 \text{ N/mm}^2$$

Design bending strength, $f_{m,g,d}$:

$$k_h = \min(1.1, (600/h)^{0.1}) = \min(1.1, (600/550)^{0.1}) = 1.01 \qquad (\textit{Equation 3.2})$$

$k_{mod} = 0.8$

$\gamma_M = 1.25$

So

$f_{m.g.d} = k_{mod}k_h f_{m,g,k}/\gamma_M = 0.8 \times 1.01 \times 28/1.25 = 18.1\ \text{N/mm}^2$

Validation requirement (*Equation 6.33*):

$\sigma_{m,g,d}/k_{crit}f_{m.g.d} = 12.17/0.96 \times 18.1 = 0.7 < 1$

which is OK.

6.4. Design of cross-sections in members with varying cross-section or a curved shape

6.4.1 General

The design rules for tapered beams apply to beams profiled from a solid cross-section (generally glulam or LVL) with the taper formed by cutting across the grain direction. The slope normally comes within the range 0–10°.

Clause 6.4.2
Clause 6.4.3
Clause 6.2
Clause 6.3

Rules for these types of beams are given in *clauses 6.4.2* and *6.4.3*, but they also have to satisfy the requirements of the relevant parts of *clauses 6.2* and *6.3*.

Clause 6.2.4
Clause 6.3.2
Clause 6.3.3

When subjected to an axial design force, the design stress is assumed to be uniform over the depth of the section, and the greatest value will occur where the section area is smallest. For combined bending and axial compression, the relevant relationships in *clauses 6.2.4, 6.3.2* and *6.3.3* – as discussed in Sections 6.2.4, 6.3.2 and 6.3.3 in this guide – will apply.

6.4.2 Single tapered beams

Because of the taper, when this type of beam is subjected to bending, the bending stress distribution across any section (e.g. A–A in Figure 6.10) varies non-linearly, and the bending stress parallel to the tapered face at the tapered face surface, $\sigma_{m,\alpha,d}$, will be smaller than the maximum bending stress at the horizontal face, $\sigma_{m,0,d}$. As the range of slope associated with these beams is normally small, in EN 1995-1-1 this difference in stress is ignored, and conventional bending theory is taken to apply. For a tapered beam as shown in Figure 6.10, of width b and subjected to a design moment M_d at any position x along the beam length, the design bending stress at the taper face, $\sigma_{m,\alpha,d}$, is taken to act horizontally and to equate in value to the design bending stress at the extreme horizontal face, $\sigma_{m,0,d}$. The design value of each stress is obtained from *Equation 6.37*:

$$\sigma_{m,\alpha,d} = \sigma_{m,0,d} = 6M_d/bh^2 \qquad (6.37)$$

where b is the beam width and h is the depth of the beam at the stress position.

Figure 6.10. Assumed stress distribution in a single tapered beam

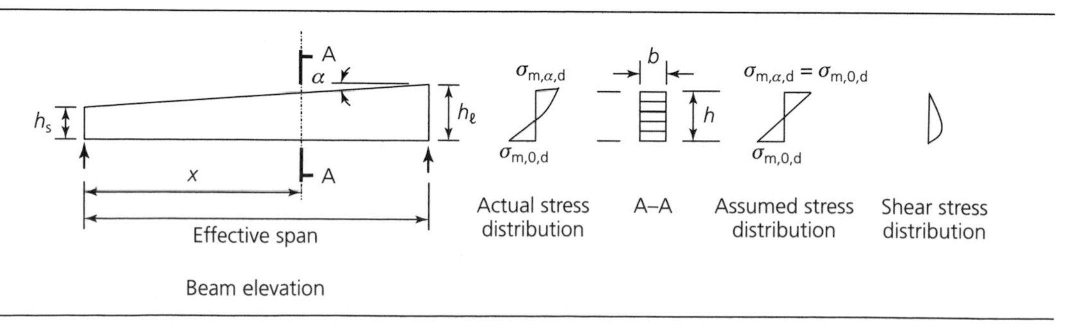

The position of maximum bending stress is dependent on the loading configuration and the beam profile, and for other than simple loading arrangements it is best derived by trial-and-error analysis along the beam length. For cases where the beam is simply supported and loaded by a uniformly distributed load or a point load, expressions for the maximum stress are given in *Structural Timber Design to Eurocode 5* (Porteous and Kermani, 2007).

The design condition to be satisfied in bending is given in *Equation 6.38*, and is

$$\sigma_{m,\alpha,d} \leq k_{m,\alpha} f_{m,d} \qquad (6.38)$$

where $f_{m,d}$ is the design bending strength, and is obtained as defined in Section 6.1.6 of this guide.

Because bending in tapered beams also induces shear stresses and stresses perpendicular to the grain, the strength criterion takes these into account through a reduction factor, $k_{m,\alpha}$, which is derived from *Equations 6.39* and *6.40*. If the stress parallel to the tapered edge is in tension, *Equation 6.39* applies, and if it is in compression, *Equation 6.40* applies. The design strengths for tension perpendicular to the grain ($f_{t,90,d}$), compression perpendicular to the grain ($f_{c,90,d}$) and shear ($f_{v,d}$) referred to in these equations are obtained as described in Sections 6.1.3, 6.1.5 and 6.1.7, respectively, in this guide.

Equation 6.38 only applies when lateral torsional instability will not arise. If lateral torsional instability can occur (i.e. $\lambda_{rel,m} > 0.75$), the instability factor, k_{crit}, referred to in Section 6.3.3, is applied as follows:

$$\sigma_{m,\alpha,d} \leq k_{crit} k_{m,\alpha} f_{m,d} \qquad (D6.20)$$

Where the beam is restrained laterally along its full length, then $k_{crit} = 1$, and where restraint is by spaced bracing members, a conservative value can be obtained by assuming that the beam has a uniform depth based on the deepest end h_ℓ.

The shear stress will vary across the beam depth and along its length, but, for design purposes, under normal loading conditions the maximum shear stress will be at the shallowest end, h_s, and the verification requirements will follow the rules given in Section 6.1.7 in this guide. Should there be a need to check the shear stress at a position along the beam length, the methodology proposed by Maki and Keunzi (1965) can be used.

6.4.3 Double-tapered, curved and pitched cambered beams

The design of tapered, curved and pitched cambered beams is referred to in *clause 6.4.3*, and only applies to beams made from glulam and LVL.

Clause 6.4.3

Where the beams are made from single-taper sections connected by a central apex zone, the design of the taper section must comply with the requirements of *clause 6.4.2*. The rules in *clause 6.4.3* apply to verification of the design within the apex zone, which is the hatched area shown for each beam type in *Figure 6.9*.

Clause 6.4.2
Clause 6.4.3

The design condition will arise when positive bending, as shown in Figure 6.11, is applied to the section, and the fibre direction is parallel to the lower edge of the beam. Under this condition, bending and radial stresses will be generated across the apex zone, and will be greatest at the

Figure 6.11. Curved, double-tapered and pitched cambered beams subjected to positive bending

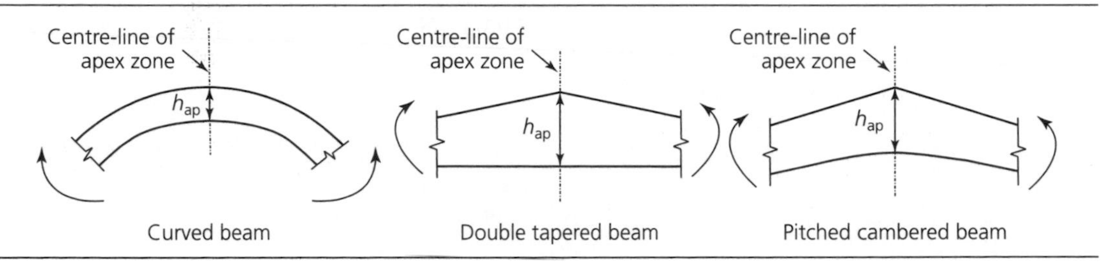

Table 6.6. Stress and strength equations for double-tapered, pitched cambered and curved beams

Clause 6.4.3

	Double tapered and pitched cambered beams (equations in *clause 6.4.3*)	Curved beams (equations in *clause 6.4.3*)
Bending stress	*Equation 6.42*	*Equation 6.42* with $\alpha_{ap} = 0^b$
Radial tensile stress	*Equation 6.54*[a]	*Equation 6.54*[a] with $\alpha_{ap} = 0^b$
Bending strength condition	*Equation 6.41* and: $f_{m,d}$ is derived as described in Section 6.1.6 in this guide; k_r will be: 1 for double tapered beams; obtained from *Equation 6.49* for pitched cambered beams	*Equation 6.41*, and k_r will be obtained from *Equation 6.49*
Tensile strength condition	*Equation 6.50* and: $k_{dis} = 1.4$ for double tapered beams; $k_{dis} = 1.7$ for pitched cambered beams; $V_0 = 0.01\ \text{m}^3$; $V =$ min (volume of the apex zone) or 2 ((volume of the beam)/3)	*Equation 6.50*, and: $k_{dis} = 1.4$ for curved beams; $V_0 = 0.01\ \text{m}^3$; $V =$ min (volume of the apex zone) or 2 ((volume of the beam)/3)

Clause NA.2.4

[a] In accordance with the requirements of *clause NA.2.4* in the National Annex to EN 1995-1-1
[b] Where angle α_{app} is as defined in *Figure 6.9*

Clause 6.4.3

centre-line position. They are defined by the equations given in *clause 6.4.3* and summarised in Table 6.6.

As with single tapered beams, under normal loading conditions the maximum shear stress will be at the support position and will be determined by following the rules given in 6.1.7 in this guide. Where, however, there are shear stresses in the apex zone there will be an interaction with the tensile stress perpendicular to the grain and the combined strength relationship given in *Equation (6.53)* must be complied with.

Example 6.5: Calculation of a curved beam in the apex zone

A 200 mm (b) thick by 1200 mm deep (h) glued laminated beam comprises two straight lengths and a curved central apex zone, all of uniform section along its length. It forms part of a roof structure, and is laterally braced along its compression flange. It is strength class GL 32h in accordance with BS EN 1194, made from 40 mm thick laminations (t) and supports a design load due to permanent and short-term snow loading, including its self-weight, $q_d = 6.50$ kN/m, as shown in Figure 6.12. The bearing length at each end is 290 mm. For the design loading condition given below, which includes an allowance for the self-weight of the beam, confirm that the beam will comply with the design rules in EN 1995-1-1 at the ultimate limit states. Service class 2 conditions apply.

Shear at support: the maximum shear stress will be at the beam end.

Clause 6.1.7(3)

Design shear force, V_d (ignoring load reduction allowed in *clause 6.1.7(3)*):

$$V_d = q_d L \cos\beta/2 = 6.50 \times 25 \times \cos(10^\circ)/2 = 80.02\ \text{kN}$$

Design shear stress at each end of the beam, $\tau_{v,d}$:

Clause 6.1.7(2)

$$k_{cr} = 0.67 \qquad (\textit{clause 6.1.7(2)})$$

$$\tau_{v,d} = 1.5V_d/k_{cr}bh = 1.5 \times 80.02 \times 10^3/0.67 \times 200 \times 1200 = 0.75\ \text{mm}^{-2}$$

Figure 6.12. Roof structure in Example 6.5

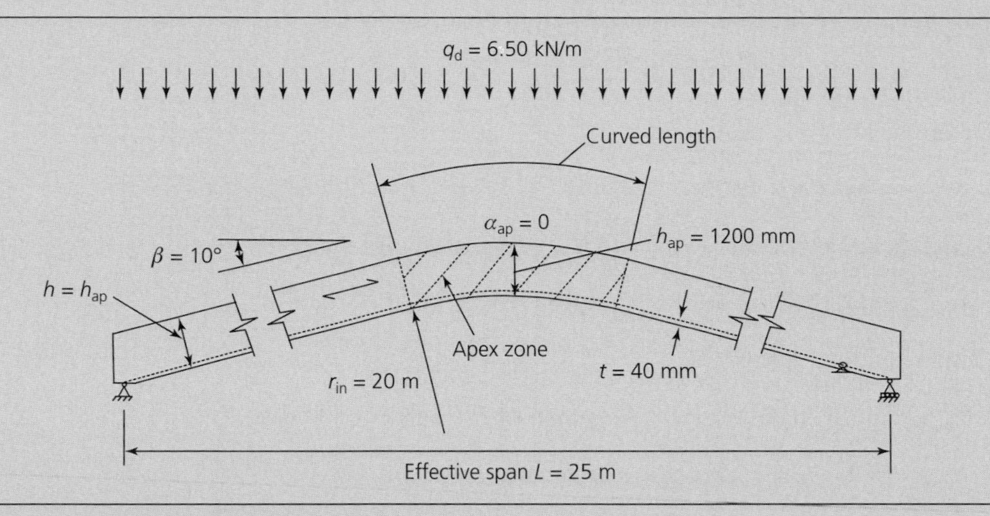

Design shear strength, $f_{v,g,d}$:

$$k_{mod} = 0.9,\ \gamma_M = 1.25,\ f_{v,g,k} = 3.2\ \text{N/mm}^2$$

$$f_{v,g,d} = k_{mod} f_{v,g,k}/\gamma_M = 0.9 \times 3.2/1.25 = 2.30\ \text{N/mm}^2$$

So the beam is OK in shear.

$$\tau_{v,d}/f_{v,g,d} = 0.75/2.3 = 0.32 < 1$$

which is OK.

Bearing at support: bearing length, $b_l = 290$ mm; end reaction $= V_d$.

Ignoring any enhancement of the bearing area permitted by *clause 6.1.5(1)*:

Clause 6.1.5(1)

Bearing stress:

$$\sigma_{c,\beta,d} = V_d/bb_l = 80.02 \times 10^3/200 \times 290 = 1.38\ \text{N/mm}^2$$

Bearing strength, $f_{c,\beta,g,d}$, will be a function of the bearing strength parallel and perpendicular to the grain (*Equation 6.16*): $f_{c,0.g.k} = 26.5\ \text{N/mm}^2$; $f_{c,90.g.k} = 3.0\ \text{N/mm}^2$; and taking $k_{c,90} = 1$.

Compression strength parallel to grain:

$$f_{c,0,g,d} = k_{mod} f_{c,0,g,k}/\gamma_M = 0.9 \times 26.5/1.25 = 19.08\ \text{N mm}^2$$

Bearing strength perpendicular to grain, $f_{c,90,g,d} = k_{mod} f_{c,90,g,k}/\gamma_M$:

$$f_{c,90,g,d} = 0.9 \times 3.0/1.25 = 2.16\ \text{N mm}^2$$

Design bearing strength, $f_{c,\beta,g,d}$:

$$f_{c,\beta,g,d} = f_{c,0,g,d}/[(f_{c,0,g,d}/k_{c,90} f_{c,90,g,d})\sin^2(90^\circ - \beta) + \cos^2(90^\circ - \beta)]$$

$$= 19.08/[(19.08/1.0 \times 2.16)\sin^2(80^\circ) + \cos^2(80^\circ)] = 2.22$$

$$\sigma_{c,\beta,d}/f_{c,\beta,g,d} = 1.38/2.22 = 0.62 < 1$$

which is OK.

Bending strength in the apex zone: from *Equations 6.44–6.48*, $k_1 = 1$, $k_2 = 0.35$, $k_3 = 0.6$, $k_4 = 0$, $r = r_{in} + 0.5h_{ap} = 20.6$ m; and from *Equation 6.43*

$$k_\ell = k_1 + k_2(h_{ap}/r) + k_3(h_{ap}/r)^2 + k_4(h_{ap}/r)^3 = 1 + 0.35(1.2/20.6) + 0.6(1.2/20.6)^2 = 1.022$$

Design bending moment:

$$M_{ap,d} = q_d L^2/8 = 6.50 \times 25^2/8 = 507.81 \text{ kN m}$$

Design bending stress, $\sigma_{m,g,d} = k_\ell 6M_{ap,d}/bh_{ap}^2$ (*Equation 6.42*):

$$\sigma_{m,g,d} = 1.022 \times 6 \times 507.81 \times 10^6/200 \times 1200^2 = 10.81 \text{ N/mm}^2$$

Design bending strength, $f_{m,g,d}$:

$$k_h = \min(1.1, (600/h)^{0.1}) = 1 \text{ (\textit{Equation (3.2)} when } h > 600 \text{ mm)}$$

$$k_{mod} = 0.9,\ \gamma_M = 1.25,\ f_{m,g,k} = 28 \text{ N/mm}^2$$

$$f_{m.g.d} = k_{mod} k_h f_{m,g,k}/\gamma_M = 0.9 \times 1 \times 28/1.25 = 20.16 \text{ N/mm}^2$$

Validation requirement (*Equation 6.41*): from *Equation 6.49*, as $r_{in}/t = 20 \times 1000/40 = 500$, $k_r = 1$ and the beam is laterally restrained along its length, $k_{crit} = 1$,

$$\sigma_{m,g,d}/k_{crit} k_r f_{m.g.d} = 10.81/20.16 = 0.54 < 1$$

which is OK.

Radial strength in the apex zone: from *Equations 6.57–6.59*, $k_5 = 0$, $k_6 = 0.25$ and $k_7 = 0$; and from *Equation 6.56*

$$k_p = k_5 + k_6(h_{ap}/r) + k_7(h_{ap}/r)^2 = 0.25(1.2/20.6) = 0.01456$$

Design tensile stress perpendicular to the grain due to $M_{ap,d}$, $\sigma_{t,g,90,d} = k_p 6M_{ap,d}/bh_{ap}^2$ (*Equation 6.54*):

$$\sigma_{t,g,90,d} = 0.01456 \times 6 \times 507.81 \times 10^6/200 \times 1200^2 = 0.154 \text{ N/mm}^2$$

Design tensile strength perpendicular to the grain, $f_{t,g,90,d}$:

$$k_{mod} = 0.9;\ \gamma_M = 1.25;\ f_{t,g,90,k} = 0.45 \text{ N/mm}^2$$

$$f_{t.g.90,d} = k_{mod} f_{t,g,90,k}/\gamma_M = 0.9 \times 0.45/1.25 = 0.32 \text{ N/mm}^2$$

From *Equations 6.51* and *6.52*: reference volume, $V_0 = 0.01$ m^3.

Volume of the apex zone of the beam (see Porteous and Kermani, 2007), $V = (\beta/180)\pi b(h_{ap}^2 + 2r_{in}h_{ap})$:

$$V = (10/180) \times \pi \times 0.2(1.2^2 + 2 \times 20 \times 1.2) = 1.73 \text{ m}^2$$

Volume of the beam (approximate), $V_b = V + 2bh_{ap}(L/2 - (r_{in} + h_{ap}/2)\sin\beta))/\cos\beta$:

$$V_b = 1.73 + 2 \times 0.2 \times 1.2[25/2 - (20 + 1.2/2) \times \sin(10°)]/\cos(10°) = 6.07 \text{ m}^3$$

Ratio $V/V_b = 1.73/6.07 = 0.28$, which is less than 0.67 and thus OK. From *Equations 6.51* and *6.52*, $k_{vol} = (V_0/V)^{0.2} = (0.01/1.73)^{0.2} = 0.36$ and $k_{dis} = 1.4$.

Validation requirement (*Equation 6.50*):

$$\sigma_{t,g,90,d}/k_{dis} k_{vol} f_{t.g.90,d} = 0.154/1.4 \times 0.36 \times 0.32 = 0.95 < 1$$

which is OK.

6.5. Notched beams

6.5.1 General

When a member with a notch formed within its length is subjected to bending, stress concentrations will occur at the notch position. If the notch is subjected to bending tension stresses (Figure 6.13(a)) and the slope of the taper at the notch is $<1:10$, or it is subjected to bending compression stresses (Figure 6.13(b)), the effect can be ignored.

Where it is subjected to bending tension stresses and the slope of the taper at the notch is $\geq 1:10$ the effect must be taken into account to prevent failure due to the initiation of tension fracture at the inset corner of the notch. Although the verification equation (*Equation 6.60*) for this condition is given in terms of shear strength, the critical failure condition will change from shear to tension perpendicular to the grain, and this is achieved in the equation by the k_v factor referred to in Section 6.5.2.

6.5.2 Beams with a notch at the support

The design procedure for a notched beam only applies when the beam is rectangular in section and the grain runs essentially parallel to its length.

For this condition, the design is validated against the shear stress calculated using the effective depth, h_{ef}, shown in Figure 6.14. Although not referred to, the k_{cr} in *clause 6.1.7* will apply (see Appendix A), and, under a design shear force, V_d, the design shear stress, τ_d, in a beam of width b will be obtained from

Clause 6.1.7

$$\tau_d = 1.5V_d/k_{cr}bh_{ef} = 1.5V_d/b_{ef}h_{ef} \tag{D6.21}$$

and the verification requirement for this condition is

$$\tau_d \leq k_v f_{v,d} \tag{6.60}$$

Figure 6.13. Bending at a notch

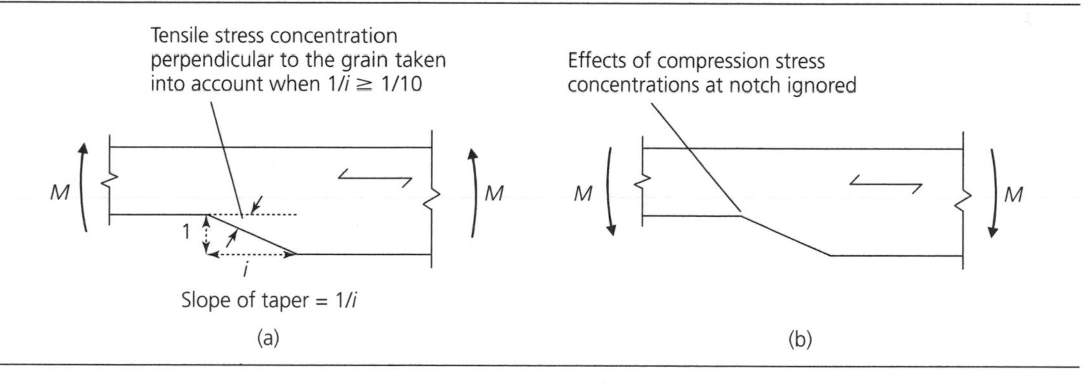

Figure 6.14. End notched beams. (Based on EN 1995-1-1 (*Figure 6.11*). Reproduced with permission from BS EN 1995-1-1, © British Standards Institute, 2004)

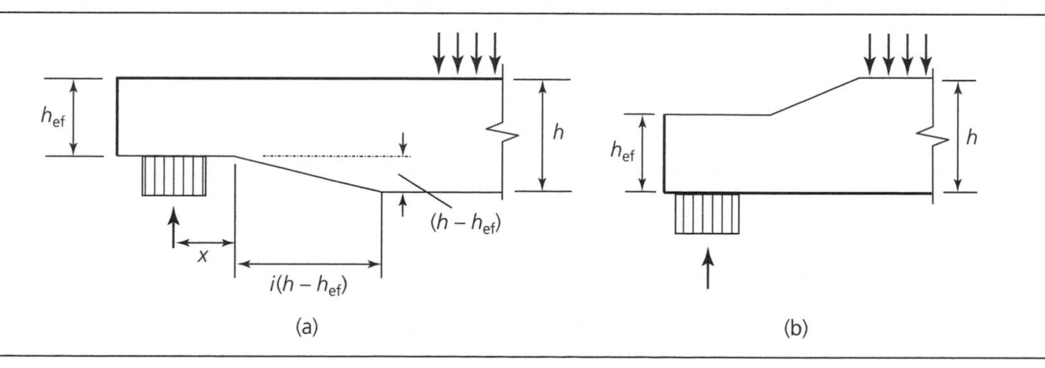

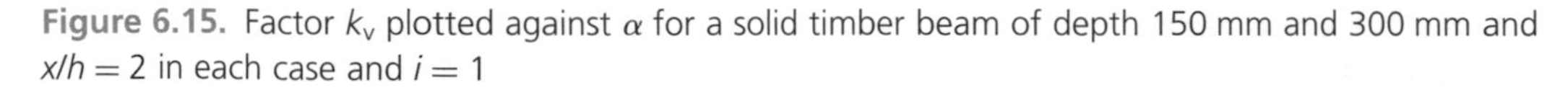

Figure 6.15. Factor k_v plotted against α for a solid timber beam of depth 150 mm and 300 mm and $x/h = 2$ in each case and $i = 1$

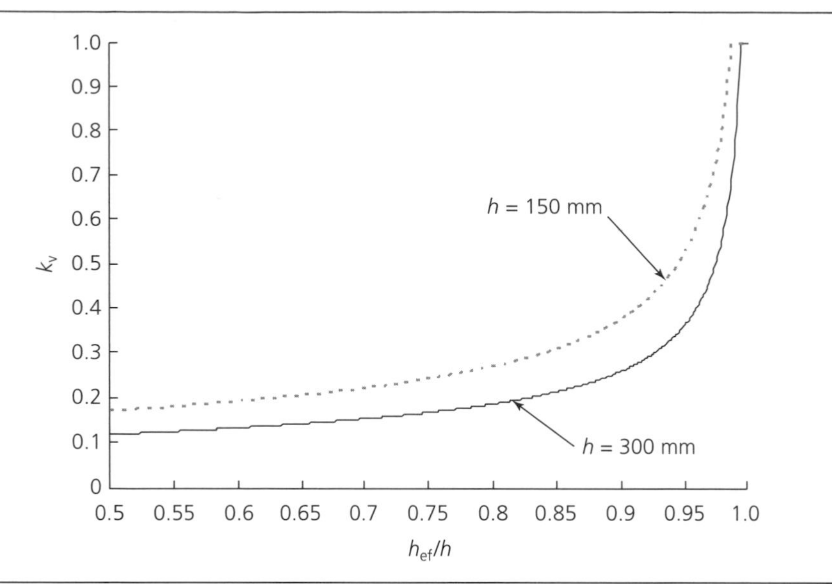

where $f_{v,d}$ is the design shear strength of the material, defined in Section 6.1.7 in this guide, and b_{ef} is as defined in Section 6.1.7 in this guide.

The factor k_v is a reduction factor that takes into account which side the beam is notched relative to the beam support and the geometry of the beam in the area of the notch. It is also a function of the fracture strength of the beam material, and the value to be used in design is:

- for beams notched on the side opposite to the support (see Figure 6.14(b)), $k_v = 1.0$
- for beams of depth h notched on the same side as the support (see Figure 6.14a)

$$k_v = \min(1, k_n(1 + 1.1i^{1.5}/h^{0.5})/\{h^{0.5}[(\alpha(1-\alpha))^{0.5} + 0.8x(1/\alpha - \alpha^2)^{0.5}/h]\}) \qquad \text{(D6.22)}$$

where $\alpha = h_{ef}/h$; x is the distance from the centroid of the support reaction to the notch corner, in mm; i is the notch inclination; and $k_n = 4.5$ for LVL, 5.0 for solid timber and 6.5 for glulam.

As structured, *Equation 6.60* gives the impression that failure will always be a shear condition relating solely to the shear strength, $f_{v,d}$. However, shear failure will only occur when $k_v = 1$. When k_v is less than 1, failure will be due to tension fracture initiated at the corner of the notch, and for this condition the fracture energy in tension perpendicular to the grain is included within the k_n factor. The k_v factor is greatly influenced by the value of α, and a plot of k_v against α for $i = 1$, $x/h = 2$, $h = 300$ mm and 150 mm for a solid timber beam is shown in Figure 6.15.

Values for k_v for solid timber beams with different notch depths on the same side as the support, with $i = 0$ and with different values of x/h, are given in Table 6.7. If glulam or LVL is used, the k_v value from the table should be multiplied by 1.3 and 0.9, respectively.

Table 6.7. Values of k_v for solid timber beams of $h = 150$–300 mm and $x/h = 0.5$ to 2.0 with $i = 0$

$x/h = 0.5$	$\alpha = 0.5$	$\alpha = 0.6$	$\alpha = 0.7$	$\alpha = 0.8$	$\alpha = 0.9$	$\alpha = 1.0$
h						
150	0.397	0.431	0.483	0.573	0.786	1.0
175	0.367	0.399	0.447	0.531	0.728	1.0
200	0.344	0.373	0.418	0.496	0.681	1.0
225	0.324	0.352	0.394	0.468	0.642	1.0
250	0.307	0.334	0.374	0.444	0.609	1.0
275	0.293	0.318	0.356	0.423	0.580	1.0
300	0.280	0.305	0.341	0.405	0.556	1.0
$x/h = 1.0$	$\alpha = 0.5$	$\alpha = 0.6$	$\alpha = 0.7$	$\alpha = 0.8$	$\alpha = 0.9$	$\alpha = 1.0$
h						
150	0.262	0.291	0.331	0.398	0.552	1.0
175	0.243	0.269	0.306	0.369	0.511	1.0
200	0.227	0.252	0.287	0.345	0.478	1.0
225	0.214	0.237	0.270	0.325	0.451	1.0
250	0.203	0.225	0.265	0.309	0.428	1.0
275	0.193	0.215	0.244	0.294	0.408	1.0
300	0.185	0.206	0.234	0.282	0.391	1.0
$x/h = 1.5$	$\alpha = 0.5$	$\alpha = 0.6$	$\alpha = 0.7$	$\alpha = 0.8$	$\alpha = 0.9$	$\alpha = 1.0$
h						
150	0.196	0.219	0.252	0.305	0.426	1.0
175	0.181	0.203	0.233	0.283	0.394	1.0
200	0.169	0.190	0.218	0.264	0.369	1.0
225	0.160	0.179	0.206	0.249	0.348	1.0
250	0.151	0.170	0.195	0.236	0.330	1.0
275	0.144	0.162	0.186	0.225	0.315	1.0
300	0.138	0.155	0.178	0.216	0.301	1.0
$x/h = 2.0$	$\alpha = 0.5$	$\alpha = 0.6$	$\alpha = 0.7$	$\alpha = 0.8$	$\alpha = 0.9$	$\alpha = 1.0$
h						
150	0.156	0.176	0.203	0.247	0.347	1.0
175	0.144	0.163	0.188	0.229	0.321	1.0
200	0.135	0.152	0.176	0.214	0.300	1.0
225	0.127	0.144	0.166	0.202	0.283	1.0
250	0.121	0.136	0.157	0.192	0.268	1.0
275	0.115	0.130	0.150	0.183	0.256	1.0
300	0.110	0.124	0.144	0.175	0.245	1.0

Example 6.6: calculation of a notched beam in a floor structure

Confirm that the shear strength at the end of the 50 mm wide (b) solid timber beam shown in Figure 6.16 will comply with the ultimate limit state requirements of EN 1995-1-1. The beam is one of a number in a floor system, spaced 400 mm apart, and is strength class C24 in accordance with BS EN 338: 2009. The design shear force in the beam at the notch position is $V_d = 3.09$ kN due to permanent and medium-term variable loading. Service class 2 applies.

Design shear force at support:

$V_d = 3.09$ kN

Design shear stress in notch area, $\tau_{v,d}$:

Clause 6.1.7(2)

$k_{cr} = 0.67$ (*clause 6.1.7(2)*), $h_{ef} = 215$ mm

$\tau_{v,d} = 1.5V_d/k_{cr}bh_{ef} = 1.5 \times 3.09 \times 10^3/0.67 \times 50 \times 215 = 0.64$ mm^{-2}

Figure 6.16. Beam in the floor system of Example 6.6

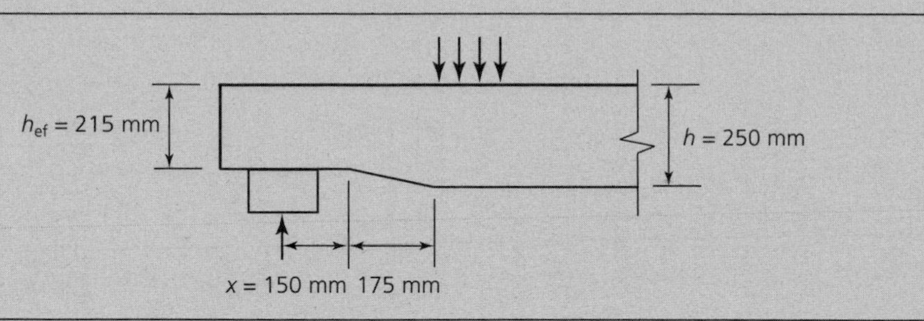

Design shear strength, $f_{v,d}$:

$k_{mod} = 0.8$, $k_{sys} = 1.1$, $\gamma_M = 1.3$, $f_{v,k} = 4.0$ N/mm^2

$f_{v,d} = k_{mod} k_{sys} f_{v,k}/\gamma_M = 0.8 \times 1.1 \times 4.0/1.3 = 2.71$ N/mm^2

which is OK in shear.

From *Equation 6.62*:

$i = 175/(250 - 215) = 5$, $\alpha = h_{ef}/h = 215/250 = 0.86$, $k_n = 5$

$$k_v = \min(1, k_n[1 + 1.1i^{1.5}/(h)^{0.5}]/((h)^{0.5}\{[\alpha(1 - \alpha)]^{0.5} + 0.8(x/h)(1/\alpha - \alpha^2)^{0.5}\})$$

$$k_v = \min(1, 5(1 + 1.1 \times 5^{1.5}/(250)^{0.5}\{[0.86(1 - 0.86)]^{0.5} + 0.8(150/250)(1/0.86 - 0.86^2)^{0.5}\})$$

$$= 0.85$$

Validation requirement (*Equation 6.60*):

$\tau_{v,d}/k_v f_{v,g,d} = 0.64/0.85 \times 2.71 = 0.28 < 1$

which is OK.

(If the notch were vertical, i.e. $i = 0$, $k_v = 0.48$ and shear stress/shear strength = 0.49, i.e. it would still be OK.)

6.6. System strength

Where there are several equally spaced similar members, components or assemblies connected by a continuous load distribution system, the member strength properties may be increased by multiplying by a system factor, k_{sys}.

The argument used to justify the factor is that the probability that every member connected by the load distribution system will have the minimum strength (i.e. the fifth percentile value), and stiffness (the mean value) can be considered to be outside the design basis. One member may have these properties, but the others will have greater values. In this type of situation, when the minimum strength member takes its share of the system loading, it will deform by a greater amount than the adjacent members, but the stiffness of the load distribution system will restrain this movement and transfer loading to adjacent stronger and stiffer members. On this basis, the load on the weaker member will be reduced, and to reach the failure condition a greater load can be applied to the system. In EN 1995-1-1 this is achieved by increasing the affected strength properties through the application of the system factor k_{sys}. The properties affected will be the bending, shear, bearing, compression and tension strengths of the connected members in the system.

The strength of the load distribution system must be validated, and *clause 6.6* requires that this is carried out assuming the loads being supported are of short-term duration.

Clause 6.6

In *clause 6.6*, $k_{sys} = 1.1$ when the requirements are met. From experience in the UK, where there is a continuous load distribution system this factor can be applied to floor or wall or roof systems where there are at least four members connected by the distribution system. The joints in the load distribution system must be staggered, and the members must be no greater than 610 mm apart. With roof trusses, *clause 6.6* also permits the factor to be used with trusses that are up to 1.2 m apart when supporting tiled roofs, providing the load distribution members are continuous over at least two spans and any joints are staggered.

Clause 6.6

Clause 6.6

Where the spacing is more than 610 mm or there are fewer than the number of members referred to above or the distribution system is not continuous or fixed in accordance with design requirements, the factor cannot be applied. Common situations in design where this condition will apply are:

- the sole plate of a stud wall – as the sole plate is the only member in the system
- the flanges in thin flanged beams – as the same flange panel will connect several webs.

Where laminated timber decks are to be used, which is more common in timber bridge structures than in buildings, the value used for k_{sys} will be obtained from *Figure 6.12*.

Table 6.8(a). Values of k_{cy} or k_{cz} for softwoods compliant with the strength classes C14 to C20 in EN 338, based on *Equations 6.25–6.29* inclusive in EN 1995-1-1 with $\beta_c = 0.2$ (based on slenderness ratios, λ, up to 240)

C14 $f_{c,0,k} = 16$ N/mm^2 $E_{0.05} = 4.7$ kN/mm^2		C16 $f_{c,0,k} = 17$ N/mm^2 $E_{0.05} = 5.4$ kN/mm^2		C18 $f_{c,0,k} = 18$ N/mm^2 $E_{0.05} = 6.0$ kN/mm^2		C20 $f_{c,0,k} = 19$ N/mm^2 $E_{0.05} = 6.4$ kN/mm^2	
λ	k_{cy} (k_{cz})	λ	k_{cy} (k_{cz})	λ	k_{cy} (k_{cz})	λ	k_{cy} (k_{cz})
16.153	1.000	16.797	1.000	17.207	1.000	17.298	1.000
20	0.984	20	0.987	20	0.989	20	0.989
25	0.960	25	0.965	25	0.968	25	0.968
30	0.932	30	0.939	30	0.943	30	0.944
35	0.899	35	0.908	35	0.914	35	0.915
40	0.856	40	0.870	40	0.878	40	0.880
45	0.804	45	0.823	45	0.834	45	0.836
50	0.741	50	0.766	50	0.781	50	0.784
55	0.673	55	0.702	55	0.720	55	0.723
60	0.605	60	0.636	60	0.655	60	0.659
65	0.540	65	0.572	65	0.591	65	0.595
70	0.482	70	0.512	70	0.531	70	0.535
75	0.430	75	0.459	75	0.477	75	0.481
80	0.387	80	0.412	80	0.429	80	0.433
85	0.347	85	0.371	85	0.387	85	0.391
90	0.313	90	0.336	90	0.351	90	0.354
95	0.284	95	0.305	95	0.318	95	0.321
100	0.258	100	0.278	100	0.290	100	0.293
105	0.236	105	0.254	105	0.265	105	0.268
110	0.217	110	0.233	110	0.244	110	0.246
115	0.199	115	0.214	115	0.224	115	0.226
120	0.184	120	0.198	120	0.207	120	0.209
125	0.170	125	0.183	125	0.192	125	0.194
130	0.158	130	0.170	130	0.178	130	0.180
135	0.147	135	0.158	135	0.166	135	0.167
140	0.137	140	0.148	140	0.155	140	0.156
145	0.128	145	0.138	145	0.145	145	0.146
150	0.120	150	0.129	150	0.136	150	0.137
155	0.113	155	0.122	155	0.127	155	0.129
160	0.106	160	0.114	160	0.120	160	0.121
165	0.100	165	0.108	165	0.113	165	0.114
170	0.094	170	0.102	170	0.107	170	0.108
175	0.089	175	0.096	175	0.101	175	0.102
180	0.084	180	0.091	180	0.095	180	0.096
185	0.08	185	0.086	185	0.091	185	0.091
190	0.076	190	0.082	190	0.086	190	0.087
195	0.072	195	0.078	195	0.082	195	0.083
200	0.069	200	0.074	200	0.078	200	0.079
205	0.066	205	0.071	205	0.074	205	0.075
210	0.063	210	0.068	210	0.071	210	0.072
215	0.06	215	0.065	215	0.068	215	0.068
220	0.057	220	0.062	220	0.065	220	0.065
225	0.055	225	0.059	225	0.062	225	0.063
230	0.052	230	0.057	230	0.059	230	0.06
235	0.05	235	0.054	235	0.057	235	0.057
240	0.048	240	0.052	240	0.055	240	0.055

Table 6.8(b). Values of k_{cy} or k_{cz} for softwoods compliant with the strength classes C22 to C30 in EN 338, based on *Equations 6.25–6.29* inclusive in EN 1995-1-1 with $\beta_c = 0.2$ (based on slenderness ratios, λ, up to 240)

C22 $f_{c,0,k} = 20$ N/mm^2 $E_{0.05} = 6.7$ kN/mm^2		C24 $f_{c,0,k} = 21$ N/mm^2 $E_{0.05} = 7.4$ kN/mm^2		C27 $f_{c,0,k} = 22$ N/mm^2 $E_{0.05} = 7.7$ kN/mm^2		C30 $f_{c,0,k} = 23$ N/mm^2 $E_{0.05} = 8.0$ kN/mm^2	
λ	k_{cy} (k_{cz})	λ	k_{cy} (k_{cz})	λ	k_{cy} (k_{cz})	λ	k_{cy} (k_{cz})
17.250	1.000	17.692	1.000	17.632	1.000	17.577	1.000
20	0.989	20	0.991	20	0.991	20	0.991
25	0.968	25	0.971	25	0.970	25	0.97
30	0.944	30	0.948	30	0.947	30	0.947
35	0.915	35	0.92	35	0.919	35	0.919
40	0.879	40	0.887	40	0.886	40	0.885
45	0.835	45	0.846	45	0.844	45	0.843
50	0.782	50	0.796	50	0.794	50	0.793
55	0.721	55	0.739	55	0.736	55	0.734
60	0.657	60	0.676	60	0.674	60	0.671
65	0.593	65	0.614	65	0.611	65	0.608
70	0.533	70	0.554	70	0.551	70	0.548
75	0.479	75	0.499	75	0.496	75	0.494
80	0.431	80	0.450	80	0.447	80	0.445
85	0.389	85	0.406	85	0.404	85	0.402
90	0.352	90	0.368	90	0.366	90	0.364
95	0.320	95	0.335	95	0.333	95	0.331
100	0.291	100	0.305	100	0.303	100	0.302
105	0.267	105	0.279	105	0.278	105	0.276
110	0.245	110	0.256	110	0.255	110	0.253
115	0.225	115	0.236	115	0.235	115	0.233
120	0.208	120	0.218	120	0.217	120	0.216
125	0.193	125	0.202	125	0.201	125	0.200
130	0.179	130	0.188	130	0.186	130	0.185
135	0.167	135	0.175	135	0.174	135	0.173
140	0.155	140	0.163	140	0.162	140	0.161
145	0.145	145	0.153	145	0.152	145	0.151
150	0.136	150	0.143	150	0.142	150	0.141
155	0.128	155	0.134	155	0.133	155	0.133
160	0.120	160	0.126	160	0.126	160	0.125
165	0.113	165	0.119	165	0.118	165	0.118
170	0.107	170	0.112	170	0.112	170	0.111
175	0.101	175	0.106	175	0.106	175	0105
180	0.096	180	0.101	180	0.100	180	0.099
185	0.091	185	0.096	185	0.095	185	0.094
190	0.086	190	0.091	190	0.09	190	0.09
195	0.082	195	0.086	195	0.086	195	0.085
200	0.078	200	0.082	200	0.082	200	0.081
205	0.075	205	0.078	205	0.078	205	0.077
210	0.071	210	0.075	210	0.074	210	0.074
215	0.068	215	0.071	215	0.071	215	0.07
220	0.065	220	0.068	220	0.068	220	0.067
225	0.062	225	0.065	225	0.065	225	0.065
230	0.06	230	0.063	230	0.062	230	0.062
235	0.057	235	0.06	235	0.06	235	0.059
240	0.055	240	0.058	240	0.057	240	0.057

Table 6.8(c). Values of k_{cy} or k_{cz} for softwoods compliant with the strength classes C35 to C50 in EN 338, based on *Equations 6.25–6.29* inclusive in EN 1995-1-1 with $\beta_c = 0.2$ (based on slenderness ratios, λ, up to 240)

C35 $f_{c,0,k} = 25$ N/mm^2 $E_{0.05} = 8.7$ kN/mm^2		C40 $f_{c,0,k} = 26$ N/mm^2 $E_{0.05} = 9.4$ kN/mm^2		C45 $f_{c,0,k} = 27$ N/mm^2 $E_{0.05} = 10.0$ kN/mm^2		C50 $f_{c,0,k} = 29$ N/mm^2 $E_{0.05} = 10.7$ kN/mm^2	
λ	k_{cy} (k_{cz})	λ	k_{cy} (k_{cz})	λ	k_{cy} (k_{cz})	λ	k_{cy} (k_{cz})
17.582	1.000	17.92	1.000	18.138	1.000	18.104	1.000
20	0.991	20	0.992	20	0.993	20	0.993
25	0.970	25	0.972	25	0.973	25	0.973
30	0.947	30	0.950	30	0.951	30	0.951
35	0.919	35	0.923	35	0.925	35	0.925
40	0.885	40	0.890	40	0.894	40	0.893
45	0.843	45	0.851	45	0.855	45	0.855
50	0.793	50	0.803	50	0.809	50	0.808
55	0.734	55	0.747	55	0.755	55	0.754
60	0.672	60	0.686	60	0.695	60	0.694
65	0.608	65	0.624	65	0.633	65	0.632
70	0.549	70	0.564	70	0.574	70	0.572
75	0.494	75	0.509	75	0.518	75	0.517
80	0.445	80	0.459	80	0.468	80	0.467
85	0.402	85	0.415	85	0.424	85	0.422
90	0.364	90	0.376	90	0.384	90	0.383
95	0.331	95	0.342	95	0.350	95	0.348
100	0.302	100	0.312	100	0.319	100	0.318
105	0.276	105	0.286	105	0.292	105	0.291
110	0.253	110	0.263	110	0.268	110	0.267
115	0.233	115	0.242	115	0.247	115	0.246
120	0.216	120	0.223	120	0.229	120	0.228
125	0.200	125	0.207	125	0.212	125	0.211
130	0.185	130	0.192	130	0.197	130	0.196
135	0.173	135	0.179	135	0.183	135	0.183
140	0.161	140	0.167	140	0.171	140	0.170
145	0.151	145	0.156	145	0.160	145	0.159
150	0.141	150	0.147	150	0.150	150	0.149
155	0.133	155	0.138	155	0.141	155	0.140
160	0.125	160	0.130	160	0.133	160	0.132
165	0.118	165	0.122	165	0.125	165	0.124
170	0.111	170	0.115	170	0.118	170	0.118
175	0.105	175	0.109	175	0.112	175	0.111
180	0.100	180	0.103	180	0.106	180	0.105
185	0.094	185	0.098	185	0.1	185	0.1
190	0.09	190	0.093	190	0.095	190	0.095
195	0.085	195	0.088	195	0.091	195	0.09
200	0.081	200	0.084	200	0.086	200	0.086
205	0.077	205	0.08	205	0.082	205	0.082
210	0.074	210	0.077	210	0.078	210	0.078
215	0.071	215	0.073	215	0.075	215	0.075
220	0.067	220	0.07	220	0.072	220	0.071
225	0.065	225	0.067	225	0.069	225	0.068
230	0.062	230	0.064	230	0.066	230	0.065
235	0.059	235	0.062	235	0.063	235	0.063
240	0.057	240	0.059	240	0.06	240	0.06

Table 6.9. Values of k_{cy} or k_{cz} for LVL, based on *Equations 6.25–6.29* inclusive in EN 1995-1-1 with $\beta_c = 0.1$ (based on slenderness ratios, λ, up to 240)

Kerto-S $f_{c,0,k} = 35$ N/mm^2 $E_{0.05} = 11.6$ kN/mm^2		Kerto-Q $f_{c,0,k} = 26$ N/mm^2 $E_{0.05} = 8.8$ kN/mm^2	
λ	k_{cy} (k_{cz})	λ	k_{cy} (k_{cz})
17.158	1	17.339	1
20	0.994	20	0.995
25	0.983	25	0.984
30	0.970	30	0.971
35	0.954	35	0.955
40	0.932	40	0.934
45	0.901	45	0.904
50	0.857	50	0.863
55	0.798	55	0.806
60	0.727	60	0.736
65	0.653	65	0.663
70	0.582	70	0.592
75	0.518	75	0.528
80	0.463	80	0.471
85	0.415	85	0.423
90	0.373	90	0.381
95	0.337	95	0.344
100	0.306	100	0.312
105	0.279	105	0.285
110	0.255	110	0.260
115	0.234	115	0.239
120	0.216	120	0.220
125	0.200	125	0.204
130	0.185	130	0.189
135	0.172	135	0.175
140	0.160	140	0.163
145	0.149	145	0.153
150	0.140	150	0.143
155	0.131	155	0.134
160	0.123	160	0.126
165	0.116	165	0.118
170	0.109	170	0.112
175	0.103	175	0.106
180	0.098	180	0.100
185	0.093	185	0.095
190	0.088	190	0.090
195	0.084	195	0.085
200	0.080	200	0.081
205	0.076	205	0.077
210	0.072	210	0.074
215	0.069	215	0.070
220	0.066	220	0.067
225	0.063	225	0.064
230	0.060	230	0.062
235	0.058	235	0.059
240	0.055	240	0.057

Table 6.10(a). Values of k_{cy} or k_{cz} for glued laminated timber compliant with EN 1194, based on *Equations 6.25–6.29* inclusive in EN 1995-1-1 with $\beta_c = 0.1$ (based on slenderness ratios, λ, up to 240)

GL 24h $f_{c,0,g,k} = 24$ N/mm^2 $E_{0,g,0.5} = 9.4$ kN/mm^2		GL 28h $f_{c,0,g,k} = 26.5$ N/mm^2 $E_{0,g,0.5} = 10.2$ kN/mm^2		GL 32h $f_{c,0,g,k} = 29$ N/mm^2 $E_{0,g,0.5} = 11.1$ kN/mm^2		GL 36h $f_{c,0,g,k} = 31$ N/mm^2 $E_{0,g,0.5} = 11.9$ kN/mm^2	
λ	k_{cy} (k_{cz})	λ	k_{cy} (k_{cz})	Λ	k_{cy} (k_{cz})	λ	k_{cy} (k_{cz})
18.652	1	18.49	1	18.439	1.000	18.466	1.000
20	0.998	20	0.997	20	0.997	20	0.997
25	0.988	25	0.988	25	0.987	25	0.987
30	0.977	30	0.976	30	0.976	30	0.976
35	0.964	35	0.963	35	0.962	35	0.962
40	0.947	40	0.945	40	0.945	40	0.945
45	0.924	45	0.922	45	0.921	45	0.922
50	0.893	50	0.890	50	0.889	50	0.98
55	0.851	55	0.846	55	0.845	55	0.846
60	0.796	60	0.789	60	0.787	60	0.788
65	0.730	65	0.722	65	0.720	65	0.721
70	0.662	70	0.653	70	0.651	70	0.652
75	0.596	75	0.587	75	0.585	75	0.586
80	0.535	80	0.527	80	0.525	80	0.526
85	0.482	85	0.475	85	0.472	85	0.473
90	0.435	90	0.428	90	0.426	90	0.427
95	0.394	95	0.388	95	0.386	95	0.387
100	0.358	100	0.353	100	0.351	100	0.352
105	0.327	105	0.322	105	0.32	105	0.321
110	0.299	110	0.294	110	0.293	110	0.294
115	0.275	115	0.271	115	0.269	115	0.27
120	0.254	120	0.249	120	0.248	120	0.249
125	0.234	125	0.231	125	0.229	125	0.230
130	0.217	130	0.214	130	0.213	130	0.213
135	0.202	135	0.199	135	0.198	135	0.198
140	0.188	140	0.185	140	0.184	140	0.185
145	0.176	145	0.173	145	0.172	145	0.172
150	0.165	150	0.162	150	0.161	150	0.161
155	0.154	155	0.152	155	0.151	155	0.151
160	0.145	160	0.143	160	0.142	160	0.142
165	0.137	165	0.134	165	0.134	165	0.139
170	0.129	170	0.127	170	0.126	170	0.126
175	0.122	175	0.120	175	0119	175	0.119
180	0.115	180	0.113	180	0.113	180	0.113
185	0.109	185	0.107	185	0.107	185	0.107
190	0.104	190	0.102	190	0.101	190	0.102
195	0.099	195	0.097	195	0.096	195	0.097
200	0.094	200	0.092	200	0.092	200	0.092
205	0.089	205	0.088	205	0.087	205	0.088
210	0.085	210	0.084	210	0.083	210	0.083
215	0.081	215	0.080	215	0.079	215	0.080
220	0.078	220	0.076	220	0.076	220	0.076
225	0.074	225	0.073	225	0.073	225	0.073
230	0.071	230	0.070	230	0.070	230	0.070
235	0.068	235	0.067	235	0.067	235	0.067
240	0.065	240	0.064	240	0.064	240	0.064

Table 6.10(b). Values of k_{cy} or k_{cz} for glued laminated timber compliant with EN 1194, based on *Equations 6.25–6.29* inclusive in EN 1995-1-1 with $\beta_c = 0.1$ (based on slenderness ratios, λ, up to 240)

GL 24c $f_{c,0,g,k} = 21.0$ N/mm^2 $E_{0,g,0.5} = 9.4$ kN/mm^2		GL 28c $f_{c,0,g,k} = 24.0$ N/mm^2 $E_{0,g,0.5} = 10.2$ kN/mm^2		GL 32c $f_{c,0,g,k} = 26.5$ N/mm^2 $E_{0,g,0.5} = 11.1$ kN/mm^2		GL 36c $f_{c,0,g,k} = 29.0$ N/mm^2 $E_{0,g,0.5} = 11.9$ kN/mm^2	
λ	k_{cy} (k_{cz})	λ	k_{cy} (k_{cz})	Λ	k_{cy} (k_{cz})	λ	k_{cy} (k_{cz})
19.94	1	19.43	1	19.289	1	19.092	1
		20	0.999	20	0.999	20	0.998
25	0.991	25	0.990	25	0.990	25	0.989
30	0.981	30	0.980	30	0.979	30	0.979
35	0.970	35	0.968	35	0.967	35	0.966
40	0.956	40	0.952	40	0.951	40	0.950
45	0.938	45	0.933	45	0.931	45	0.929
50	0.914	50	0.907	50	0.905	50	0.901
55	0.882	55	0.871	55	0.868	55	0.863
60	0.840	60	0.824	60	0.819	60	0.812
65	0.786	65	0.765	65	0.759	65	0.751
70	0.724	70	0.700	70	0.693	70	0.684
75	0.659	75	0.635	75	0.628	75	0.618
80	0.598	80	0.573	80	0.566	80	0.557
85	0.541	85	0.518	85	0.511	85	0.502
90	0.490	90	0.468	90	0.462	90	0.454
95	0.445	95	0.425	95	0.419	95	0.411
100	0.405	100	0.387	100	0.381	100	0.374
105	0.371	105	0.353	105	0348	105	0.342
110	0.340	110	0.323	110	0.319	110	0.313
115	0.312	115	0.297	115	0.293	115	0.288
120	0.288	120	0.274	120	0.270	120	0.265
125	0.267	125	0.254	125	0.250	125	0.245
130	0.247	130	0.235	130	0.232	130	0.227
135	0.230	135	0.219	135	0.216	135	0.211
140	0.214	140	0.204	140	0.201	140	0.197
145	0.200	145	0.190	145	0.188	145	0.184
150	0.187	150	0.178	150	0.176	150	0.172
155	0.176	155	0.167	155	0.165	155	0.162
160	0.165	160	0.157	160	0.155	160	0.152
165	0.156	165	0.148	165	0.146	165	0.143
170	0.147	170	0.140	170	0.138	170	0.135
175	0.139	175	0.132	175	0.130	175	0.128
180	0.131	180	0.125	180	0.123	180	0.121
185	0.125	185	0.118	185	0.117	185	0.114
190	0.118	190	0.112	190	0.111	190	0.109
195	0.112	195	0.107	195	0.105	195	0.103
200	0.107	200	0.102	200	0.100	200	0.098
205	0.102	205	0.097	205	0.095	205	0.093
210	0.097	210	0.092	210	0.091	210	0.089
215	0.093	215	0.088	215	0.087	215	0.085
220	0.089	220	0.084	220	0.083	220	0.081
225	0.085	225	0.081	225	0.079	225	0.078
230	0.081	230	0.077	230	0.076	230	0.075
235	0.078	235	0.074	235	0.073	235	0.071
240	0.075	240	0.071	240	0.070	240	0.069

REFERENCES

Aune P (1995) Shear and torsion. In *Timber Engineering: STEP 1* (Blass HJ, Aune P, Choo BS *et al.* (eds)). Centrum Hout, Almere, lecture B4.

Blass HJ and Gorlacher R (2004) Compression perpendicular to the grain. *Proceedings of the World Conference on Timber Engineering*, Lahti, vol. II, pp. 435–440.

BSI (2004) BS EN 13986: 2004. Wood-based panels for use in construction. Characteristics, evaluation of conformity and marking. BSI, London.

BSI (2010) BS EN 408: 2010. Timber structures. Structural timber and glued laminated timber – Determination of some physical and mechanical properties. BSI, London.

Hankinson RL (1921) Investigation of crushing strength of spruce at varying angles to the grain. *US Air Service Information Circular*, vol. 3. No. 259 (Material Section Paper No. 130).

Maki AC and Keunzi EW (1965) *Deflection and Stresses of Tapered Wood Beams. Research Paper FPL 34*. US Forest Service, Madison, WI.

Porteous J and Kermani A (2007) *Structural Timber Design to Eurocode 5*. Blackwell, Oxford.

Timoshenko S and Goodier JN (1951) *Theory of Elasticity,* 2nd edn. McGraw-Hill, New York.

Designers' Guide to Eurocode 5: Design of Timber Buildings
ISBN 978-0-7277-3162-3

http://dx.doi.org/10.1680/dtb.31623.079

Chapter 7
Serviceability limit states

This chapter is concerned with the requirements for serviceability limit states. It covers the material in *Section 7* of EN 1995-1-1, and addresses the following clauses:

- Joint slip *Clause 7.1*
- Limiting values for deflection of beams *Clause 7.2*
- Vibrations *Clause 7.3*

Serviceability limit states are states that concern the functioning and appearance of the structure as well as the comfort of the users, and are defined in clause 3.4 of EN 1990. The requirements for verification of these states are fragmented throughout EN 1995-1-1, and the information provided in this chapter of the guide attempts to bring together the relevant information under the above headings.

Serviceability limit states can be either irreversible or reversible, and the loading requirements used to calculate deformations will differ for each type. Irreversible limit states are those that will be exceeded even when the design loading has been removed (e.g. cracking to plastered ceilings). Reversible limit states are states that will be exceeded under the design loading but when removed there will be no irreversible damage (e.g. deflections exceeding design limits resulting in visual impact but causing no permanent damage).

Reversible limit states are permitted in EN 1990, providing the designer can agree the design criteria with the client, but in EN 1995-1-1 only irreversible limit states are permitted in the design rules. Although EN 1990 sets the framework for EN 1995-1-1, and reversible limit states are able to be used if agreed, the rules in the code are written in such a way that the designer is directed to use irreversible limit states to determine deformations in timber structures.

The loading combinations for irreversible limit states are the characteristic and the quasi-permanent, referred to in EN 1990 and in Chapter 2 in this guide. These combinations apply permanent and variable actions, and the design loading condition will be the one that results in the greatest deformation effect in the structural element being assessed. When the creep behaviour is different throughout the structure, the loading requirements of *clauses 2.2.3(3)* and *2.2.3(4)* are unclear, and are to be revised as stated in Appendix A and as explained in Section 2.2.3 and Chapter 5 in this guide. Also, as suggested in this guide, a simpler but more conservative option for this condition will be to apply the characteristic loading combination, and proposals for this are also given in Section 2.2.3 and Chapter 5.

Clause 2.2.3(3)
Clause 2.2.3(4)

7.1. Joint slip

When a connection is subjected to lateral loading, because of yielding of the timber or wood product in the connection, possible deformation of the fastener type being used and movement due to the take up of any tolerance gaps, slip will occur. The amount of slip will depend on the type of fastener being used, and the load duration and typical load–slip curves for a nailed or screwed fastener connection and a bolted connection subjected to test loading conditions are shown in Figure 7.1.

The ratio of the serviceability load on a fastener, F_{ser}, divided by the instantaneous fastener slip under this load, u_{inst}, is defined as the slip modulus, K_{ser}, and modulus values for the different type of fastener referred to in EN 1995-1-1 are given in Table 7.1. K_{ser} is in N/mm/shear

Figure 7.1. Typical load–slip curves for a nail/screw and a bolted connection subjected to lateral loading

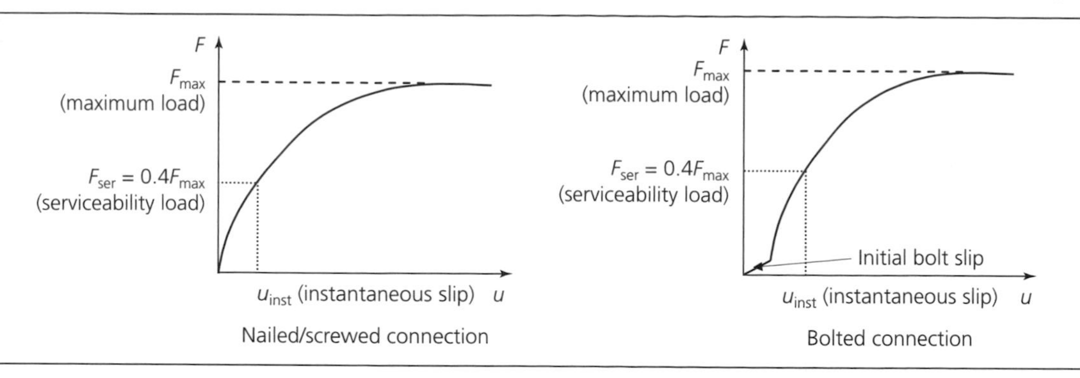

plane/fastener for common fastener types, noting that no value is provided for punched metal plate fasteners as it is dependent on the detail of the punched metal plate being used.

To prevent creep slip from occurring in tests used to derive K_{ser}, the load duration allowed in the relevant European standards is less than 2 minutes. On this basis the slip, u_{inst}, is taken to be elastic and referred to as the instantaneous slip of the fastener.

When a connection is made from timber or wood-based products having different densities, the mean density used in the expressions in Table 7.1 is obtained from *Equation 7.1*:

$$\rho_m = (\rho_{m1}\rho_{m2})^{0.5} \qquad (7.1)$$

where ρ_{m1} and ρ_{m2} are the mean densities in kg/m^3 of the respective members in the connection.

Clause 7.1(3)

With timber-to-steel or timber-to-concrete connections, *clause 7.1(3)* states that K_{ser} may be multiplied by 2. This assumes that there is no slip in the steel or concrete element of the

Table 7.1. Values of K_{ser} for fasteners and connectors in timber–timber and wood-based panel–timber connections (K_{ser} is in N/mm/shear plane/fastener)

Fastener type	K_{ser}^{c}	Clearance allowance, c
Dowels	$\rho_m^{1.5}d/23$	None
Bolts with a clearance[a]		a
Bolts without a clearance		None
Screws		None
Nails (with predrilling)		None
Nails (without predrilling)	$\rho_m^{1.5}d^{0.8}/30$	None
Staples	$\rho_m^{1.5}d^{0.8}/80$	None
Split-ring connectors type A according to EN 912 (BSI, 2011)	$\rho_m d_c/2$	b
Shear-plate connectors type B according to EN 912		b
Toothed-plate connectors:		
■ connectors types C1 to C9 according to EN 912	$1.5\rho_m d_c/4$	b
■ connectors types C10 and C11 according to EN 912	$\rho_m d_c/2$	b

Data from EN 1995-1-1

Clause 10.4.3

[a]The clearance allowance should be the difference between the predrilled hole size and the bolt diameter – *see clause 10.4.3*. It should be added separately to the deformation

[b]There is no stated requirement in EN 1995-1-1. This may result in an underestimate of slip (e.g. with toothed-plate connectors), and it is suggested in this guide that an allowance is made for clearance

[c]ρ_m is the mean density of the connection members in kg/m^3; d is the fastener diameter in mm and will be the dowel diameter, the nominal diameter for nails and screws (as defined in EN 14592) or the bolt solid shank diameter. d_c will be as defined in EN 912 for the appropriate connector (in mm)

Figure 7.2. Single- and double-shear laterally loaded connections: (a–c) single shear; (d) double shear

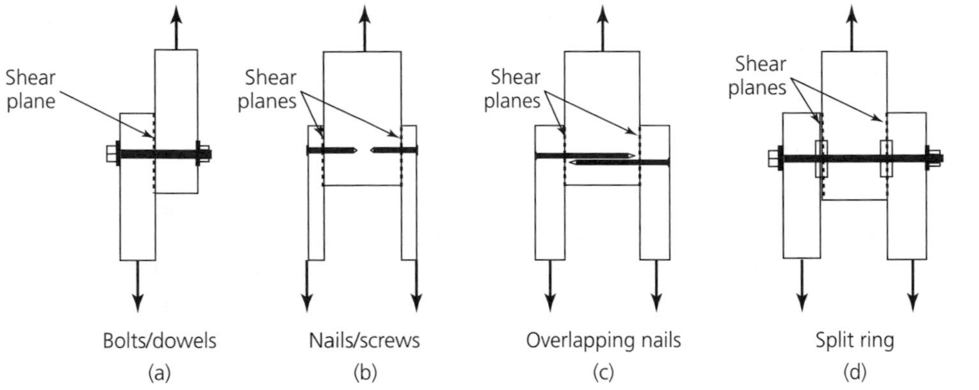

Shear plane | Shear planes | Shear planes | Shear planes

Bolts/dowels (a) | Nails/screws (b) | Overlapping nails (c) | Split ring (d)

connection, however, because of the effects of tolerance and yielding, particularly in steel connections and where fasteners are being used, multiplying by 2 may significantly overestimate the connection stiffness. It is suggested in this guide that a value less than 2 may be more appropriate when using fasteners, and, when dealing with timber-to-concrete connections, guidance on the stiffness behaviour of such joints may be obtained from research undertaken by Dias *et al.* (2010).

Connections in timber are generally single or double shear, and examples of each are shown in Figure 7.2. In single-shear connections there is a single shear plane per fastener, and in double-shear connections there are two shear planes per fastener.

To obtain the stiffness of a connection at the serviceability limit state, the slip modulus of the fastener type being used must be multiplied by the number of fasteners in the shear plane and the number of shear planes in the connection. When single-sided connectors are used in a back-to-back configuration in a shear plane (e.g. shear plate or single-sided toothed-plate connectors), the shear plane stiffness will be obtained by multiplying the K_{ser} value for the connector by the number of bolts in the shear plane, as shown in Figure 7.3.

The instantaneous slip will occur under the serviceability limit states design load, and guidance on the loading combination to be used in this analysis is given in Chapter 2 and in Section 5.4.2 in this guide.

When a connection is subjected to serviceability limit states loading over a period of time, creep slip will also occur, and the combined instantaneous and creep slip is referred to as the final deformation of the connection. If a structure consists of members, components and connections

Figure 7.3. Shear plane stiffness of a single row of type B2 shear plate connectors in a timber-to-timber connection

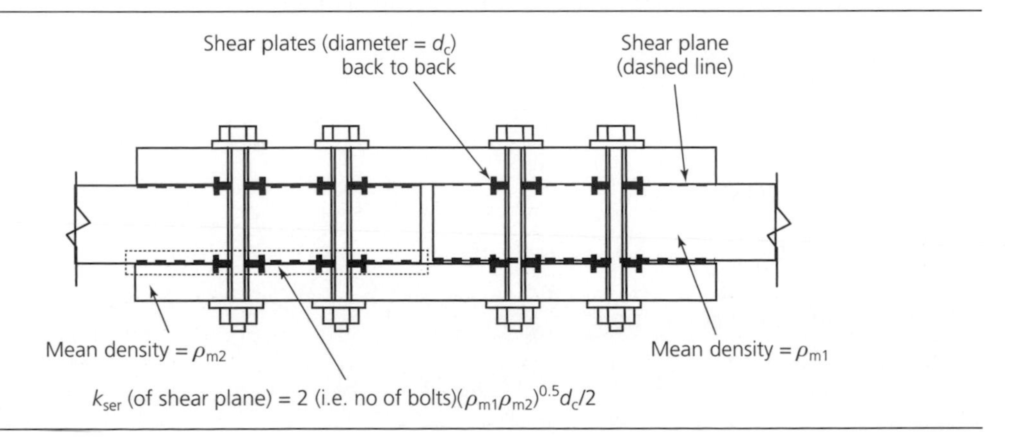

Clause 2.3.2.2(3)
Clause 2.3.2.2(4)

having the same creep behaviour (i.e. the same value of k_{def} as referred to in Chapters 2 and 3 in this guide and taking the requirements of *clauses 2.3.2.2(3)* and *2.3.2.2(4)* into account), the final deformation can be obtained from an analysis using mean stiffness properties, and the slip modulus for the fastener type being used for the connections will be the relevant value of K_{ser} given in Table 7.1. The serviceability limit states design loading will be derived from the combination of the characteristic loading and the quasi-permanent loading combination, and is referred to in Chapters 2 and 5 in this guide.

Clause 2.3.2.2(1)
Clause 2.3.2.2(3)
Clause 2.3.2.2(4)

Where a structure consists of members, components and connections having different creep behaviour, *clause 2.3.2.2(1)* requires the final deformation analysis to be carried out using final mean values of stiffness properties. *Clauses 2.3.2.2(3)* and *2.3.2.2(4)* state that for a connection the deformation factor, k_{def}, of each of the connection materials must be doubled, and so, unless steel gusset plates are used (where it can be argued that only k_{def} will apply), the creep properties of connections will always differ from those of the members, requiring final mean stiffness values to be used. For a fastener in a connection, the final mean slip modulus, $K_{ser,fin}$, is derived from *Equation 2.9*:

$$K_{ser,fin} = K_{ser}/(1 + k_{def}) \tag{2.9}$$

where K_{ser} is the slip modulus obtained from Table 7.1 and:

- for connections formed from timber elements having the same time-dependent properties, k_{def} = twice the value of the deformation factor from *Table 3.2*
- for connections formed from two wood-based elements having different time-dependent properties, $k_{def} = 2(k_{def,1}k_{def,2})^{0.5}$, where $k_{def,1}$ and $k_{def,2}$ are the deformation factors of the respective elements, obtained from *Table 3.2*.

Clauses 2.2.3(3)
Clause 2.2.3(4)

As stated in Section 2.2.3, the statements in *clauses 2.2.3(3)* and *2.2.3(4)* are to be revised as given in Appendix A, to clarify the loading condition to be used for this situation, and the new requirements are referred to in Section 2.2.3 and Chapter 5 in this guide. Where there is a clearance allowance, *c*, this will result in a slip in the connection at the commencement of loading, and this must be added to the sum of the instantaneous and creep slip, to obtain the final deformation.

Example 7.1: the stiffness, instantaneous slip and final slip in a connection formed with bolts

A timber–OSB/3 gusset plate joint in a structure is subjected to a design lateral permanent action $F_{d,G} = 2.5$ kN, a design lateral variable action $F_{d,Q} = 4.0$ kN with an associated Ψ_2 factor = 0.3, at the serviceability limit state (Figure 7.4). The gusset plates are 12 mm thick, and have a mean density of 650 kg/m^3; the timber is strength class C18, 50 mm thick, and the joint functions under service class 1 conditions. There are six No. 8 mm diameter bolts per connection acting in double shear, and spaced as shown in Figure 7.4, and the tolerance per bolt hole is 1 mm. Determine the lateral stiffness, instantaneous slip and the final slip of the joint at the serviceability limit states.

Mean density of the timber:

$$\rho_{t,m} = 380 \text{ kg/m}^3$$

Mean density of OSB/3:

$$\rho_{osb,m} = 650 \text{ kg/m}^3$$

Mean density of the connection:

$$\rho_m = \sqrt{\rho_{t,m}\rho_{osb,m}} = \sqrt{380 \times 650} = 497 \text{ kg/m}^3$$

Figure 7.4. Timber–OSB/3 connection

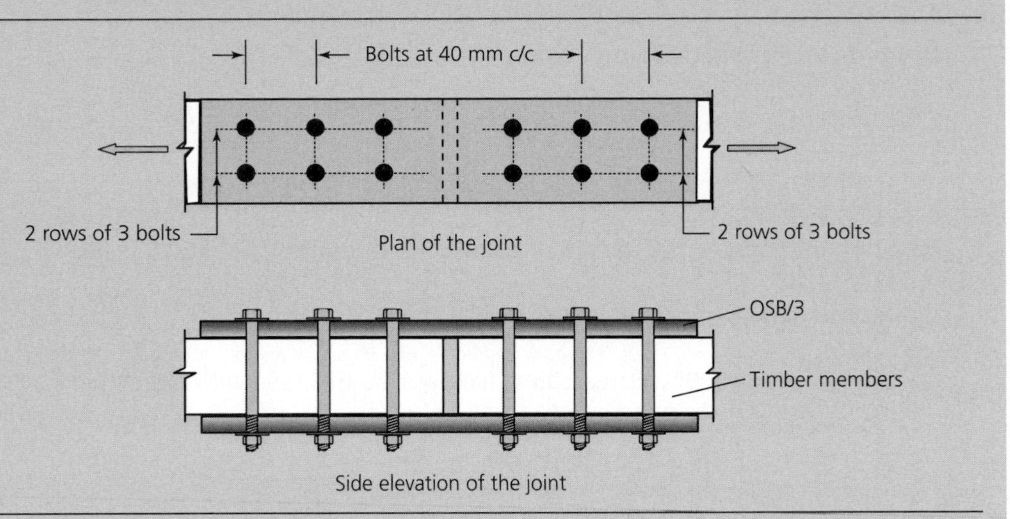

Slip modulus/shear plane/bolt:

$$K_{ser} = \rho_m^{1.5} d/23 = (497^{1.5} \times 8)/23 = 3.8538 \text{ kN/mm/sp/bolt}$$

Lateral stiffness per connection:

$$2 \times 6 \times 3.8538 = 46.246 \text{ kN/mm}$$

Lateral stiffness of the joint:

$$K_{ser,joint} = 46.246/2 = 23.123 \text{ kN/mm}$$

Characteristic lateral load on the joint:

$$F_{d.sls} = F_{d.G} + F_{d.Q} = 2.5 + 4.0 = 6.5 \text{ kN}$$

Taking the bolt clearance to be 1 mm, the instantaneous slip at the serviceability limit states, u_{inst} is

$$u_{inst} = F_{d.sls}/K_{ser,joint} + 2 \text{ mm} = 0.28 + 2 = 2.28 \text{ mm}$$

Deformation factor for the timber:

$$k_{t,def} = 0.6$$

Deformation factor for OSB/3:

$$k_{osb,def} = 1.5$$

The deformation factor for each connection:

$$k_{def} = 2\sqrt{k_{t,def} k_{osb,def}} = 2\sqrt{0.6 \times 1.5} = 1.897$$

Joint stiffness at final slip condition:

$$K_{ser,fin} = K_{ser,joint}/(1 + k_{def}) = 23.123/2.897 = 7.982 \text{ kN/mm}$$

The final slip of the joint, u_{fin}:

1 Based on the revised method in Appendix A:

$$u_{fin} = (F_{d.sls} - (F_{d.G} + \Psi_2 \times F_{d.Q})/K_{ser,joint}) + (F_{d.G} + \Psi_2 \times F_{d.Q})/K_{ser,fin} + 2\ \text{mm}$$

$$= [6.5 - (2.5 + 0.3 \times 4.0)]/23.123 + (2.5 + 0.3 \times 4.0)/7.982 + 2 = 2.58\ \text{mm}$$

2 Based on the conservative approach referred to in Section 2.2.3 and Chapter 5:

$$u_{fin} = F_{d.sls}/K_{ser,fin} + 2\ \text{mm} = 6.5/7.982 + 2 = 2.81\ \text{mm}$$

(i.e. approximately a 9% increase in value over the EN 1995-1-1 method).

7.2. Limiting values for deflection of beams

In EN 1995-1-1, 'deformation' is the generic term used to cover all forms of displacement (i.e. deflection, joint slip, joint rotation, etc.). It is made up from that element of the displacement arising immediately the design loading is applied, referred to as the instantaneous deformation, u_{inst}, and that arising over the design life of the building due to creep, referred to as the creep deformation, u_{creep}. The summation of the instantaneous and the creep deformation is the final deformation, u_{fin}.

The loading requirements for calculating deformations are referred to in Sections 2.2.3, 2.3.1 and 5.4.2 in this guide, and to prevent the occurrence of unacceptable damage due to excessive deformations as well as to meet functional and visual requirements, limits for each project have to be agreed with the client. For beams, guidance on limits that are deemed to be acceptable is given in *Table 7.2* in EN 1995-1-1 and in the UK National Annex to EN 1995-1-1. The components of the deflection arising under the design loading condition are shown in *Figure 7.1* (reproduced as Figure 7.5), and are defined as follows:

w_c is the value of any pre-camber set for the beam (where applied)
w_{inst} is the instantaneous deflection (measured from the pre-cambered position where used)
w_{creep} is the creep deflection
w_{fin} is the final deflection (i.e. $w_{fin} = w_{inst} + w_{creep}$)
$w_{net,fin}$ is the net final deflection (i.e. $w_{net,fin} = w_{fin} - w_c$).

In the National Annex to EN 1995-1-1, the validation requirement relates solely to the final deformation condition, and is

$$u_{net,fin} \leq w_{net,fin} \qquad \text{(D7.1)}$$

Figure 7.5. Components of deflection. (Based on EN 1995-1-1 (*Figure 7.1*). Reproduced with permission from EN 1995-1-1, © British Standards Institute, 2004)

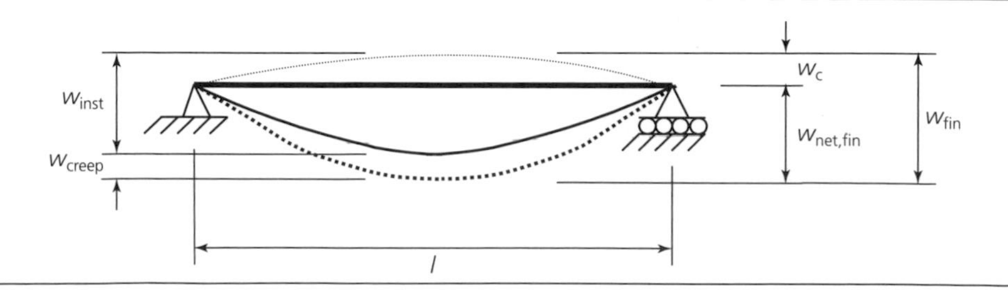

where $u_{net,fin}$ is the net final deflection of the beam and $w_{net,fin}$ is obtained from the National Annex or is any alternative limit agreed between the designer and the client.

In timber structures, because the ratio of the shear modulus to the bending modulus is high, being approximately eight times higher than that for steel, in addition to bending deformation, shear deformation must also be taken into account when calculating deformations, and the limits referred to above are for combined shear and bending conditions.

In timber construction, beams are commonly simply supported over single-span conditions, and the associated deflection formulae for typical loading arrangements at the instantaneous deformation condition for solid rectangular sections and for built-up I beams of the type referred to in Chapter 9 are given in Table 7.2.

Table 7.2. Combined instantaneous bending and shear deflection for simply supported and cantilever beams of (a) solid rectangular cross-section and (b) glued I section

(a) Solid rectangular beams of width b (mm), depth h (mm) and span, l (mm); $E_{0,mean}$ is the mean modulus (kN/mm^2) and $G_{0,mean}$ is the mean shear modulus (kN/mm^2)

Load case	Combined bending and shear deflection: mm
Uniformly distributed instantaneous design load (total value), V kN, along the length of a simply supported beam of span l	$u_{mid\ span} = \frac{5Vl^3}{32E_{0,mean}bh^3} + \frac{Vl}{6.67G_{0,mean}bh}$
A point load V kN, at the mid-span of a simply supported beam of span l	$u_{mid\ span} = \frac{Vl^3}{4E_{0,mean}bh^3} + \frac{Vl}{3.333G_{0,mean}bh}$
Point loads kN at the quarter and three-quarter points of a simply supported beam of span l	$u_{mid\ span} = \frac{11Vl^3}{32E_{0,mean}bh^3} + \frac{Vl}{3.332G_{0,mean}bh}$
A point load V kN, at the end of a cantilever of length l	$u_{end\ of\ cantilever} = \frac{4Vl^3}{E_{0,mean}bh^3} + \frac{1.2Vl}{G_{0,mean}bh}$

(b) Glued I beams as described in Section 9.1.1 in this guide; $E_{f,mean}$ is the mean modulus of the flange material (kN/mm^2); $E_{w,mean}$ is the mean modulus of the web material (kN/mm^2); $G_{w,mean}$ is the mean shear modulus of the web material (kN/mm^2) and $I_{ef} = I_f + (E_{w,mean}/E_{f,mean})\, I_w$ is the effective section modulus (mm^4), where I_f and I_w are as described in Section 9.1.1 in this guide; b is the web thickness (mm) and h_w is the clear distance (mm) between the flanges as shown in Figure 9.1.

Load case	Combined bending and shear deflection (mm)[a]
Uniformly distributed instantaneous design load (total value), V kN, along the length of a simply supported beam of span l (mm)	$u_{mid\ span} = \frac{5Vl^3}{384E_{0,mean}I_{ef}} + \frac{Vl}{8G_{0,mean}bh_w}$
A point load V kN, at the mid-span of a simply supported beam of span l (mm)	$u_{mid\ span} = \frac{Vl^3}{48E_{0,mean}I_{ef}} + \frac{Vl}{4G_{0,mean}bh_w}$
Point loads V kN at the quarter and three-quarter points of a simply supported beam of span l (mm)	$u_{mid\ span} = \frac{11Vl^3}{32E_{0,mean}I_{ef}} + \frac{Vl}{4G_{0,mean}bh_w}$
A point load V kN, at the end of a cantilever of length l (mm)	$u_{end\ of\ cantilever} = \frac{Vl^3}{3E_{0,mean}I_{ef}} + \frac{Vl}{G_{0,mean}bh_w}$

[a]The shear deformation is based on an approximate solution: if a more accurate value is required, guidance is given in *Roark's Formulas for Stress and Strain* (Young, 1989)

Example 7.2: the deflection of a box beam

A simply supported ply-webbed box beam in an office floor is laterally supported along the length of its compression flange and has an effective span of 6.75 m. Including the self-weight of the structure, the beam supports permanent design loading, $F_{d,G} = 1.04$ kN/m and variable medium duration loading, $F_{d,Q} = 2.7$ kN/m. The timber used for the flanges is class C24, and each web is Canadian plywood, 12.5 mm thick with the face ply parallel to the direction of span. The cross-section of the beam is shown in Figure 7.6, and it functions in service class 2 conditions. Show that the instantaneous and the final deflection of the box beam will not exceed span/300 and span/150, respectively. The value of the Ψ_2 factor associated with the variable loading is 0.3.

Figure 7.6. Beam cross-section

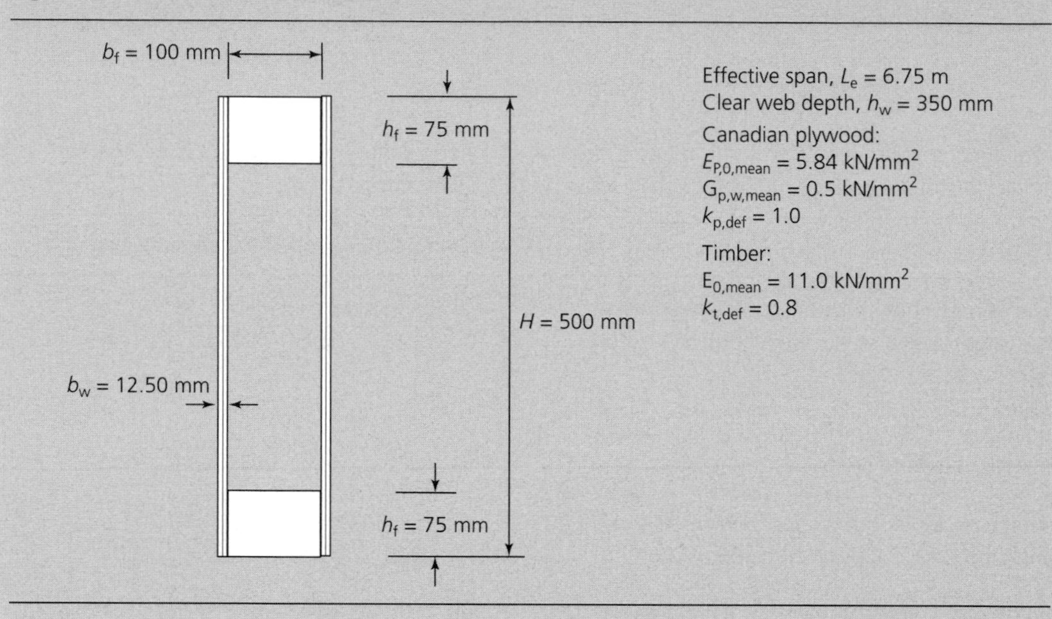

Characteristic load on the beam:

$$F_{d,sls} = (F_{d,G} + F_{d,Q})L_e = (1.04 + 2.7)6.75 = 25.245 \text{ kN}$$

The second moment of area of the section at the instantaneous condition, I_{ef}, and the final deformation condition, $I_{ef,fin}$ (based on the section transformed into timber) are calculated. For the instantaneous condition, the transformed web thickness is

$$b_{w,tfd} = b_w(E_{p,0,mean}/E_{0,mean}) = 12.5 \times 5.84/11.0 = 6.636 \text{ mm}$$

and

$$I_{ef} = 2b_{w,tfd}\mathrm{H}^3/12 + b_f(H^3 - h_w^3)/12$$

$$= 2 \times 6.636 \times 500^3/12 + 100(500^3 - 350^3)/12 = 8.2263 \times 10^8 \text{ mm}^4$$

For the final condition, the transformed web thickness is

$$b_{cw,tfd} = b_{wtfd}(1 + k_{t,def})/(1 + k_{p,def}) = 6.636(1 + 0.8)/(1 + 1.0) = 5.973 \text{ mm}$$

and

$$I_{c,ef} = 2b_{cw,tfd}H^3/12 + b_f(H^3 - h_w^3)/12$$

$$= 2 \times 5.973 \times 500^3/12 + 100(500^3 - 350^3)/12 = 8.088 \times 10^8 \text{ mm}^4$$

Effective area of the web for shear:

$$A_{ef} = 2 \times b \times h_w = 2 \times 12.5 \times 350 = 8.75 \times 10^3 \text{ mm}^2$$

The instantaneous deformation of the beam, u_{inst} is calculated.

Combined bending deflection (u_m) and shear deflection (u_v) of the beam at mid-span under uniformly distributed loading at the instantaneous condition, (u_{inst}):

$$u_m = \frac{5}{384} \frac{F_{d,sls} L_e^3}{E_{0,mean} I_{ef}}$$

$$u_v = \frac{1}{8} \frac{F_{d,sls} L_e}{G_w A_{ef}}$$

$$u_{inst} = u_m + u_v$$

$$u_{inst} = \frac{5}{384} \frac{F_{d,sls} L_e^3}{E_{0,mean} I_{ef}} + \frac{1}{8} \frac{F_{d,sls}}{G_w A_{ef}} = \frac{5}{384} \frac{(25.245) \times 6750^3}{11 \times 8.2263 \times 10^8} + \frac{25.245 \times 6750}{8 \times 0.5 \times 8.75 \times 10^3}$$

$$= 16.04 \text{ mm}$$

The design limit is 6750/300 = 22.5 mm, and is therefore OK.

The final deformation of the beam, u_{fin}:

1 Based on the revised method in Appendix A:

$$u_{m.fin} = \frac{5}{384} \frac{[F_{d,sls} - (F_{d.G} + \varphi_2 F_{d.Q}) L_e] L_e^3}{E_{0,mean} I_{ef}} + \frac{5}{384} \frac{[(F_{d.G} + \varphi_2 F_{d.Q}) L_e] L_e^3 (1 + k_{def,f})}{E_{0,mean} I_{c,ef}}$$

$$= \frac{5}{384} \frac{[25.245 - (1.04 + 0.3 \times 2.7) \times 6.75] \times 6750^3}{11 \times 8.2263 \times 10^8}$$

$$+ \frac{5}{384} \frac{(1.04 + 0.3 \times 2.7) \times 6.75 \times 6750^3 \times 1.8}{11 \times 8.088 \times 10^8} = 5.65 + 10.12 = 15.77 \text{ m}$$

$$u_{v.fin} = \frac{1}{8} \frac{[F_{d,sls} - (F_{d.G} + \varphi_2 F_{d.Q}) L_e] L_{ef}}{G_w A_{ef}} + \frac{1}{8} \frac{(F_{d.G} + \varphi_2 F_{d.Q}) L_{ef}^2 (1 + k_{def,w})}{G_w A_{ef}}$$

$$= \frac{1}{8} \frac{[25.245 - (1.04 + 0.3 \times 2.7) \times 6.75] \times 6750}{0.5 \times 8.75 \times 10^3}$$

$$+ \frac{1}{8} \frac{(1.04 + 0.3 \times 2.7) \times 6.75 \times 6750 \times 2}{0.5 \times 8.75 \times 10^3} = 2.46 + 4.82 = 7.28 \text{ mm}$$

$$u_{fin} = u_{m.fin} + u_{v.fin} = 15.77 + 7.28 = 23.05 \text{ mm}$$

2 Based on the conservative approach referred to in 2.2.3 and Chapter 5:

$$u_{fin} = \frac{5}{384} \frac{(F_{d,sls}) L_e^3 (1 + k_{def,f})}{E_{0,mean} I_{c,ef}} + \frac{1}{8} \frac{(F_{d,sls}) L_{ef} (1 + k_{def,w})}{G_w A_{ef}}$$

$$= \frac{5}{384} \frac{25.245 \times 6750^3 \times 1.8}{11 \times 8.008 \times 10^8} + \frac{1}{8} \frac{25.245 \times 6750 \times 2}{0.5 \times 8.75 \times 10^3} = 20.45 + 9.74 = 30.19 \text{ mm}$$

(i.e. approximately a 31% increase in value over EN 1995-1-1 method).

The design limit is 6750/150 = 45 mm O.K. (for both options).

7.3. Vibrations

7.3.1 General

The vibration of timber members, components and structures must be kept within levels that ensure there will be no adverse impact on the ability of the structure to fulfil its functional requirements and that there will be an acceptable level of comfort for the user.

Vibration behaviour is largely controlled by ensuring that the fundamental frequencies of the structural elements are kept above minimum values, and the particular problem associated with timber structures is the vibration behaviour caused by footfall-induced vibrations on residential floors. Guidance on the specific requirements to be met in floors in residential accommodation is given in *clause 7.3.3*.

Clause 7.3.3

Residential floor structures in Europe are generally detailed and designed to span two ways, and based on research by Ohlsson (1982), it has been concluded that a modal damping ratio, ζ, of 1% should be used for these structures, and has been adopted in *clause 7.3.1*. However, where an alternative value can be shown to be appropriate, it can be used, and in the UK it has been decided that for UK floors, which are normally designed to span one way, a value of 2% should be used. This is confirmed in *Table NA.6* in the National Annex to EN 1995-1-1.

Clause 7.3.1

7.3.2 Vibrations from machinery

Where machinery is to be supported by the structure and can cause steady vibrations, the levels of vibration against which the structure and its elements must be validated should be based on unfavourable combinations of the permanent and variable load. For floor structures, acceptable limits for continuous vibration can be obtained from Figure 5a in Appendix A of ISO 2631-2: 2003 (ISO, 2003).

These types of problem are common in structures built from any type of structural material, and are normally analysed using a dynamic analysis. Where the acceptable limits are exceeded, the options open to the designer are to detune the vibration effect by the use of anti-vibration mountings or to isolate the machinery from the structure.

7.3.3 Residential floors

For floor structures in residential accommodation, the critical loading condition is caused by footfall-induced vibrations, and under this loading, unless the fundamental frequency of the floor structure, f_1, is high, dynamic resonant response and 'bouncy' floor behaviour can arise. Floors in which $f_1 \leq 8$ Hz can be classed as low frequency and those where $f_1 > 8$ Hz can be classed as high frequency, and in EN 1995-1-1 a threshold level, $f_1 > 8$ Hz, has been taken as the level at which the design rules in *clause 7.3.3* will apply. With low-frequency floor structures, different design principles to those used for high-frequency behaviour have to be applied, but no guidance is given on how this should be done. The design rules dictate that all floor structures must have a fundamental frequency >8 Hz.

Clause 7.3.3

Clause NA.2.7

Taking into account the requirements of *clause NA.2.7* in the National Annex to EN 1995-1-1, for a floor structure of length ℓ and width b, simply supported on its four sides and comprising timber floor beams having an effective span ℓ, the fundamental frequency of the floor structure, f_1, is approximately obtained from *Equation (7.5)* as follows:

$$f_1 = \frac{\pi}{2\ell^2}\sqrt{\frac{(EI)_\ell}{m}} \tag{7.5}$$

ℓ is the design span of the floor beams (in m). $(EI)_\ell$ is the equivalent flexural rigidity of the floor structure about the axis perpendicular to the direction of span (in $\mathrm{Nm^2/m}$). Unless the floor decking has been glued to the beams in accordance with the requirements of *clauses 3.6* and *10.3*, and designed in accordance with the rules in *clause 9.1.2*, no allowance can be made for composite action. m is the mass per unit area of the floor (in $\mathrm{kg/m^2}$), based solely on permanent actions and excluding loads from partitions.

Clause 3.6
Clause 10.3
Clause 9.1.2

Clause 7.3.3

Although not stated in *clause 7.3.3*, *Equation 7.5* also applies to timber floors supported along two edges as well as to simply supported beams carrying uniformly distributed loading.

In the National Annex to EN 1995-1-1, an expression is also given for calculating the fundamental frequency of a girder joist (i.e. a single member or multiple member joist that directly supports either other joists or another girder joist).

The fundamental frequency relationships referred to above assume that the end supports of the floor structure beams (or of the girder joists) are rigid, assuming no interaction between supporting structures. Where there is concern that the interaction between the floor structure and its supporting beams could result in a fundamental frequency less than 8 Hz, the approximate approach proposed by Wyatt (1989) could be used to determine the fundamental frequency of the combined system, using Dunkerly's method. In this method, the estimated frequency of the combined floor system, f_0, is obtained from

$$\frac{1}{f_0^2} = \frac{1}{f_1^2} + \frac{1}{f_2^2} \qquad \text{(D7.2)}$$

where f_1 is the fundamental frequency of the floor structure assuming rigid beam supports and f_2 is the fundamental frequency of each beam (assuming loading from the floor).

Example 7.3: the fundamental frequency of the simply supported floor structure in a domestic building, formed with solid timber beams and OSB flooring

The floor beams are 47 mm by 190 mm strength class C18, spaced at 400 mm c/c with an effective span of 3.85 m (Figure 7.7). OSB/3 flooring, 18 mm thick, is fixed to the floor beams but does not form a composite structure with the beams; 12.5 mm-thick plasterboard is fixed on the underside; and the floor mass is 40 g/m^2.

Figure 7.7. Cross-section of the floor

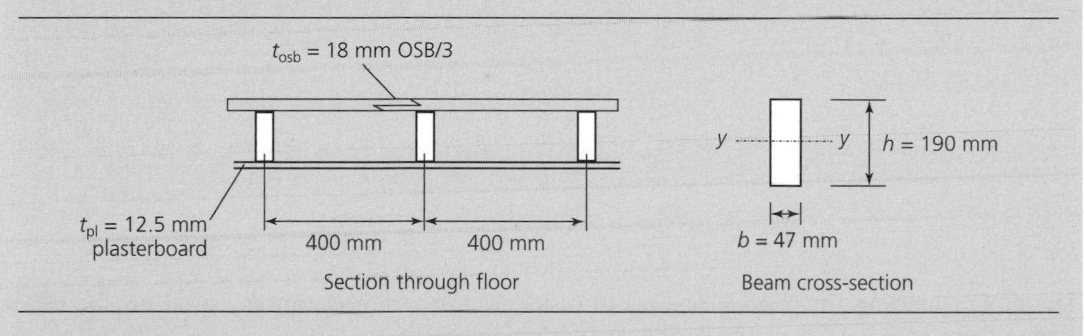

Mean modulus of elasticity of the timber beams:

$$E_{0,\text{mean}} = 9\ \text{kN/mm}^2$$

Mean modulus of elasticity of the OSB parallel to the beam span:

$$E_{\text{osb},90,\text{mean}} = 1.98\ \text{kN/mm}^2$$

Mean modulus of elasticity of the OSB perpendicular to the beam span:

$$E_{\text{osb},0,\text{mean}} = 4.93\ \text{kN/mm}^2$$

Mean modulus of elasticity of the plasterboard:

$$E_{\text{pl},0,\text{mean}} = 2.00\ \text{kN/mm}^2 \qquad (\textit{clause NA.2.7})$$

Clause NA.2.7

Effective span:

$$L_{\text{ef}} = 3.85\ \text{m}$$

Floor mass:

$$m = 40\ \mathrm{kg/m^2}$$

Beam spacing:

$$b_s = 400\ \mathrm{mm}$$

I of the beam about its y–y axis:

$$I_y = bh^3/12 = 47 \times 190^3/12 = 2.686 \times 10^7\ \mathrm{mm^4}$$

I of the OSB about its y–y axis:

$$I_{osb,y} = b_S t_{osb}^3/12 = 400 \times 18^3/12 = 1.944 \times 10^5\ \mathrm{mm^4}$$

I of the plasterboard about its y–y axis:

$$I_{pl,y} = b_S t_{pl}^3/12 = 400 \times 12.5^3/12 = 6.51 \times 10^4\ \mathrm{mm^4}$$

The fundamental frequency of the floor, f_1, where EI values are in $\mathrm{Nm^2/m}$, based on *Equation 7.5*:

$$f_1 = \frac{\pi}{2L_{ef}^2}\sqrt{\frac{E_{0,mean}I_y + E_{osb,90,mean}I_{osb,y} + E_{pl,0,mean}I_{pl,y}}{b_s m}}$$

$$= \frac{\pi}{2 \times 3.85^2}\sqrt{\frac{9 \times 2.686 \times 10^4 + 1.98 \times 1.944 \times 10^2 + 2 \times 6.51 \times 10}{0.4 \times 40}}$$

$$= 13.04\ \mathrm{Hz}$$

(Note: if the contribution of the flooring and the plasterboard are ignored, $f_1 = 13.027$ Hz.)

The design loading for floor structures in residential accommodation is based on the effects arising from walking across the floor structure, and comprises:

(*a*) a low-frequency force arising from the weight of the person walking over the floor
(*b*) a high-frequency force arising from heel impact.

Because the enforcing step frequency is lower than the fundamental frequency of the floor structure, the force generated by (a) can be considered to be a static vertical load, F_d (kN), and, allowing for load distribution within the floor structure the resulting maximum instantaneous vertical deflection, w, must comply with the requirements of *clause 7.3.3*, *Equation 7.3*:

Clause 7.3.3

$$w/F_d \leq a\ (\mathrm{mm/kN}) \tag{7.3}$$

The ratio w/F_d is equivalent to the instantaneous vertical deflection of the floor structure under a 1 kN vertical force, with the largest deflection being obtained by placing this force at the centre of the floor. The expression to be used to calculate the deflection is given in *clause NA.2.7.2* in the National Annex to EN 1995-1-1, and is

Clause NA.2.7.2

$$w/F_d = 1000 k_{dist} \ell_{eq}^3 k_{amp}/48(EI)_{joist} \leq a \tag{D7.3}$$

where the functions are as described in the National Annex.

The limiting value of a in *Equation 7.3* is given in *Table NA.6* in the National Annex, and for a floor of span ℓ mm is

$a \leq 1.8$ mm for floor spans $\leq$4000 mm

$a \leq 16\,500/\ell^{1.1}$ mm for floor spans >4000 mm

When designing for the high-frequency heel impact effect, the loading from (b) above is taken to be a vertical unit impact of 1.0 N s applied at the floor position giving the maximum vibration response. This generates vertical vibrations in the floor, and the vertical floor vibration velocity response under this impulse, v, has to be shown to be less than the limiting value obtained from *Equation 7.4* as follows:

$$v \leq b^{(f_1\zeta - 1)} \qquad (7.4)$$

v is dependent on the natural frequencies excited and the modal mass. From research it has been concluded that only natural frequencies less than 40 Hz need be considered, and the modal mass will be the mass of the floor (as used to derive the fundamental frequency, f_1) plus an allowance of 50 kg from the person on the floor being disturbed by this vibration. Taking these factors into account, v is derived from *Equation 7.6*. b is a constant that depends on the value of a, and by setting $a = w/F_d$ (in mm) is obtained from the equations given in *Table NA.6* as follows:

$$b = 180 - 60a \quad \text{when } a \leq 1 \text{ mm} \qquad (7.4)$$

$$b = 160 - 40a \quad \text{when } a > 1 \text{ mm} \qquad (7.5)$$

f_1 is the fundamental frequency of the floor structure (in Hz). ζ is the modal damping ratio, and, as stated in *Table NA.6* in the National Annex to EN 1995-1-1 and noted in Section 7.3.1, is to be taken as 2%.

Because the UK has adopted such a high value for the damping ratio, it is extremely unlikely that unit impulse velocity response will ever be a limiting factor in floor design.

Example 7.4: check on floor deflection and unit impulse velocity response of the floor structure in Example 7.3

The floor width, w, is 4.5 m, and the fundamental frequency, f_1, will be 13.04 Hz. The mean modulus of elasticity of the timber beams, $E_{0,\text{mean}} = 9$ kN/mm^2; the mean modulus of elasticity of the OSB perpendicular to beam span, $E_{\text{osb},0,\text{mean}} = 4.93$ kN/mm^2; the mean modulus of elasticity of the plasterboard, $E_{\text{pl},0,\text{mean}} = 2.0$ kN/mm^2 (*clause NA.2.7*); I of a beam about its y–y axis, $I_y = 2.686 \times 10^7$ mm^4; and the effective beam span, $L_{\text{ef}} = 3850$ mm.

Clause NA.2.7

1 Check the static floor deflection based on the requirements of the National Annex to EN 1995-1-1.

Maximum allowable deflection under 1 kN:

$a_p = 1.8$ mm (*Table NA.6*)

k_{strut} factor = 1.0 (*clause NA.2.7.2*)

Clause NA.2.7.2

$k_{\text{amp}} = 1.05$ (for solid timber beams)

$$EI_{\text{joist}} = E_{0,\text{mean}} \times I_y = 9 \times 2.686 \times 10^{10} = 2.418 \times 10^{11} \text{ N mm}^2$$

The k_{dist} factor over a width $b_t = 1000$ mm and beam spacing $b_s = 400$ mm is now calculated.

I of the OSB flange perpendicular to the beam:

$$I_{osb,b} = b_t t_{osb}^3/12 = 10^3 \times 18^3/12 = 4.86 \times 10^5\ \mathrm{mm}^4$$

I of the plasterboard perpendicular to the beam:

$$I_{pl} = b_t t_{pl}^3/12 = 10^3 \times 12.5^3/12 = 1.628 \times 10^5\ \mathrm{mm}^4$$

$$EI_b = E_{osb,0,mean} I_{osb,b} + E_{pl,0,mean} I_{pl} = (4.93 \times 4.86 + 2.0 \times 1.628) \times 10^8$$

$$= 2.722 \times 10^9\ \mathrm{N\,mm^2/m}$$

$$k_{dist} = \max\{k_{strut} \times [0.38 - 0.08 \ln(14 \times EI_b/b_s^4)], 0.3\} = 0.348$$

$$\delta = (10^3 k_{dist} L_{ef}^3 k_{amp})/(48 EI_{joist}) = (10^3 \times 0.348 \times 3850^3 \times 1.05)/(48 \times 2.418 \times 10^{11})$$

$$= 1.8\ \mathrm{mm}$$

The value is the same as a_p, so it is just OK.

2 Check the unit impulse velocity response of the floor based on the combined requirements of EN 1995-1-1 and its National Annex.

Constant for the control of unit impulse response:

$$b = 160 - 40a \qquad (\textit{Table NA.6})$$

$$= (160 - 40 \times 1.8) = 88$$

Modal damping ratio:

Clause NA.2.7

$$\zeta = 0.02 \qquad (\textit{clause NA.2.7})$$

Allowable unit impulse velocity response:

$$b^{(f_1 \zeta - 1)} = 88^{(13.04 \times 0.02 - 1)} = 0.037\ \mathrm{m/(N\,s^2)} \qquad (\textit{Equation 7.4})$$

Equivalent plate bending stiffness of the floor about an axis parallel to the beam direction (see Example 7.3):

$$EI_\ell = \frac{(E_{0,mean} I_y + E_{osb,0,mean} I_{osb,y} + E_{pl,0,mean} I_{pl,y})}{J_s} = 6.057 \times 10^5\ \mathrm{Nm^2/m}$$

Number of first-order modes with natural frequencies up to 40 Hz:

$$n_{40} = \left\{\left[\left(\frac{40}{f_1}\right)^2\right] - 1\left(\frac{w}{L_{ef}}\right)^4 \frac{(EI)_\ell}{(EI)_b}\right\}^{0.25}$$

$$= \left\{\left[\left(\frac{40}{13.04}\right)^2\right] - 1\left(\frac{4.5}{3.85}\right)^4 \frac{6.057 \times 10^5}{2.722 \times 10^3}\right\}^{0.25} = 7.69$$

Unit impulse velocity response:

$$v = \frac{4(0.4 + 0.6 n_{40})}{mw\ell + 200} = \frac{4(0.4 + 0.6 \times 7.69)}{40 \times 4.5 \times 3.85 + 200} = 0.022\ \mathrm{m/(N\,s^2)}$$

which is less than 0.037 (*Equation 7.4*), and therefore OK.

REFERENCES

BSI (2011) BS EN 912: 2011. Timber fasteners. Specifications for connections for timber. BSI, London.

Dias AMPG, Cruz HMP, Lopes SMR and van de Kuilen JW (2010) Stiffness of dowel-type fasteners in timber–concrete joints. *Proceedings of the ICE: Structures and Buildings* **163(584)**: 257–266.

ISO (2003) ISO 2631-2: 2003. Mechanical vibration and shock – Evaluation of human exposure to whole-body vibration – Part 2: Vibrations in buildings. International Organization for Standardization, Geneva.

Ohlsson S (1982) Floor vibrations and human discomfort. PhD thesis, Chalmers University of Technology, Gothenburg.

Wyatt TA (1989) *Design Guide on the Vibration of Floors*. SCI Publication 076. The Steel Construction Institute, Ascott.

Young WC (1989) *Roark's Formulas for Stress and Strain*, 6th edn. McGraw-Hill, New York.

Designers' Guide to Eurocode 5: Design of Timber Buildings
ISBN 978-0-7277-3162-3

http://dx.doi.org/10.1680/dtb.31623.095

Chapter 8
Connections with metal fasteners

This chapter is concerned with the requirements for connections formed using metal fasteners. It covers the material in *Section 8* of EN 1995-1-1, and addresses the following clauses:

- General *Clause 8.1*
- Lateral load-carrying capacity of metal dowel-type fasteners *Clause 8.2*
- Nailed connections *Clause 8.3*
- Stapled connections *Clause 8.4*
- Bolted connections *Clause 8.5*
- Dowelled connections *Clause 8.6*
- Screwed connections *Clause 8.7*
- Connections made with punched metal plate fasteners *Clause 8.8*
- Split ring and shear plate connectors *Clause 8.9*
- Toothed-plate connectors *Clause 8.10*

8.1. General

In most timber structures, the strength, stiffness and spatial requirements of the fasteners in connections will generally determine the strength and stiffness behaviour of the structure, as well as the size of members that can be used. *Section 8* of EN 1995-1-1 covers connection strength behaviour, and connection stiffness requirements are addressed in *Section 7*. Where rules for calculating the characteristic load-carrying capacity and the stiffness of the connection are not given, values must be derived by testing.

Where reference is made in *Section 8* to timber members, unless otherwise stated the rules will apply to members made from softwood, hardwood, glued laminated timber and laminated veneer lumber.

EN 1995-1-1 only gives design rules for metal-based fasteners, and the types covered are nails, staples, bolts, dowels, screws, punched metal plate fasteners and connectors (split ring, shear plate and toothed-plate).

Connection strength is a function of several factors, and a comparison of the characteristic lateral load-carrying capacity of a timber-to-timber connection as shown in Figure 8.1, formed using C24 timber to EN 338 (BSI, 2009a) with different types and sizes of fastener, loaded parallel to the grain is given in Table 8.1. The values in the table are for a single fastener, but derived on the basis that there is more than one in the connection. Using the slip modulus at the serviceability limit state given in *Table 7.1*, the stiffness of each fastener type is also included in the table.

Connections can be formed with fasteners laterally loaded in single or double shear, and examples of the different types are shown in Figure 8.2. The plane between connected members is defined as a shear plane, there being one shear plane in a single-shear connection, and two in a double-shear connection.

Connections will fail in either a ductile or a brittle manner when subjected to lateral loading. Ductile failure occurs when the connection yields without significant strength loss and brittle failure when the timber splits or tears or the fastener shears and there is sudden failure without warning. The rules in EN 1995-1-1 are written to try to ensure that ductile failure will

Figure 8.1. Laterally loaded connection

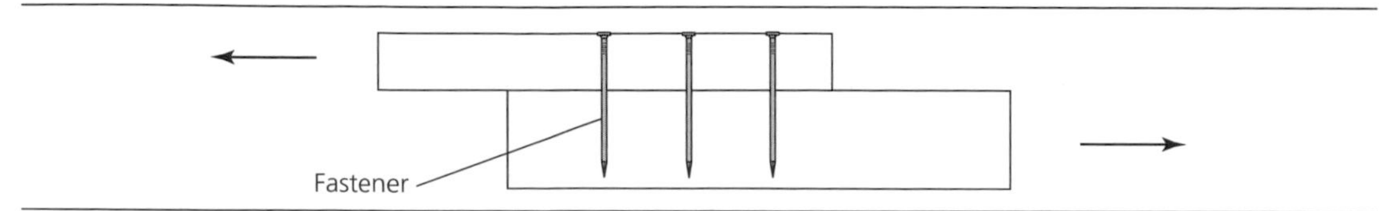

always arise, and in situations where both ductile and brittle types are possible it is good practice to try to ensure that the design condition is based on the ductile failure mechanism.

Clause 2.1.4

In *Section 8*, the rules for fasteners are generally given as a **characteristic load-carrying capacity per shear plane per fastener**, and to convert to the design value, the rules in *clause 2.1.4* will apply. For example, to obtain the design load-carrying capacity of a nail loaded laterally, $F_{v,Rd}$, from its characteristic capacity, $F_{v,Rk}$:

$$F_{v,Rd} = \frac{k_{mod}}{\gamma_M} F_{v,Rk} \qquad \text{(D8.1)}$$

Clause 3.1.3

where k_{mod} is the modification factor referred to in *clause 3.1.3*, γ_M is the partial factor for the connection given in the National Annex to EN 1995-1-1; both factors are also referred to in Chapters 2 and 3 in this guide.

Clause 5.2

The rules in *Section 8* cover the design requirements of the fastener as well as any brittle failure condition that can arise in the connection. There is also the need to validate the strength of the members in the connection in accordance with the relevant rules in *Section 6*, taking into account the reduction in cross-sectional area arising from fitting the fastener(s), and design rules for area loss and for connections with multiple fasteners are given in *clause 5.2*.

Unless otherwise stated in EN 1995-1-1, the strength rules for fasteners loaded laterally assume that all fasteners are driven/fitted **perpendicular to the grain direction**.

Clause 10.3

No guidance is given in EN 1995-1-1 on the design requirements of glued connections, but rules for the design of overlapping glued joints are given in PD 6693-1, with quality control requirements being as referred to in *clause 10.3* and in Section 10.3 of this guide.

8.1.1 Fastener requirements

Clause 8.1.1

Where no design rules are given in *Section 8* to enable the stiffness and the characteristic load-carrying capacity of a connection to be calculated, they have to be determined by tests, and the standards to be used are given in *clause 8.1.1*.

Table 8.1. Strength and stiffness of different types of fastener

Nominal diameter (d) of the fastener: mm	Characteristic lateral capacity of the fastener[a]: kN					Stiffness: kN/mm (based on *Table 7.1*)
	Smooth nails:[b] $f_u = 600$ N/mm^2	Screws:[b] $f_u = 500$ N/mm^2	Bolts/dowels:[c] $f_{uk} = 500$ N/mm^2	Split ring (EN 912, Type A2)	Toothed-plate (EN 912, Type C1)	
3.0	0.72	0.42	–	–	–	0.69 (nail); 1.12 (screw)
6.0	2.25	1.33	2.6	–	–	1.20 (nail); 2.25 (screw) 2.25 (bolt)
20.0	–	–	14.3	–	–	7.48
64	–	–	–	21.4	–	15.12
62 (with 20 mm diameter bolt)	–	–	–	–	21.07	9.77

[a]The strength excludes the rope effect and is the value per shear plane
[b]Nails and screws are not pre-drilled and the screw strength is derived for a condition where $d_{ef} = 1.1(0.7d)$
[c]The member thicknesses used is the same for both bolt sizes

Figure 8.2. Laterally loaded fasteners: (a–c) single shear; (d) double shear

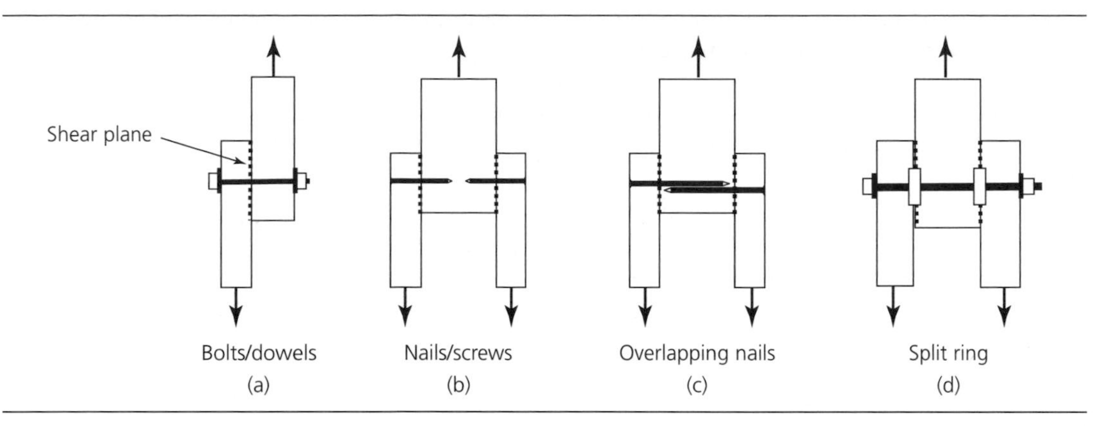

8.1.2 Multiple fastener connections

In most connections, there will be several fasteners, particularly when using nails, screws and bolts, and it is normal practice not to vary the type and size of the fastener in the connection. The rules in EN 1995-1-1 have been derived on this understanding. If different types and sizes are to be used, the compatibility of the design rules used must be verified by the designer.

Connections can be subjected to lateral forces with and without moments, to axial forces and to combinations of these actions. Where it is certain that connection failure will be based solely on ductile behaviour, it is possible to determine the forces on the fasteners using pure plastic theory. However, because there is the risk of brittle failure modes in connections (e.g. connections loaded at an angle to the grain or where steel gusset plates are used), it is suggested in this guide that the design forces in fasteners are derived assuming a linear elastic approach. The basic rules of statics are used, taking into account the effect of any eccentricity of the loading relative to the geometric centre of the fastener group in the connection.

When a connection is formed using dowel-type fasteners and subjected to an in-plane moment, the assumption is made in this guide that the members will rotate as rigid bodies and that all movement in the connection will be due to lateral displacement of the fasteners. Each fastener will move by an amount proportional to its distance from the centre of rotation of the fastener group. For the generalised arrangement shown in Figure 8.3, in which the connection has n fasteners in a shear plane and is subjected to a design moment M_d per shear plane, the force due to the moment in fastener i, $F_{md,i}$, at a radius r_i from the centre of rotation, will be

$$F_{md,i} = M_d \frac{r_i}{\sum_{i=1}^{n} r_i^2} \tag{D8.2}$$

Figure 8.3. Connection with dowel fasteners subjected to an in-plane moment and a direct force

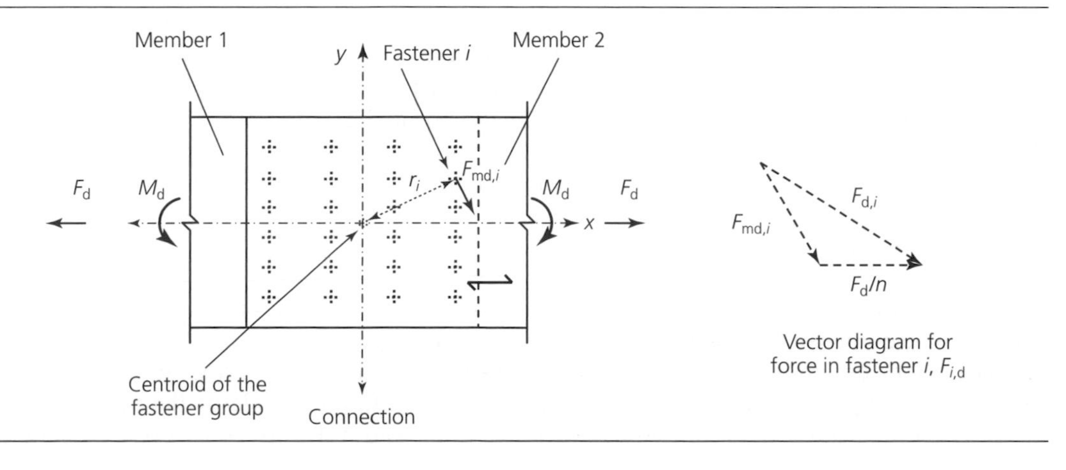

The fastener having the largest radius will be subjected to the greatest design force. If each shear plane is also subjected to lateral design force, F_d, the resultant force on fastener i will be the vector sum of the forces, as shown in Figure 8.3.

The value of the force in a fastener per shear plane in some generalised arrangements under differing loading regimes can be obtained from the relationships given in Figure 8.4.

When a number of fasteners, n, are aligned in a row running parallel to the grain of the timber, the effective characteristic load carrying capacity of the row per shear plane when loaded laterally along the row direction, $F_{v,ef,Rk}$, will be:

$$F_{v,ef,Rk} = n_{ef}F_{v,Rk} \quad (8.1)$$

where $F_{v,Rk}$ is the characteristic load-carrying capacity of each fastener per shear plane loaded parallel to the grain, and n_{ef} is the effective number of fasteners in the row. If there are r rows in the connection, the characteristic load-carrying capacity per shear plane will be $rF_{v,ef,Rk}$.

Clause 8.3.1.1(8)
Clause 8.5.1.1(4)
Clause 8.9(12)

The value of n_{ef} is dependent on the type of fastener being used, and is defined in *clause 8.3.1.1(8)* for nails, screws (6 mm or less in diameter) and staples; *clause 8.5.1.1(4)* for bolts, dowels and screws >6 mm in diameter; and in *clause 8.9(12)* for ring and shear plate connectors. When using nails, screws and bolts and staples, the effective number is also dependent on the spacing between fasteners in the grain direction, and examples of the calculation of n_{ef} in connections formed with bolts/dowels where the grain direction of all of the members in the connection is not aligned or where panel or steel gusset plates are being used are shown in Figure 8.5.

Clause 8.1.4

When the force in the shear plane in a connection is at an angle to the row of fasteners parallel to the grain, the force component parallel to the row must be shown not to exceed the effective load-carrying capacity of the row calculated using *Equation 8.1*. The force component perpendicular to the grain of the timber member must also be checked to ensure that the splitting strength referred to in *clause 8.1.4* is not exceeded.

8.1.3 Multiple shear plane connections

Where there are more than two shear planes in a connection and the members are at varying angles to each other, the connection behaviour is highly redundant and cannot be solved by the simple statics approach used to design single and double shear connections. This is a common situation in truss structures, and an example of a typical five-member connection is shown in Figure 8.6.

Clause 8.1.3(1)

For these types of connection, *clause 8.1.3(1)* states that the capacity of each shear plane is derived by analysing the connection assuming that each shear plane is part of a series of three-member connections.

The three-member connections are formed by working from one side of the connection (the left side being used in this guide), forming a series of symmetrical three-member connections such that the central member of each is the actual joint member and the outer members are its adjacent members in the original connection. Where the adjacent members have different properties (e.g. material, cross-sectional or directional), the member on the right side of the connection is replaced by the left-hand member to form the symmetrical connection. This is shown in Figure 8.6 where connections (a), (b) and (c) are formed from members 1, 2, 1, 2, 3, 2 and 3, 2, 3. For each three-member connection, assuming the fastener behaves as a rigid member, the force in each shear plane together with its direction relative to the member grain direction can be derived by statics.

Clause 8.2

Knowing the directions of the shear plane forces, using the design rules in *clause 8.2*, the strength of each connection can be derived, and for each three-member connection it must be shown that the shear plane design strength is not less than the shear plane design force. When deriving the connection strength, EN 1995-1-1 requires the governing failure mode of the fasteners in the respective shear planes to be compatible with each other and must not consist of a combination

Figure 8.4. Value of the force in a fastener per shear plane for different connections

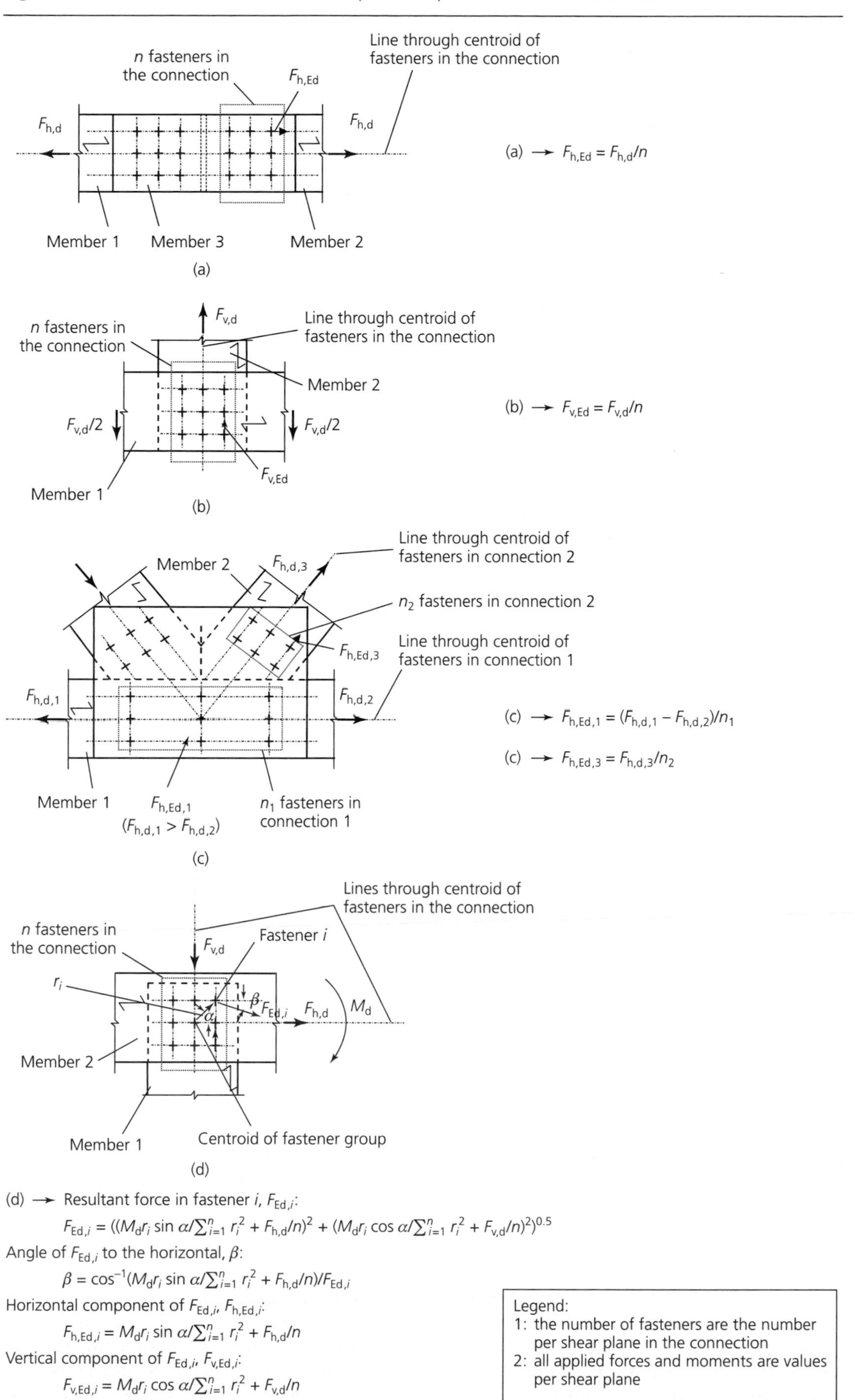

Figure 8.5. Effective number of fasteners (bolts/dowels) in connections

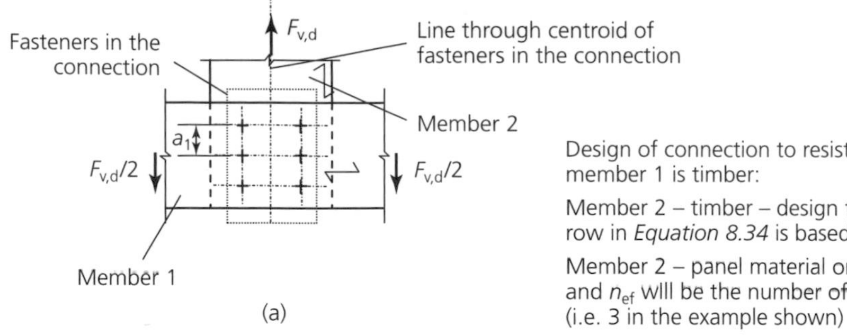

Design of connection to resist $F_{v,d}$, and member 1 is timber:

Member 2 – timber – design for 2 rows and n_{ef} per row in *Equation 8.34* is based on the fastener spacing a_1

Member 2 – panel material or steel – design for 2 rows and n_{ef} will be the number of fasteners in each row (i.e. 3 in the example shown)

Design of connection to resist $F_{v,d}$, and member 1 is timber

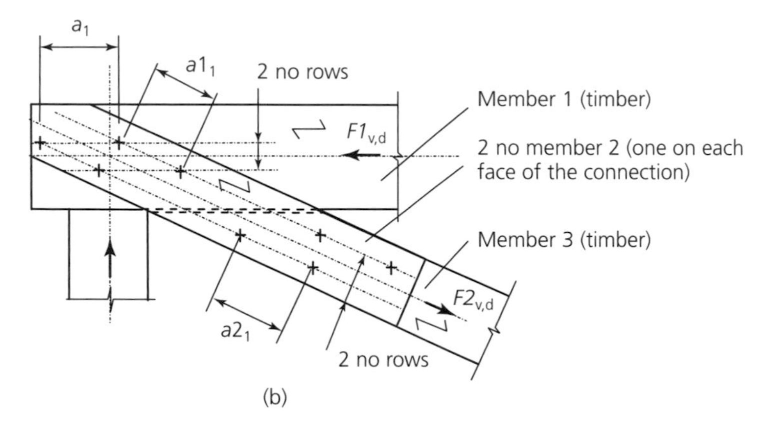

Design of connection to resist connection component force $F1_{v,d}$:

Member 2 – timber, panel material or steel – design for 2 rows and n_{ef} per row derived from *Equation 8.34* is based on fastener spacing a_1

Design of connections to resist connection shear plane force $F2_{v,d}$:

Member 2 – timber – design for 2 rows and n_{ef} per row derived from *Equation 8.34* for the connection in member 1 is based on the fastener spacing $a1_1$ and for the connection with member 3, on fastener spacing $a2_1$

Clause 8.5.1.1(4)

Member 2 – panel material or steel – design for 2 rows and n_{ef} per row for the connection in member 1 may be determined using the requirements of the final paragraph in *clause 8.5.1.1(4)*, i.e. by linear interpolation between *Equations 8.34* and *8.35*, where the fastener spacing in *Equation 8.34* will be a_1. For the connection with member 3, n_{ef} per row will be derived from *Equation 8.34* and the fastener spacing will be $a2_1$

of type 1 failure modes (i.e. failure modes (a), (b), (g) and (h) from *Figure 8.2* or modes (c), (f) and (j/l) from *Figure 8.3*) with the other failure modes.

8.1.4 Connection forces at an angle to the grain

In a connection involving timber members, if any member is loaded at an angle to its grain, the component of the force at right angles to the grain can result in tension splitting of the member, requiring the member splitting strength to be validated. Examples of this type of situation are shown in *Figure 8.1*, and reproduced in Figure 8.7.

Clause 8.1.4
Clause 8.1.4(3)

Consider, for example, connection (a) in Figure 8.7. The design force in the inclined members, F_{Ed}, arises from actions on the horizontal member, and from statics the vertical component of this force, $F_{ed} \sin \alpha$, will equal the sum of the shear forces $F_{v,Ed,1}$ and $F_{v,Ed,2}$. This force arrangement can cause tension splitting in the horizontal member, as shown in Figure 8.7(d), and the design rules in *clause 8.1.4* require that the larger of $F_{v,Ed,1}$ and $F_{v,Ed,2}$ does not exceed the design splitting capacity of the arrangement. The design rules in *clause 8.1.4(3)* **only apply** to softwood members and there are not rows of fasteners in the member being split.

Figure 8.6. Multiple shear plane connection

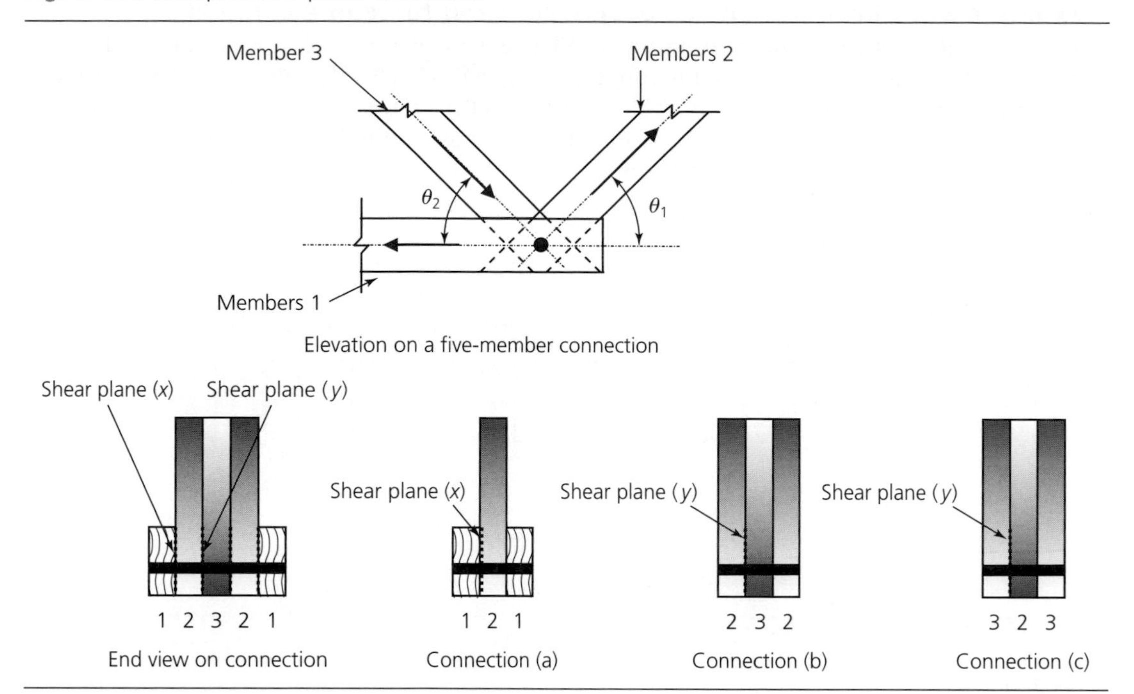

The characteristic splitting strength of softwoods, $F_{90,\mathrm{Rk}}$ is obtained from *Equation 8.4*, and the design splitting strength, $F_{90,\mathrm{Rd}}$, will be

$$F_{90,\mathrm{Rd}} = \frac{k_{\mathrm{mod}}}{\gamma_{\mathrm{M}}} F_{90,\mathrm{Rk}} \tag{D8.3}$$

where k_{mod} and γ_{M} are defined in *Equation D8.1*, and

$$F_{90,\mathrm{Rk}} = 14bw\sqrt{\frac{h_{\mathrm{e}}}{1 - h_{\mathrm{e}}/h}} \tag{8.4}$$

The symbols are as defined for *Equation 8.4* and shown in Figure 8.7, and w is a modification factor = 1 for all types of fastener except punched metal plate fasteners.

The design condition to be verified is

$$\mathrm{Max}(F_{\mathrm{v,Ed,1}}, F_{\mathrm{v,Ed,2}}) \leq F_{90,\mathrm{Rd}} \tag{D8.4}$$

Figure 8.7. Tension splitting situation in a connection

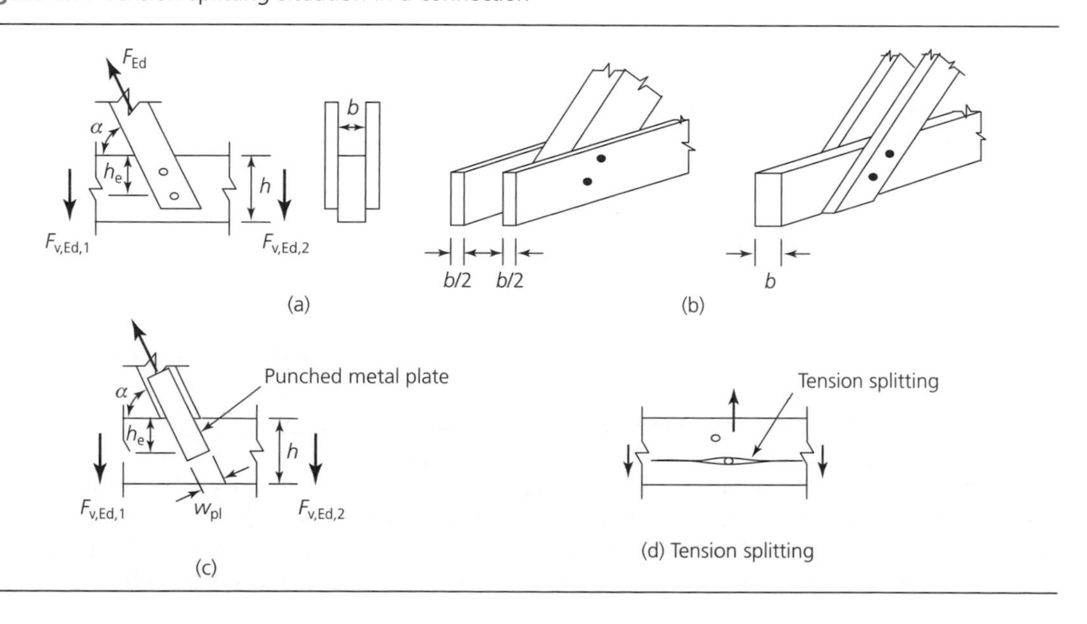

Example 8.1: calculation of the design splitting force in a connection

The corner joint of a truss made from softwood members is supported at each end, as shown in Figure 8.8. The design value of the support reaction is 8.9 kN ($V_{,d}$), due to a combination of permanent and medium-term variable loading, and the truss functions in service class 2 conditions. Check that the splitting strength of the horizontal 50 mm (t) by 120 mm (h) members complies with the requirements of EN 1995-1-1.

Figure 8.8. Corner joint of a truss

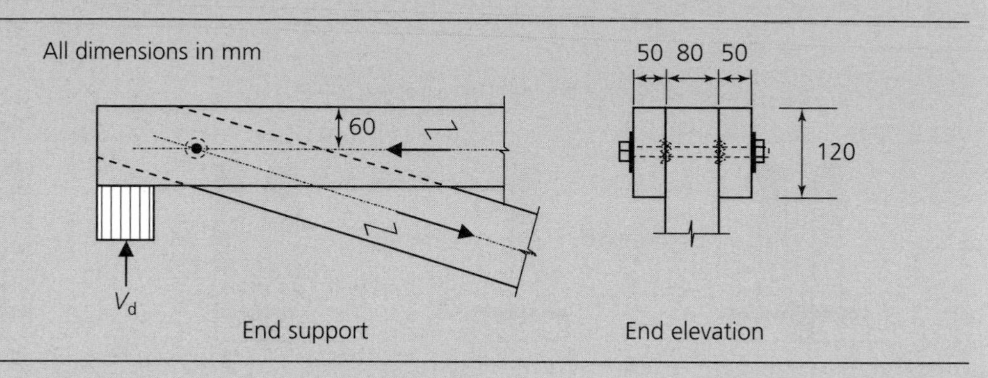

Design shear force:

$$V_d = 8.9 \text{ kN}$$

Characteristic splitting capacity of the connection, $F_{90,Rk}$ (*Equation 8.4*), with loaded edge distance $h_e = 60$ mm, member thickness $t = 50$ mm, $b = 2 \times 50$ mm and $w = 1$:

$$F_{90,Rk} = 14bw[h_e/(1 - h_e/h)]^{0.5} = 14 \times (2 \times 50) \times (1) \times [60/(1 - 60/120)]^{0.5} = 15.34 \text{ kN}$$

Design splitting capacity of the connection, $F_{90,Rd}$ (*Equation 2.14*), with $k_{mod} = 0.8$ and $\gamma_M = 1.3$:

$$F_{90,Rd} = k_{mod}F_{90,Rk}/\gamma_M = 0.8 \times 15.34/1.3 = 9.44 \text{ kN}$$

$$V_d/F_{90,Rd} = 8.9/9.44 = 0.94 < 1$$

Therefore OK.

8.1.5 Alternating connection forces

The strength of a connection must be reduced when any of its members are subjected to alternating forces caused by long- or medium-term actions. Where member design forces alternate between a tensile value, $F_{t,Ed}$, and a compressive value, $F_{c,Ed}$, the strength has to be validated against the following design forces:

$$\text{design force in tension} = (F_{t,Ed} + 0.5|F_{c,Ed}|) \qquad \text{(D8.5)}$$

$$\text{design force in compression} = (F_{c,Ed} + 0.5|F_{t,Ed}|) \qquad \text{(D8.6)}$$

8.2. Lateral load-carrying capacity of metal dowel-type fasteners

Clause 8.2

Clause 8.2 covers the design equations required to validate the strength of connections formed with dowel-type fasteners (i.e. nails, staples, screws, bolts and dowels) when subjected to lateral loading. Lateral loading occurs when a connection is subjected to actions acting in the plane of the connection members. The equations have been derived from the application of pure plastic theory to connection behaviour, assuming failure will be in a ductile yielding manner rather than a sudden brittle manner, as mentioned in Section 8.1 in this guide. The basic strength equations for timber connections were originally developed by Johansen (1949), and the equations given in *clauses 8.2.2* and *8.2.3* have been developed from his work, taking into account refinements and subsequent research findings.

Clause 8.2.2
Clause 8.2.3

To prevent spitting, rules for minimum spacings, edge and end distances, and, where necessary, for material thickness, have been derived for each dowel type, and, by complying with these requirements, ductile rather than brittle failure can be assumed. However, where connection members are loaded at an angle to the grain or a block and plug shear failure can occur, these brittle modes, which are referred to in Sections 8.1.4, 8.2.3 and 11.2 in this guide, respectively, also have to be validated.

Strength equations are given for timber-to-timber and panel-to-timber connections in *clause 8.2.2*, and for steel-to-timber connections in *clause 8.2.3*.

Clause 8.2.2
Clause 8.2.3

8.2.1 General

When calculating the characteristic load-carrying capacity of connections formed with metal dowel-type fasteners, the contributions from the yield moment, the embedment strength and the withdrawal strength of the fastener have to be taken into account.

8.2.2 Timber-to-timber and panel-to-timber connections

Taking into account the comments in Section 8.2 in this guide, when timber-to-timber and panel-to-timber connections are loaded laterally, alternative ductile failure modes can arise, and are shown in Figure 8.9 for single and double shear connections.

Modes (a) to (f) apply to single-shear connections, and (g) to (k) to double-shear connections. In modes (a), (b), (c), (g) and (h), failure is solely by crushing/yielding of the timber or panel material, referred to as embedment failure, and these modes are commonly defined as type 1 modes. Where there is embedment failure and the fastener also has a hinge failure in one member in a single-shear connection or two hinge failures in the central member of a double-shear connection and the dowels remain straight in the other member(s), as in the case of (d), (e), and (j), these modes are defined as type 2 modes. Where there is a combination of embedment failure and the dowel yields in all members, as shown in (f) and (k), these are called type 3 modes. Where the fastener being used is slender (i.e. its length-to-diameter ratio is high), generally the lowest strength will be associated with a type 3 mode of failure.

A strength equation has been derived for each failure mode **per shear plane per fastener**, and is given in *Equations 8.6* and *8.7* in *clause 8.2.2*. The lowest value from *Equations 8.6a* to *8.6f* for a single-shear connection and from *Equations 8.7g* to *8.7k* for a double-shear connection is the characteristic load-carrying capacity per shear plane per fastener, $F_{v,Rk}$. The assumption for double-shear connections is that the material used for the outer members is the same in all respects (i.e. material, thickness, grain alignment and density). If this is not the case and the outer members are made from materials m_1 and m_2, a safe approximate value is obtained by

Clause 8.2.2

Figure 8.9. Failure modes for timber and panel connections. The suffix to *t* refers to the member number. (Based on EN 1995-1-1 (*Figure 8.2*). Reproduced with permission from EN 1995-1-1, © British Standards Institute, 2004)

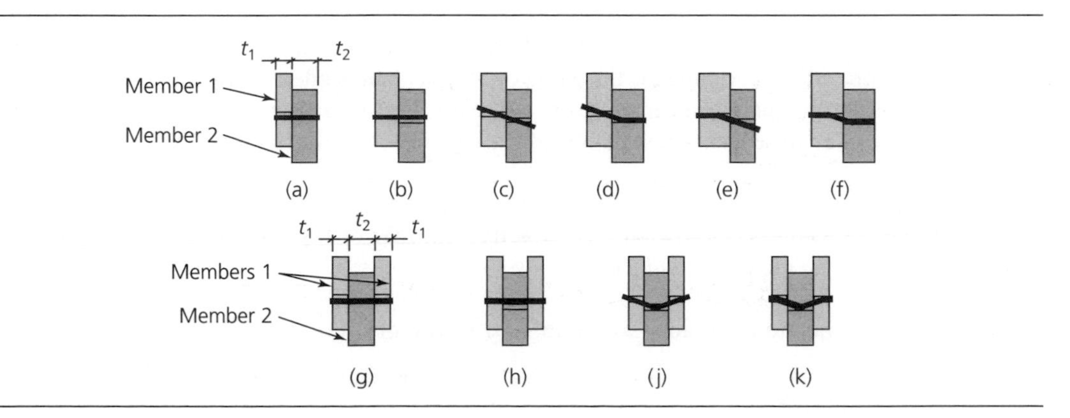

analysing two connections, one formed using m_1 for the outer members and the other using m_2. The lowest value from *Equations 8.7* for both set-ups can be taken to be $F_{v,Rk}$.

$F_{v,Rk}$ is a function of several variables that are dependent on the type of fastener being used, and where the suffix i is referred to, it means either member 1 or member 2, as shown in Figure 8.9. The variables are:

Clause 8.3.1.1
Clause 8.5.1.1

- t_i the timber or panel thickness or penetration depth in member i (it is defined in *clause 8.3.1.1* for nails, screws and staples and in *clause 8.5.1.1* for bolts)
- $f_{h,i,k}$ the characteristic embedment strength of member i (i.e. for member 1 it will be $f_{h,1,k}$ and for member 2, $f_{h,2,k}$)
- d the nominal diameter of the fastener, but in the case of screws it is the effective diameter
- $M_{y,Rk}$ the characteristic fastener yield moment
- β the ratio of the embedment strength of member 2 to member 1
- $F_{ax,Rk}$ the characteristic withdrawal capacity of the fastener.

Clause 8.3
Clause 8.4
Clause 8.5
Clause 8.6
Clause 8.7

These variables are defined in *clauses 8.3* for nails, *8.4* for staples, *8.5* for bolts, *8.6* for dowels and *8.7* for screws, and $F_{ax,Rk}$ for each fastener type is also derived from the rules in these clauses.

The strength equations are structured in two elements, as follows:

Clause 8.2.2(1)

(i) the '**Johansen**' element, which has been derived from yield theory, and is defined in *clause 8.2.2(1)* as the first part of the equation
(ii) the contribution from the '**rope effect**', which is the second part of each equation, and is derived from $F_{ax,Rk}/4$. The rope effect will generally only apply in modes where the fastener yields (i.e. type 2 and 3 modes).

An example of the two parts is shown using *Equation 8.7j*:

$$F_{v,Rk} = 1.05\frac{f_{h,1,k}t_1 d}{2+\beta}\left[\sqrt{2\beta(1+\beta)+\frac{4\beta(2+\beta)M_{y,Rk}}{f_{h,i,k}dt_1^2}}-\beta\right]+\frac{F_{ax,Rk}}{4} \qquad (8.7j)$$

first part – Johansen term   second part – rope effect term

Clause 8.2.2(2)

The contribution from the rope effect is limited for each type of fastener, and for those equations incorporating a rope effect, $F_{ax,Rk}/4$ as a percentage of the Johansen part of the equation must not exceed the values given in *clause 8.2.2(2)*. With staples, no rope effect is permitted.

Clause 10.4.3

For fasteners in single shear, $F_{ax,Rk}$ will be the lower of the capacities of the two members, and, with bolts, the resistance provided by the washer based on the maximum size permitted in *clause 10.4.3* may be taken into account. The values of $F_{ax,Rk}$ are derived from the rules discussed in this guide in Sections 8.3.2 for nails, 8.5.2 for bolts and 8.7.2 for screws. Where $F_{ax,Rk}$ is not known $F_{ax,Rk}/4$ must be taken to be 0.

Where the fastener in the connection is a nail of any diameter (i.e. up to 8 mm maximum diameter) or a screw $\leq$6 mm effective diameter, the value of $F_{v,Rk}$ is **independent** of the direction of the shear plane force relative to the grain direction. If bolts are used or screws with an effective diameter >6 mm, $F_{v,Rk}$ will be **dependent** on the angle of the shear plane force relative to the grain direction.

The characteristic load-carrying capacity of the connection, $Fc_{v,Rk}$, in which there are n_{sp} shear planes, r rows, n fasteners per row and the effective number of fasteners in a row loaded parallel to the grain is n_{ef}, is calculated as follows:

(i) Where the connection is loaded at an angle α to the grain,

$$Fc_{v,Rk} = nrn_{sp}F_{v,Rk} \qquad (D8.7)$$

and $F_{v,Rk}$ is the characteristic load-carrying capacity per shear plane per fastener loaded at an angle α to the grain. When using bolts, n is replaced by n_{ef}, and is derived from the rules given in *clause 8.5.1.1(4)*, referred to in Section 8.5.1.1 in this guide.

Clause 8.5.1.1(4)

(ii) Where the connection is loaded parallel to the grain,

$$Fc_{v,Rk} = n_{ef} r n_{sp} F_{v,Rk} \tag{D8.8}$$

and $F_{v,Rk}$ is the characteristic load-carrying capacity per shear plane per fastener loaded parallel to the grain, and n_{ef} is discussed in Section 8.1.2 and other sections in this guide.

The design strength of the connection will be

$$Fc_{v,Rd} = k_{mod} \frac{Fc_{v,Rk}}{\gamma_M} \tag{D8.9}$$

where k_{mod} and γ_M are defined in *Equation D8.1*.

Where a timber member in the connection is loaded at an angle to its grain direction, the capacity of the connection must also be validated for:

(i) the load component perpendicular to the grain, as described in Section 8.1.4 in this guide
(ii) the load component parallel to the grain, as described in Section 8.1.2 in this guide.

8.2.3 Steel-to-timber connections

Taking into account the comments in Section 8.2 in this guide, when steel-to-timber connections are loaded laterally, different ductile failure modes can arise, and these are shown in Figure 8.10 for single- and double-shear connections.

Modes (a) to (e) apply to single-shear connections, and (f) to (m) to double-shear connections. For the reasons explained in Section 8.2.2, modes (a), (c), (f), and (j/l) will be type 1 modes; (b), (d), (g) and (k) type 2 modes and (e), (h) and (m) type 3 modes.

With steel gusset plate connections, the characteristic load-carrying capacity is also a function of the thickness of the steel plates. Where the plate thickness, t, is $\leq 0.5d$ (d being the nominal diameter of the fastener), it is classified as a thin plate, and where $t \geq d$ and the tolerance on the fastener hole diameter is less than $0.1d$ it is classified as a thick plate. As thin plates cannot provide adequate rigidity to the head of the fastener, type 3 modes are unable to be formed with these plates.

As with timber-to-timber connections, strength equations have been derived for each failure mode per fastener per shear plane for single-shear connections, and are categorised for thin

Figure 8.10. Modes of failure for steel-to-timber connections. The suffix against t refers to the member number. (Based on EN 1995-1-1 (*Figure 8.3*). Reproduced with permission from EN 1995-1-1, © British Standards Institute, 2004)

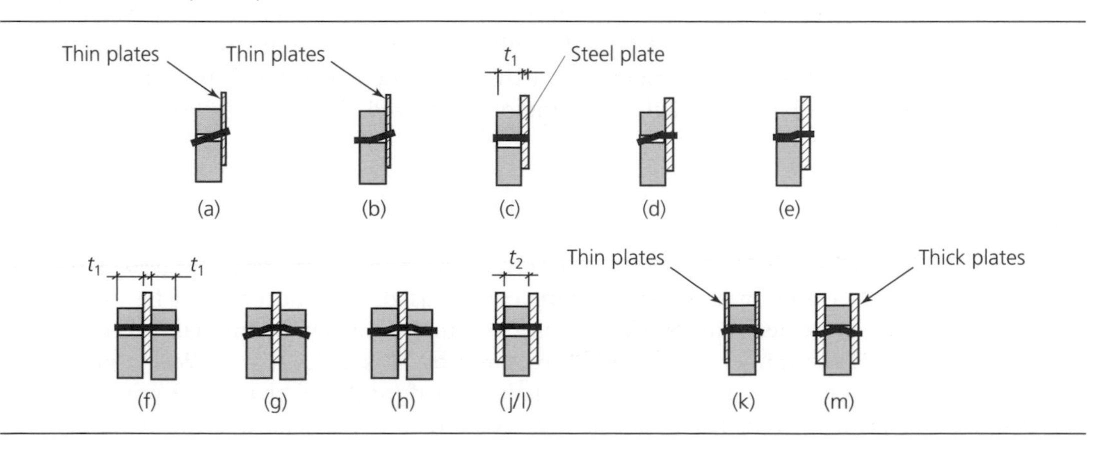

steel plates (*Equations 8.9*) and for thick steel plates (*Equations 8.10*). Double-shear connections are also categorised against thin and thick steel plates (*Equations 8.12* and *8.13*, respectively), and a category is also given for plates of any thickness when acting as the central member (*Equations 8.11*). The lowest value from the set of equations appropriate to the type of connection being considered (e.g. from *Equations 8.10c*, *8.10d* and *8.10e* for a thick steel plate single-shear connection) will be the value of the characteristic load-carrying capacity per shear plane per fastener, $F_{v,Rk}$, for that connection. If the steel plate thickness is between a thin and thick plate, the connection capacity is obtained by interpolation between the limiting thin and thick plate values.

$F_{v,Rk}$ is a function of several variables which are dependent on the type of fastener being used, and these are:

t_1	the smaller of the thickness of the timber side member or the penetration depth of the pointside of the fastener
$f_{h,k}$	the characteristic embedment strength in the timber member
d, $M_{y,Rk}$, $F_{ax,Rk}$	as referred to in Section 8.2.2 in this guide.

Clause 8.3
Clause 8.4
Clause 8.5
Clause 8.6
Clause 8.7

These variables are defined in *clauses 8.3* for nails, *8.4* for staples, *8.5* for bolts, *8.6* for dowels and *8.7* for screws, and $F_{ax,Rk}$ for each fastener type is also derived from the rules in these clauses.

Clause 8.2.2(2)

As explained in Section 8.2.2 in this guide, the strength equations comprise two parts: the Johansen part and the rope effect part. The contribution from the rope effect is limited for each type of fastener, and for those equations incorporating a rope effect, $F_{ax,Rk}/4$ as a percentage of the Johansen part of the equation must not exceed the values given in *clause 8.2.2(2)*.

Evaluation of the characteristic load-carrying capacity of the connection and the determination of the design strength are as defined for timber-to-timber connections in Section 8.2.2 in this guide.

Clause NA.3.1

Where steel-to-timber connections are formed with multiple dowel-type fasteners and there is a force component in the connection acting parallel to the grain and in a loaded end configuration (as defined in *Figure 8.7*) in addition to the ductile failure modes referred to above, there is also the risk of a brittle failure due to a block-type and/or a plug-type failure. *Clause NA.3.1* in the National Annex to EN 1995-1-1 requires these types of failure to be checked when the fastener diameter is ≤ 6 mm and there are ≥ 10 fasteners in a line or the fastener diameter exceeds 6 mm and there are five or more fasteners in a line. The design requirements for these types of failure are covered in Chapter 11 of this guide.

8.3. Nailed connections

The types of nail covered by the design rules in EN 1995-1-1 are described in EN 14592 (BSI, 2008a) and EN 10230-1 (BSI, 2000), and are referred to as:

- Plain shank nail (smooth nail) – these have a constant cross-section along its entire length (e.g. round, square or grooved nails).
- Threaded nail – the shank is profiled over a minimum length of 4.5 times its nominal diameter, and the characteristic withdrawal parameter, $f_{ax,k}$, must be at least 6 N/mm^2 (although it should be noted that EN 14592 only requires 4.5 N/mm^2) when derived under service class 1 conditions using timber with a characteristic density of 350 kg/m^3 (i.e. C24 timber). In EN 1995-1-1, these nails are commonly referred to as '*nails other than smooth nails*'.

Examples of smooth and threaded nails are shown in Figure 8.11.

Clause 8.3.1.2(6)

Nails can be driven by hand or, more commonly, using a nail gun, and the design rules in EN 1995-1-1 apply to both methods. Predrilling can be used, and indeed must be used where the timber density is greater than 500 kg/m^3, the nominal diameter of the nail exceeds 6 mm or where splitting of the timber can occur (as required by *clause 8.3.1.2(6)*). Where predrilling is used, the embedment strength of the timber is increased, and nail spacing requirements are reduced. With panel material, predrilling is not required, but, with hardboard, it reduces the

Figure 8.11. Some types of nails and their nominal diameter *d*

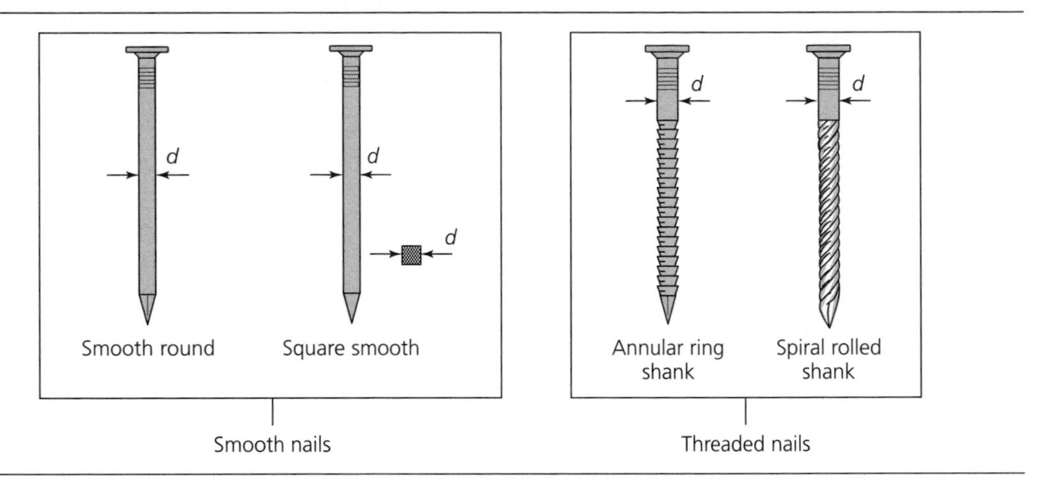

risk of physical damage on the underside when the nail is being driven. In general, predrilling will add direct and indirect costs to the project, and where it is not required to form the connection, its use is unlikely to be cost-effective.

8.3.1 Laterally loaded nails

Clause 8.3.1.1
Clause 8.3.1.2
Clause 8.3.1.3

8.3.1.1 Laterally loaded nails – general

The rules in *clauses 8.3.1.1*, *8.3.1.2* and *8.3.1.3* apply to nails that are loaded laterally, as defined in Sections 8.1 and 8.2 in this guide. As the shear strength of a nail is always greater than the capacity derived from the strength equations in *clauses 8.2.2* and *8.2.3*, there is no requirement to check the nail shear strength. Nails can, however, shear at loads lower than the shear strength when used with steel gusset plates due to high stress concentrations where the nail bears against the edge of the predrilled hole in the plate.

Clause 8.2.2
Clause 8.2.3

In a connection, the headside thickness is the thickness of the member containing the nail head, and the pointside penetration is the penetration length of the nail in the pointside member.

When referring to the nail diameter, *d*, the size to be used is the 'nominal diameter', as defined in EN 14592, and for common nail types the nominal diameter is shown in Figure 8.11. The maximum diameter of a nail is stated in EN 14592 to be 8 mm, and the design rules for nails in EN 1995-1-1 apply up to and including this size. Although EN 1995-1-1, *clause 8.3.1.1(6)* states that nails with a diameter greater than 8 mm must follow the rules for bolts to calculate the embedment strength, such nails will not comply with EN 14592, and it is to be questioned whether the rules in EN 1995-1-1 can be used.

Clause 8.3.1.1(6)

The characteristic yield moment defined in *Equations 8.14*, only applies to smooth nails. For threaded nails, the value must be obtained from the nail manufacturer or by testing in accordance with the requirements of EN 409 (BSI, 2009b).

Although the actual value of the characteristic embedment strength of timber when a nail is loaded parallel to the grain is different to the strength when loaded perpendicular to the grain, in the design rules in EN 1995-1-1 it is taken to be the same for all angles of loading, and is calculated using *Equation 8.15* when predrilling is not used, and *Equation 8.16* when predrilling is used. When predrilling is used, there will be a strength increase over the non-predrilled connection, as shown in Figure 8.12. The ratio of *Equation 8.16* to *Equation 8.15* (referred to as the 'strength factor') is plotted against the nominal diameter, and with a 3.00 mm-diameter predrilled nail there will be a 35% increase in embedment strength, increasing to over 65% when the diameter is 8 mm.

If nails are driven from the same position on opposite faces of a three-member connection, they will overlap in the central member, and to prevent splitting due to the increased splitting force in the overlapping zone the length of overlap must be limited. The criterion used is $(t - t_2) > 4d$, as

Figure 8.12. Ratio of *Equation 8.16* to *Equation 8.15* for nail diameters up to 8 mm

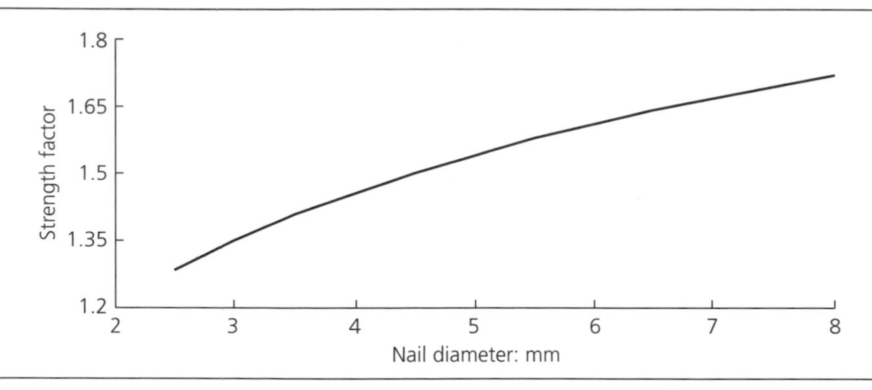

shown in *Figure 8.5*. It is possible to increase the length of the overlap, but the nail spacing and end distances must also be increased, and no rules are given for this condition in EN 1995-1-1.

Clause 8.3.1.2(5)

As previously stated, nails fixed in a line running parallel to the grain of a timber member form a row of nails. The nails can be staggered in a row by a distance of at least $1d$ and are still taken to be functioning in the row. However, if the stagger is $>1d$, separate rows will be formed, as shown in Figure 8.13, and for this condition *clause 8.3.1.2(5)* requires the minimum spacing between adjacent rows to be a_2. The value of a_2 for nailed connections is obtained from *Table 8.2*.

Where overlapping nails are used, although not stated in EN 1995-1-1, the limiting uniform stagger beyond which two rows will be formed will be $2d$.

When nails in a row are loaded parallel to the grain direction, the effective number of nails, n_{ef}, rather than the actual number in the row, n, is used in the strength calculation. n_{ef} is a function of the number of nails and the nail spacing, a_1, parallel to the grain, and for uniformly spaced nails

Figure 8.13. Staggered nails

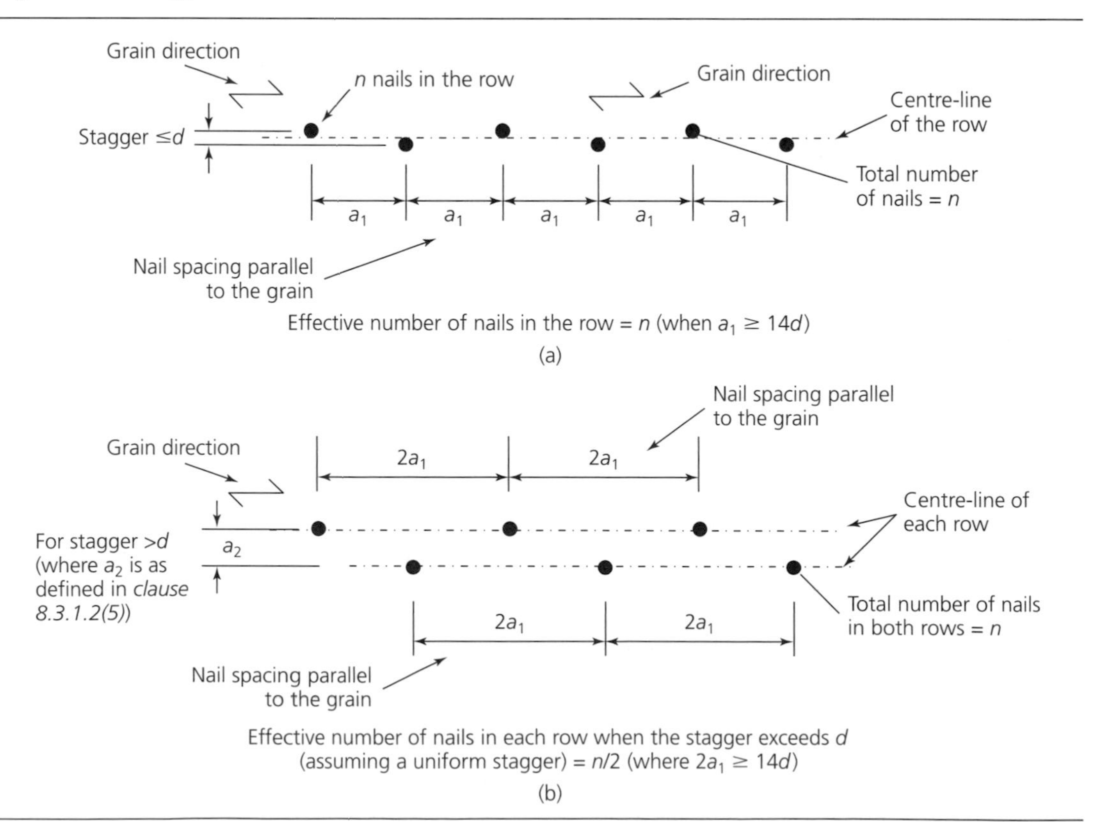

Table 8.2. Values for k_{ef}

Spacing a_1 as shown in *Figure 8.7(a)*	Factor k_{ef}	
	Whether or not pre-drilled	Only applies if pre-drilled
14*d*	1.0	
12*d*	0.925	
10*d*	0.85	
9*d*	0.8	
8*d*	0.75	
7*d*	0.7	
4*d*	–	0.5

Data taken from EN 1995-1-1
d is the nail diameter, and for spacings between the above values, linear interpolation should be used

is obtained from *Equation 8.17*:

- for single nails in single or double shear,

$$n_{ef} = n^{k_{ef}} \tag{8.17}$$

- for overlapping nails in a row,

$$n_{ef} = n_{ol}^{k_{ef}} \tag{D8.10}$$

where the value of k_{ef} is given in Table 8.2; n_{ol} is the number of nails per shear plane in the overlapping nail connection; and, with the exception of nails that have a spacing a_2 between $5d$ and $4d$ (which can only be formed if predrilling is used), predrilling will not be a factor.

To be able to form a connection that can be designed to withstand lateral loading, at least two nails must be used.

Example 8.2: calculation of the effective number of nails in a row

There are two rows of overlapping nails in a timber joint, as shown in Figure 8.14. What is the effective number of nails per row per shear plane? The nail diameter is 3 mm (d), no predrilling is used, and all dimensions are in mm.

Figure 8.14. Two rows of overlapping nails

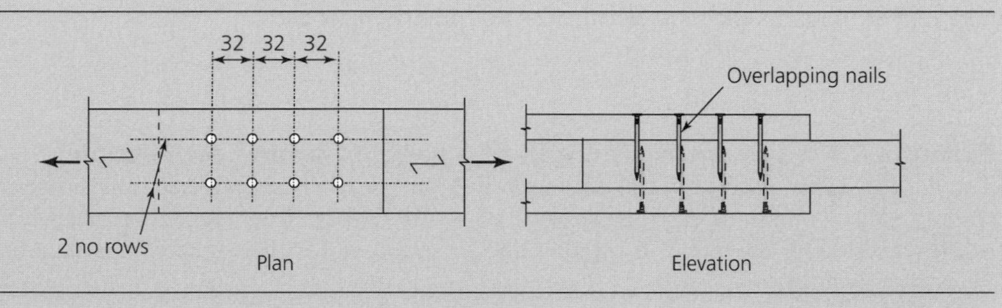

Number of nails per row per shear plane:

$$n = 4$$

Spacing parallel to the grain:

$$a_1 = 32\ \text{mm}\ \Xi\ 10.67d$$

From Table 8.2:

$$k_{ef} = [(a_1 - 10d)(1 - 0.85)/(14d - 10d)] + 0.85$$

$$= (32 - 30)(0.15)/(4 \times 3) + 0.85 = 0.875$$

Effective number of nails per row shear plane:

$$n_{ef} = n^{kef} = 4^{0.875} = 3.36$$

8.3.1.2 Laterally loaded nails – nailed timber-to-timber connections

A connection cannot be formed using smooth nails unless the pointside penetration of the nail is at least $8d$, and with threaded nails it must be at least $6d$.

Clause 8.3.1.2(3)
Clause 8.3.1.2(4)
Clause 8.3.1.2(4)

Although *clause 8.3.1.2(3)* states that nails driven into end grain are not permitted to resist lateral forces, the National Annex to EN 1995-1-1 permits the alternative ruling in *clause 8.3.1.2(4)*, allowing the use of smooth nails in secondary structures (e.g. cladding fixings to timber straps) and threaded nails in structures other than secondary structures, providing the limitations imposed in *clause 8.3.1.2(4)* are complied with.

Clause 8.2.2
Clause 8.2.3

Minimum spacings and edge and end distances for nailed connections are defined in *Table 8.2*, and unless these requirements are fully met, the lateral strength equations referred to in *clauses 8.2.2* and *8.2.3* cannot be used. Where spacings or distances include an absolute value of an angle (e.g. $|\cos \alpha|$), this means the answer of the function is always taken to be positive.

Clause 8.3.1.2(7)

In accordance with the requirements of the National Annex to EN 1995-1-1, *clause 8.3.1.2(7)*, which relates to predrilling of timber species especially sensitive to splitting, does not apply to nailed connections in UK designs.

8.3.1.3 Laterally loaded nails – nailed panel-to-timber connections

The minimum spacings in this type of connection will be those given in *Table 8.2* multiplied by a factor of 0.85. Except for plywood members, the edge and end distances will remain as stated in the table. In connections formed using plywood members, the minimum edge and end distances in the plywood member should be taken as $3d$ for an unloaded edge (or end) and $(3 + 4 \sin \alpha)d$ for a loaded edge (or end), noting that α is the angle between the direction of the load and the loaded edge (**or end**).

The embedment strength of panel material is taken to be the same for all angles of loading, and, unlike timber, remains the same whether or not predrilling is used.

Example 8.3: embedment strength of a panel to timber connection

The nailed timber–OSB gusset plate T joint in Figure 8.15 comprises 2.85 mm-diameter (d) smooth nails, 9 mm thick OSB and strength class C18 timber. What is the characteristic value of the embedment strength of the timber and the plywood for the connection in member A? Predrilling is not used.

Characteristic density of the timber:

$$\rho_k = 320 \text{ kg/m}^3$$

Characteristic density of the OSB:

$$\rho_{osb,k} = 550 \text{ kg/m}^3$$

Figure 8.15. Nailed timber–OSB gusset plate T joint

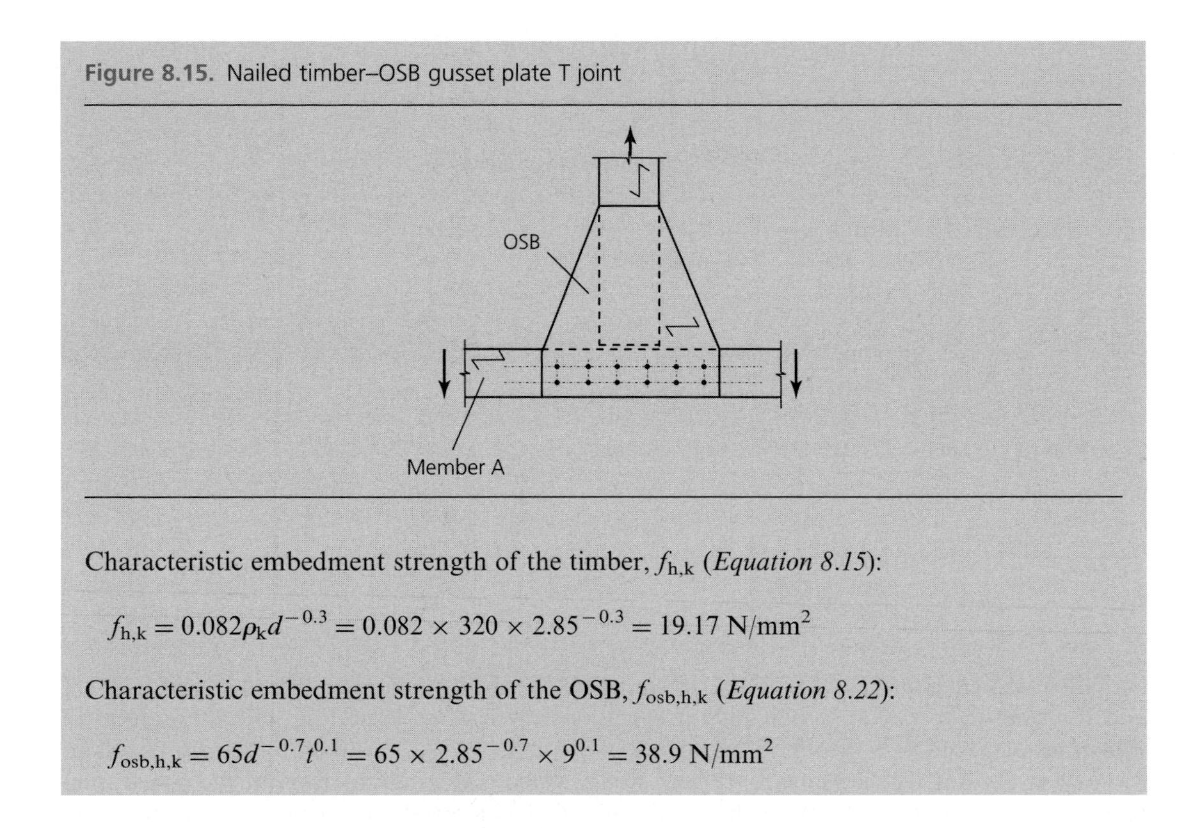

Characteristic embedment strength of the timber, $f_{h,k}$ (*Equation 8.15*):

$$f_{h,k} = 0.082\rho_k d^{-0.3} = 0.082 \times 320 \times 2.85^{-0.3} = 19.17 \text{ N/mm}^2$$

Characteristic embedment strength of the OSB, $f_{osb,h,k}$ (*Equation 8.22*):

$$f_{osb,h,k} = 65d^{-0.7}t^{0.1} = 65 \times 2.85^{-0.7} \times 9^{0.1} = 38.9 \text{ N/mm}^2$$

8.3.1.4 Laterally loaded nails – nailed steel-to-timber connections

The minimum spacings in this type of connection will be those given in *Table 8.2* multiplied by a factor of 0.7, but the edge and end distances will remain as stated in the values given in the table.

8.3.2 Axially loaded nails

The rules in this clause cover nails that are subjected to axial withdrawal, and, as the axial strength of the nail wire is always greater than the nail withdrawal strength, there is no requirement to check its tensile strength. Where the nail diameter is referred to in the design rules, the nominal diameter as defined in Section 8.3.1.1 in this guide applies.

As there is the risk of friction loss and slip over time when using smooth nails, particularly where the connection is subjected to frequent wetting and drying cycles, only threaded nails are allowed to be used to resist permanent or long-term axial loading. Smooth nails can be used for other durations of action.

The withdrawal capacity of threaded nails is dependent on the type of threaded nail being used. In the pointside member, resistance is only picked up over the **threaded** length of the nail, and in the headside member only head pull-through strength can be used. With smooth nails, in addition to the head pull-through strength, surface friction is picked up over the headside length, and for the pointside, friction is picked up over the full pointside penetration length. Nails driven in end grain cannot be used to resist axial loading.

When a connection is subjected to axial withdrawal it will fail by either the nail head being pulled through the headside member (**headside pull-through**) or by withdrawal of the nail pointside (**pointside withdrawal**), and the characteristic withdrawal capacity of the nail, $F_{ax,Rk}$, is derived from the weaker condition. It is calculated using *Equations 8.23* for threaded nails and *Equations 8.24* for smooth nails, and applies to nails driven perpendicular to the grain and also when driven in pairs at a slant, as shown in *Figure 8.8*. In *Equations 8.23*, t_{pen} is the length of the threaded part of the nail in the pointside member, and in *Equations 8.24* it is the pointside penetration length, in both cases excluding the point length (as stated in Appendix A).

$F_{ax,Rk}$ is a function of the characteristic pointside withdrawal strength, $f_{ax,k}$, and the characteristic headside pull-through strength, $f_{head,k}$. These strengths are derived from tests in accordance

Figure 8.16. Comparison of the withdrawal strength of a smooth and threaded 3.0 mm-diameter nail

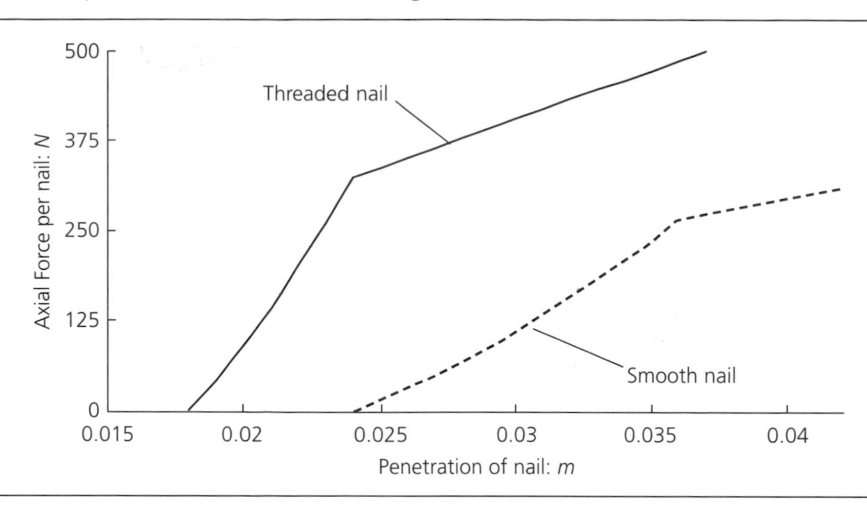

with the requirements of EN 1382 (BSI, 1999a), EN 1383 (BSI, 1999b) and EN 14358 (BSI, 2006), and for smooth nails with a pointside penetration length of at least $12d$, values are given in *Equations 8.25* and *8.26* in N/mm^2. Above $12d$ the value of $f_{ax,k}$ is taken to remain constant, but below $12d$ it is reduced linearly to zero at $8d$, which is the minimum pointside penetration permitted for smooth nails. With threaded nails, $f_{ax,k}$ is derived with a pointside penetration of $8d$, and is also reduced linearly to zero at the minimum penetration of $6d$. In the absence of a value for $f_{head,k}$ from test results for such nails, the use of $f_{head,k}$ for smooth nails will give a conservative result.

A comparison between the withdrawal capacity of a smooth and a threaded 3 mm nominal diameter nail (with $f_{ax,k} = 36.7 \times 10^{-6}\rho_k^2$ – based on the minimum value of 4.5 N/mm^2 permitted in EN 14592) having a threaded length of $6d$, both in C24 timber is given in Figure 8.16, showing that strength pick up in the threaded nail occurs over a much shorter length and has a greater rate of strength increase than the smooth nail.

Where structural timber is installed at or near its fibre saturation point and is likely to dry out under load, the values used for $f_{ax,k}$ must be multiplied by $\frac{2}{3}$.

For a connection with n nails subjected to axial withdrawal, no guidance is given on the effective number to be used in the strength connection, and in this guide it is suggested the full number of nails are used. Where the characteristic withdrawal capacity of each nail is $F_{ax,Rk}$ and there are n nails in the connection, the design value, $F_{ax,Rd}$ will be

$$F_{ax,Rd} = \frac{k_{mod}}{\gamma_M} n F_{ax,Rk} \tag{D8.11}$$

where k_{mod} and γ_M are defined in Equation D8.1.

Clause 8.3.1.1(8)

Where the connection is subjected to lateral loading and the axial withdrawal capacity is to be taken into account through the rope effect, the number of nails used to derive the lateral strength of the connection will be the effective number, n_{ef}, as defined in *clause 8.3.1.1(8)* and referred to in Section 8.3.1.1 in this guide, when the lateral loading is parallel to the grain. For any other angle, based on the rules in EN 1995-1-1, n should be used.

8.3.3 Combined axially and laterally loaded nails

For this condition, when using smooth nails a linear elastic strength approach is used, but with threaded nails the increase in axial withdrawal strength permits the use of a non-linear power function. The validation requirement is given in *Equations 8.27* and *8.28*, and the increase in strength between these relationships is shown in Figure 8.17.

Figure 8.17. Comparison between connections using smooth and threaded nails subjected to combined lateral and axial loads (*Equations 8.27* and *8.28*)

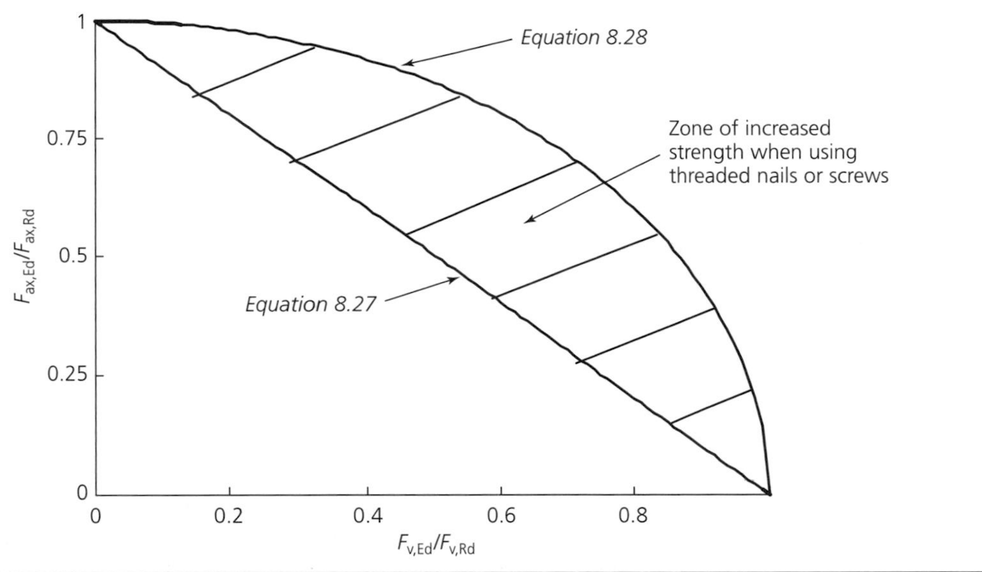

8.4. Stapled connections

Staples are generally manufactured with round, nearly round or rectangular profiles, have bevelled or symmetrical pointed legs and are always driven without predrilling. For these types of staples, the rules given in *clause 8.3* for nails driven without predrilling and up to 8 mm in diameter will apply. When deriving the characteristic yield moment, reference should be made to the content of Appendix A, where it is to be noted that *Equation 8.29* is to be revised.

Clause 8.3

Where staples are made from rectangular cross-section wire, the diameter, d, used in calculations is the square root of the product of the cross-section dimensions of the staple leg. The staple profile and pointside penetration must comply with the requirements of *clause 8.4(3)*, and in any connection there should be at least two staples.

Clause 8.4(3)

Providing the angle θ between the crown of the staple and the grain direction, as shown in *Figure 8.10*, is greater than 30°, the characteristic load-carrying capacity of each staple leg per shear plane should be calculated in accordance with the rules used to derive the characteristic load-carrying capacity of a smooth nail having the same diameter, but using *Equation 8.29* (as referred to in Appendix A) to calculate the characteristic yield moment. For $\theta \leq 30°$, the strength will be 0.7 times this value. With staples, the rope effect will not apply, and, as stated in Appendix A, in a row parallel to the grain, the effective number of staples, n_{ef}, equals the actual number of staples, n, in the row.

8.5 Bolted connections

EN 14592 states that the grades of steel that can be used for bolts in timber connections are 4.6, 4.8, 5.6 or 8.8, and their respective characteristic tensile strengths, obtained from EN 1993-1-8 (BSI, 2005) are given in Table 8.3.

Bolts shall have a minimum diameter of 6 mm and a maximum of 30 mm, and the maximum sizes for bolt holes in timber and steel as well as the requirements for bolt washers are given in *Section 10*.

Table 8.3. The characteristic tensile strength of bolts

Bolt grade	4.6	4.8	5.6	8.8
Characteristic tensile strength, $f_{u,k}$: N/mm^2	400	400	500	800

Figure 8.18. Bearing lengths of bolts when recessed

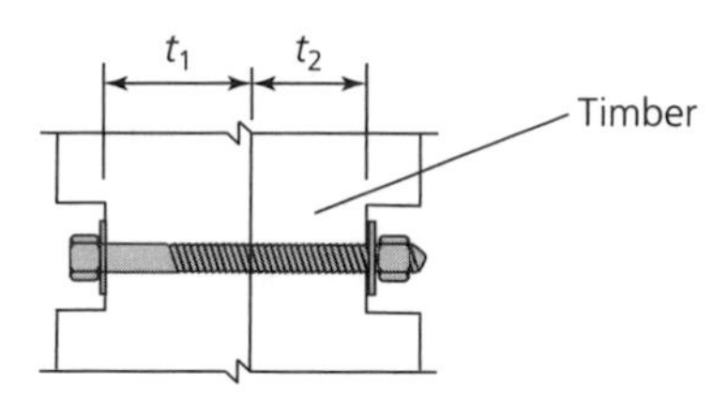

8.5.1 Laterally loaded bolts

8.5.1.1 Laterally loaded bolts – general and bolted timber-to-timber connections

Clause 8.5.1.1
Clause 8.5.1.2
Clause 8.5.1.3
Clause 8.2.2
Clause 8.2.3

The rules in *clauses 8.5.1.1, 8.5.1.2* and *8.5.1.3* apply to bolts loaded laterally, as defined in Sections 8.1 and 8.2 in this guide. As the shear strength of the bolt will always be greater than the strength derived from the strength equations in *clauses 8.2.2* and *8.2.3*, there is no requirement to check for the bolt failing in shear.

With bolts, the embedment strength of timber is dependent on the angle of loading relative to the member grain direction, and is obtained from *Equations 8.31, 8.32* and *8.33*. The factor k_{90} in *Equation 8.33* is the ratio of the characteristic embedment strength parallel to the grain divided by the value perpendicular to the grain, and (apart from 6 mm bolts in hardwood) will always be greater than 1. The largest values of k_{90} are associated with softwoods, and the smallest with hardwoods.

Clause 8.2.2
Clause 8.2.3

For bolted connections, the values used in the strength equations given in *clauses 8.2.2* or *8.2.3* to calculate the characteristic load-carrying capacity, $F_{v,Rk}$, will be:

- t_1 the bolt headside member thickness where the connection is in single shear or double shear (assuming the connection is symmetrical)
- t_2 the bolt threaded end member thickness when the connection is in single shear and the central member thickness when in double shear.

Bolts are commonly recessed into the timber, and in such situations the above values will equate to the length of the bolt that bears directly onto the timber member, as shown in Figure 8.18.

Clause 8.2.2
Clause 8.2.3

The minimum spacings and edge and end distances for bolts are given in *Table 8.4*, and, as stated in Section 8.3.1.2 of this guide, unless these requirements are fully met, the lateral strength equations referred to in *clauses 8.2.2* and *8.2.3* cannot be used. Where spacings or distances include an absolute value of an angle (e.g. $|\cos\alpha|$), this means the value of the function is always taken to be positive.

When there are n bolts in a row loaded laterally parallel to the grain, as shown in Figure 8.19, the effective number of bolts, n_{ef}, used to derive the strength of the row in that direction is obtained from *Equation 8.34*. Some values for n_{ef} using different numbers and spacings in a row are given in Table 8.4.

Figure 8.19. Bolts in a row

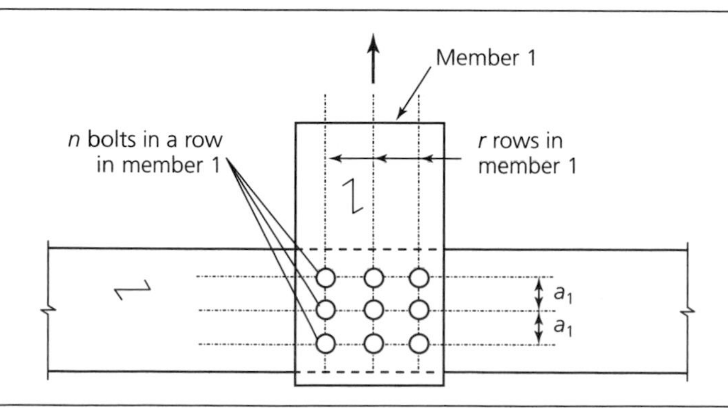

Table 8.4. The value of n_{ef} for a row of n bolts, of diameter d and spacing a_1, based on *Equation 8.34*

Number of bolts in the row, n	Bolt spacing a_1 (as defined in *Equation 8.34*)		
	5d	10d	15d
2	1.47	1.75	1.93
3	2.12	2.52	2.79
4	2.74	3.26	3.61
5	3.35	3.99	4.41

Where there are n bolts in a row loaded perpendicular to the grain, $n_{ef} = n$, and where the row is loaded at any angle α between 0 and 90° to the grain, n_{ef} may be determined by linear interpolation between n and the value from *Equation 8.34*. Some generalised cases for connections with bolts in rows are given in Example 8.4.

Example 8.4: calculation of the effective number of bolts in a connection

What is the effective number, n_{ef}, of 12 mm-diameter (d) bolts per row per shear plane in connections A and B shown in Figure 8.20?

Figure 8.20. Axial strength of a bolted connection

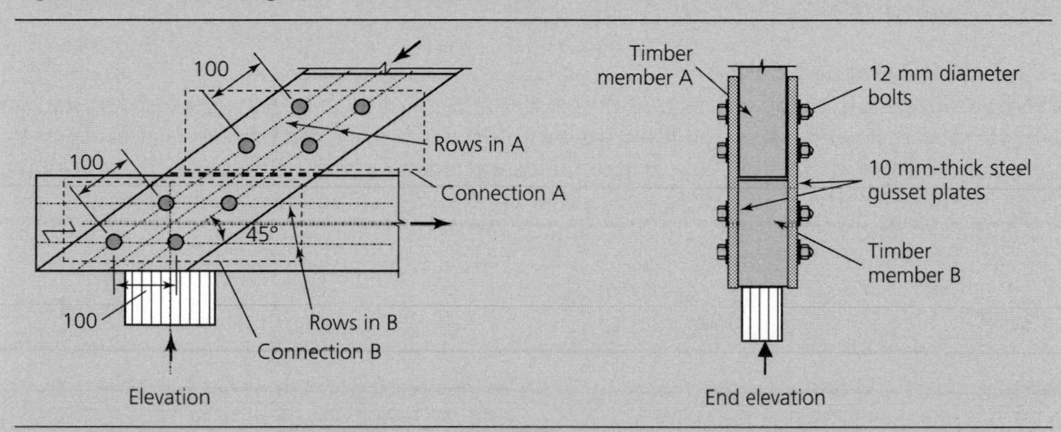

Number of bolts per row per shear plane in connection A:

$$n1 = 2$$

Spacing parallel to the grain:

$$a_1 = 100 \text{ mm}$$

Thus

$$n1_{ef} = \min[n1, n1^{0.9}(a_1/13d)^{0.25}] = \min[2, 2^{0.9} \times (100/13 \times 12)^{0.25}] = 1.67$$

Effective number of bolts per shear plane in this connection:

$$2n1_{ef} = 3.34$$

The loading parallel to the grain in member B is calculated.

Number of bolts per row per shear plane:

$$n2 = 2$$

Spacing parallel to the grain:

$a_1 = 100$ mm

So, $n2_{ef} = n1_{ef}$, and the effective number of bolts per shear plane in this connection is 3.34.

Clause 8.5.1.1(4)

For loading at 45° in member B, the final statement in *clause 8.5.1.1(4)* may be used. For this condition, the effective number of bolts when loaded perpendicular to the grain = actual number = 4 bolts per shear plane, and the effective number of bolts per shear plane parallel to the grain will be 3.34. So, the effective number per shear plane will be

$3.34 + (4 - 3.34) \times 45/90 = 3.67$ bolts

8.5.1.2 Laterally loaded bolts – bolted panel-to-timber connections

Clause 8.5.1.1

The rules in this clause apply to bolted panel-to-timber connections. When calculating the effective number of bolts in these connections the rules in *clause 8.5.1.1* for timber-to-timber connections still apply, based on the timber member.

Unlike timber, the characteristic embedment strengths of plywood and OSB are independent of the direction of load relative to the face grain of the material, and are derived from *Equations 8.36* and *8.37*, respectively.

8.5.2 Axially loaded bolts

With bolts in tension, there is a risk that the tensile strength of the bolt may be exceeded, and this strength must be validated in addition to the other strength requirements in EN 1995-1-1. Applying the rules in EN 1993-1-8, where the characteristic tensile strength is f_{uk}, in a timber design the characteristic and design tensile strengths, $F_{ax,Rk}$ and $F_{ax,Rd}$ respectively, for a normal raised head bolt will be

$$F_{ax,Rk} = \frac{0.9 f_{uk} A_s \gamma_M}{\gamma_{M2}}, \qquad F_{ax,Rd} = \frac{0.9 f_{uk} A_s}{\gamma_{M2}} \tag{D8.12}$$

where γ_{M2} (=1.25) is the partial factor for bolts in the National Annex to EN 1995-1-1; γ_{M2} (=1.25) is the partial factor for bolts in tension in the National Annex to EN 1993-1-8; and A_s is the tensile area of the bolt based on the inner thread diameter at its threaded end. If a countersunk head bolt is used, a factor of 0.63 is used rather than 0.9.

Clause 8.5.2(3)

The other strength requirement is that the load-bearing capacity of either the washer or, in steel-to-timber connections, the steel plate, is not exceeded. Guidance is given in *clause 8.5.2(3)* on the maximum area that can be taken to function as an equivalent washer where a steel plate is used, and the characteristic bearing strength of the timber over the contact area is taken to be $3.0 f_{c,90,k}$. The characteristic axial capacity of each bolt, $F_{ax,Rk}$, will be derived from the lesser of the design value of the tensile and the bearing capacity (as shown in Example 8.5).

For a connection with n bolts subjected to axial withdrawal, no guidance is given on the effective number of bolts to be used, and it is suggested in this guide that the full number of bolts will apply. Where there are n bolts in the connection, the characteristic value of the axial capacity, $F_{ax,Rk}$ will be:

$$F_{ax,Rk} = \gamma_{M2} n F_{ax,Rd} \text{ (when tension is critical), otherwise } F_{ax,Rk} = \frac{\gamma_M}{k_{mod}} n F_{ax,Rd} \tag{D8.13}$$

where k_{mod} and γ_M are defined in Equation D8.1 and γ_{M2} is stated in Equation D8.12.

Where the connection is subjected to lateral loading but the rope effect is to be taken into account, the number of bolts used to derive the lateral strength will be the effective number, n_{ef}, as defined in Section 8.5.1.1 in this guide.

Example 8.5: example calculating the axial strength of a bolted connection
Determine the characteristic axial strength of the bolts in connection A in Example 8.4 to be used for the rope effect when the timber is C22, the k_{mod} value is 0.8, the bolt is grade 4.6, the bolt hole in the timber is 1 mm larger than the bolt diameter and a minimum washer diameter is used (Figure 8.20). The steel gusset plates are 10 mm thick (t).

Tensile strength of a bolt:

$$f_{u,k} = 400\ \text{N/mm}^2$$

Bolt thread diameter:

$$d_1 = 0.7d = 0.7 \times 12 = 8.4\ \text{mm}$$

Washer diameter:

$$d_w = \min(12t,\ 4d) = 4 \times 12 = 48\ \text{mm} \qquad (\textit{clause 8.5.2(3)})$$

Clause 8.5.2(3)

Bearing strength of the timber below washer:

$$f_{c,90,k} = 3 \times 2.4 = 7.2\ \text{N/mm}^2 \qquad (\textit{clause 8.5.2(2)})$$

Clause 8.5.2(2)

Design tensile strength of a bolt:

$$Fs_{ax,Rd} = 0.9f_{u,k}(\pi d_1^2/4)/1.25 = 0.9 \times 400 \times (\pi \times 8.4^2/4)/1.25 = 15.96\ \text{kN}$$

Design bearing capacity of a bolt:

$$Fh_{ax,Rd} = f_{c,90,k}\pi[d_w^2/4 - (d + 1\ \text{mm})^2/4]k_{mod}/1.3$$

$$= 7.2 \times \pi(48^2/4 - 13^2/4)0.8/1.3 = 7.43\ \text{kN}$$

Characteristic value for design will be based on the minimum design value, i.e. bearing:

$$F_{ax,Rk} = 7.43 \times 1.3/k_{mod} = 12.07 \times 1.3/0.8 = 12.07\ \text{kN}$$

The value to be used for each fastener in the rope effect will be the smaller of

$$(F_{ax,Rk}/4)(n_{ef}/n) = (12.07/4) \times (3.34/4) = 2.52\ \text{kN}$$

and 25% of the Johansen part of the strength Equation (*clause 8.2.2(2)*).

Clause 8.2.2(2)

8.6. Dowelled connections

The rules in EN 1995-1-1 only apply to metal fasteners, and reference to dowels in the code means that they must be metal based. They must also have a diameter greater than 6 mm and less than 30 mm, and the tolerance rules for dowel diameter and for prebored holes have to comply with the requirements of *Clause 10.4.4*.

Clause 10.4.4

Dowels cannot be used in connections subjected to axial loading, and with these fixings the rope effect will be zero in joints subjected to lateral loading.

With dowels, the minimum spacings and edge and end distances shall comply with the content of *Table 8.5*, noting the changes for $a_{3,c}$ given in the Appendix A as follows:

$90° \leq \alpha < 150°$ use $a_{3,t}\,|\sin\alpha|$

$150° \leq \alpha < 210°$ use $\max(3.5d;\ 40\ \text{mm})$

$210° \leq \alpha < 270°$ use $a_{3,t}\,|\sin\alpha|$

Clause 8.5.1

Apart from the above requirements, the design rules for bolts given in *clause 8.5.1* and discussed in Section 8.5.1.1 in this guide will apply.

8.7. Screwed connections

As stated in EN 14592, screws for structural timber applications shall be formed in one of two ways:

1 By threading down the original rod, producing a screw with a smooth shank diameter equal to the outer maximum cross-sectional diameter of the threaded part. The outer diameter of the rod is referred to as the **nominal diameter** of the screw. This is how the traditional wood screw and coach screw are formed, and these types are referred to in EN 1995-1-1 as smoothed shank screws.
2 By hardening after rolling the thread. These are often referred to as self-tapping screws, and with these the outer thread diameter of the formed screw is referred to as the **nominal diameter**.

Examples of smoothed shank and self-tapping screws are shown in Figure 8.21.

All screws must be threaded over a length, l_g, ≥ 4 times the screw nominal diameter. With modern smooth shank screws, l_g is approximately two-thirds of the screw length, and the inner thread diameter (root diameter, d_1) is approximately 70% of the outer thread diameter (the original rod diameter). With coach screws, l_g is approximately 60% of the shank length.

Because self-tapping type screws are work hardened as part of the production process, they have higher strength properties than smooth shank screws, and, with this screw type, l_g is commonly the length of the screw shank.

EN 14592 states that the range of the nominal diameter for screws permitted to be used in structural connections is 2.4 mm up to a maximum of 24 mm. Also, the inner thread diameter of the screw shall not be less than 60% or more than 90% of the outer thread diameter.

For all types of screws, predrilling is required in connections formed in all hardwoods or in softwood having a density greater than 500 kg/m^3, or in connections where the outer thread diameter of the screw is greater than 6 mm and connections are formed in softwood. This is necessary to prevent splitting of the wood; torsion failure of the screw; or to enable the screw to be fully driven. Where the screw diameter is ≤ 6 times the outer thread diameter and the connection is in softwood, predrilling is not required. When predrilling is used, the lateral strength of the screw will increase, and spacings can be reduced. However, unless required by the rules in EN 1995-1-1, as with nails, predrilling will add direct and indirect costs, and where not required it is unlikely to be cost-effective.

Clause 10.4.5

The requirements for predrilling when using screws are given in *clause 10.4.5*. Where used with smooth shank screws, the predrilled hole must have the same diameter as the outer diameter of

Figure 8.21. Examples of smooth shank screws and self-tapping screws (d is the nominal diameter and d_1 is the inner thread diameter)

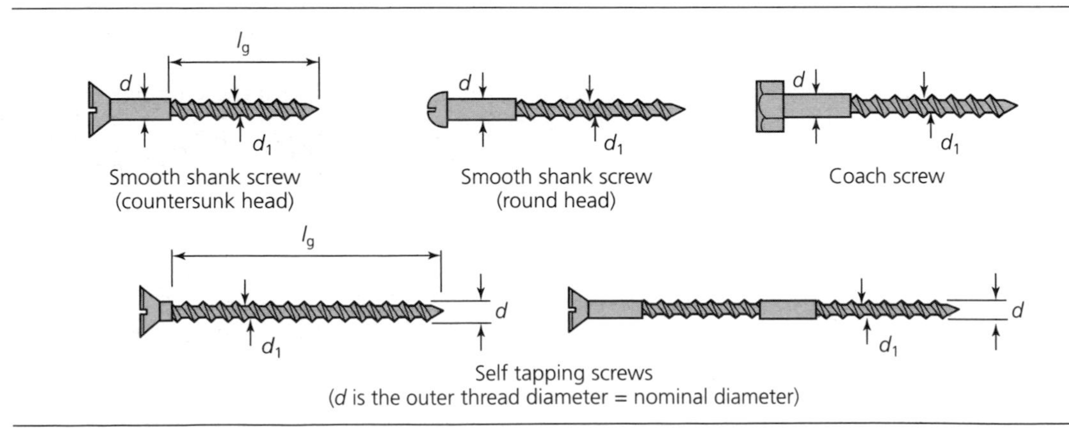

Figure 8.22. Rules for laterally loaded screws

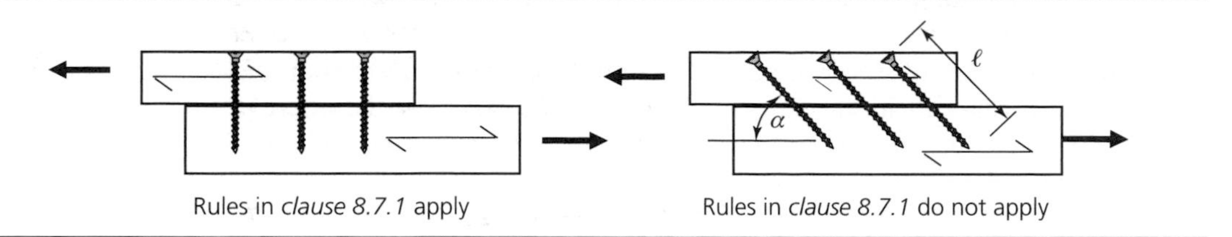

Rules in *clause 8.7.1* apply — Rules in *clause 8.7.1* do not apply

the screw for the length of the smooth shank, and over the threaded length it should be approximately 70% of the shank diameter. Where predrilling is used with self-tapping screws, the diameter of the predrilled hole should not be greater than the inner thread diameter, d_l. With any hardwood timber and where the timber density in softwoods is greater than 500 kg/m^3, tests have to be carried out to determine the diameter of the predrilled hole to be used.

Unless otherwise stated in EN 1995-1-1, the rules in *clause 8.7.1* for laterally loaded screws only apply to screws driven at right angles to the grain. Where they are fixed at an angle to the grain and subjected to lateral loading, as shown in Figure 8.22, the lateral strength design rules have to be adjusted, and a proposal for this condition is given by Bejtka and Blass (2002). An alternative is to determine a strength using the rules for axially loaded connections given in *clause 8.7.2.*

Clause 8.7.1
Clause 8.7.2

8.7.1 Laterally loaded screws

The rules in *clause 8.7.1* apply to screws that are loaded laterally, as defined in Sections 8.1 and 8.2 in this guide. As the shear strength of a screw is always greater than the capacity derived from the strength equations in *clauses 8.2.2* and *8.2.3*, there is no requirement to check its shear strength. However, as with nails, shear failure can arise when used with steel gusset plates, due to high stress concentrations where the screw bears against the edge of the predrilled hole in the plate.

Clause 8.7.1
Clause 8.2.2
Clause 8.2.3

When referring to the rules in EN 1995-1-1 to calculate the yield moment of the screw, the effective diameter of the screw, d_{ef}, as defined in EN 1995-1-1 (and in the proposed revision to *clauses 8.7.1(1)P*, *8.7.1(4)*, *8.7.1(5)* and *8.7.1(6)* in Appendix A) and also as developed further in PD 6693-1, is used, and in general is as follows:

Clause 8.7.1(1)P
Clause 8.7.1(4)
Clause 8.7.1(6)
Clause 8.7.1(7)

(*a*) with smooth shank screws where the smooth shank penetrates into the pointside member by at least 4*d*, $d_{ef} = d$
(*b*) with smooth shank screws where the shank penetrates into the pointside member by less than 4*d* and, for self-tapping screws, irrespective of the pointside penetration, $d_{ef} = 1.1d_l$, where d_l is as shown on Figure 8.21.

Based on the above requirement, the characteristic yield moment, $M_{y,Rk}$ is

$$M_{y,Rk} = 0.3 f_u d_{ef}^{2.6} \tag{D8.14}$$

where f_u is the characteristic tensile strength of the screw material. Where f_u is not able to be obtained, $M_{y,Rk}$ should be found by testing in accordance with the requirements of EN 409, also taking into account the requirements of EN 14592.

The effective diameter, d_{ef}, is also used to calculate the embedment strength, and for the nail diameter (d) in the lateral strength equations in *clauses 8.2.2* and *8.2.3*.

Clause 8.2.2
Clause 8.2.3

When the effective diameter, d_{ef}, of a smooth shank screw is $\leq$ 6 mm, the design procedure must comply with the rules in *clause 8.3.1* for nails, and when $>$ 6 mm, it must comply with the rules in *clause 8.5.1* for bolts. Although not clearly stated in EN 1995-1-1, these limits should also apply to self-tapping screws.

Clause 8.3.1
Clause 8.5.1

Where the screw is designed as a nail, the rules in *clauses 8.3.1.1*, *8.3.1.2*, *8.3.1.3* and *8.3.1.4* will apply, and if designed as a bolt, *clauses 8.5.1.1*, *8.5.1.2* and *8.5.1.3* are relevant, and the reader is also referred to the comment under these headings in this guide. However, no matter which design rules apply, where the rope effect referred to in *clause 8.2.2(2)* and in Section 8.2.2 in

Clause 8.3.1.1
Clause 8.3.1.2
Clause 8.3.1.3
Clause 8.3.1.4
Clause 8.5.1.1
Clause 8.5.1.2
Clause 8.5.1.3
Clause 8.2.2(2)

Clause 8.7.2

this guide is to be taken into account, the value of the characteristic axial withdrawal capacity, $F_{ax,Rk}$, of a screw shall be determined using the rules in *clause 8.7.2*.

Clause 8.7.1

The effective number of screws used in the connection to derive the characteristic lateral capacity shall be in accordance with the rules in *clause 8.7.1*. In other words, where $d \leq 6$ mm, the rules in Section 8.3.1.1 in this guide will apply, and where the $d > 6$ mm, the rules in Section 8.5.1.1 apply.

Example 8.6: embedment strength and yield moment of a panel-to-timber screwed connection

The tension joint in Figure 8.23 is formed from C22 timber and plywood gusset plates connected by eight 4.5 mm diameter (d), 75 mm-long self-tapping screws made from grade 5.6 steel. Calculate the characteristic embedment strength and characteristic yield moment of each screw. No predrilling is used.

Figure 8.23. Panel-to-timber screwed connection

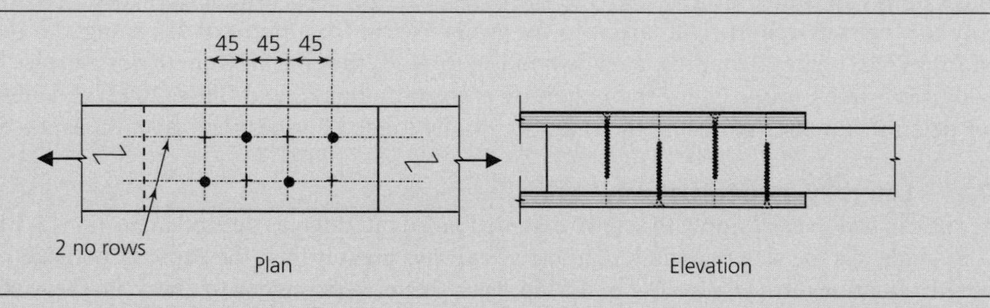

Characteristic density of timber:

$$\rho_k = 340 \text{ kg/m}^3$$

Characteristic density of plywood:

$$\rho_{p,k} = 344 \text{ kg/m}^3$$

Effective diameter of a screw:

$$1.1 \times 0.7d = 1.1 \times 0.7 \times 4.5 = 3.465 \text{ mm}$$

Characteristic embedment strength of the timber, $f_{h,k}$ (*Equation 8.15*):

$$f_{h,k} = 0.082\rho_k d^{-0.3} = 0.082 \times 340 \times 3.465^{-3} = 19.20 \text{ N/mm}^2$$

Characteristic embedment strength of the plywood, $f_{p,h,k}$ (*Equation 8.20*):

$$f_{p,h,k} = 0.11\rho_{p.k} d^{-0.3} = 0.11 \times 344 \times 3.465^{-0.3} = 26.06 \text{ N/mm}^2$$

The characteristic yield moment, $M_{y,Rk}$ (*Equation 8.14*) is calculated.

Effective diameter of a screw:

$$d_{ef} = 3.465 \text{ mm}$$

Tensile strength of the screw material:

$$f_u = 500 \text{ N/mm}^2$$

So

$$M_{y,Rk} = 0.3 f_u d_{ef}^{2.6} = 0.3 \times 500 \times 3.465^{2.6} = 3.7959 \times 10^3 \text{ N mm}$$

8.7.2 Axially loaded screws

The rules for axially loaded screws apply to connections in which screws are loaded axially or have a force component parallel to their shanks. Where the screws are fixed at an angle α to the grain and are loaded laterally, a value for the characteristic lateral load-carrying capacity of the connection can be derived from the **characteristic withdrawal capacity**, $F_{ax,\alpha,Rk}$, using the rules in *clause 8.7.2*. Where the screws are fixed at right angles to the grain and subjected to lateral loading, their lateral load carrying capacity will be derived using the rules in *clause 8.7.1*, with the rules in *clause 8.7.2* being used to calculate the characteristic axial withdrawal capacity, from which the rope effect referred to in *clause 8.2.2* will be obtained.

Clause 8.7.2
Clause 8.7.1
Clause 8.7.2
Clause 8.2.2

To determine the axial strength of a screwed connection, the following failure modes must be considered:

- the screw failing by withdrawal from the pointside member
- the screw head tearing off when used in combination with steel plates – it is a requirement that this strength exceeds the tensile strength of the screw
- pull-through of the screw head in the headside member
- the tensile strength of the screw being exceeded
- buckling failure of the screw when loaded in compression
- failure along the circumference of a group of screws used in conjunction with steel plates (block or plug shear failure modes will arise but the equations given for block and plug failure in *Annex A* in EN 1995-1-1 cannot be used as they apply to block or plug shear failure modes caused by lateral rather than axial loading).

EN 1995-1-1 draws attention to the fact that failure modes in the steel and around the screw are brittle modes, and consideration should be given to this by the designer.

To be able to resist axial load, the screw must have a minimum pointside penetration length of the threaded part of $6d$ and, provided the timber thickness in the connection is $\geq 12d$, the minimum spacings and end and edge distances in *Table 8.6* will apply.

As stated in Appendix A, the rules in *clause 8.7.2(4)* only apply to connections in softwood timber. For such connections, with screws compliant with the requirements of EN 14592 and where d is within the range 6 mm $\leq d \leq$ 12 mm and the ratio of the inner thread diameter d_1 to the outer thread diameter, d, complies with $0.6 \leq d_1/d \leq 0.75$, the characteristic withdrawal capacity, $F_{ax,\alpha,Rk}$, is obtained from *Equation 8.38*. For screws where d is less than 6 mm but still compliant with the above d_1/d ratio, the design requirement for the characteristic withdrawal strength perpendicular to the grain, $f_{ax,k}$, is given in PD 6693-1.

Clause 8.7.2(4)

In *Equation 8.38*, when screws are being fixed at a slant in both the x–z and the y–z planes (as shown in Figure 8.24), α will be the projected angle of the screw axis onto the x–z plane.

Where the ratio of d_1/d is greater than 0.75, $F_{ax,\alpha,Rk}$ is obtained from *Equation 8.40a*. If the characteristic withdrawal strength perpendicular to the grain, $f_{ax,k}$, can be derived from tests

Figure 8.24. Fixing at slanted angles in the x–y and y–z planes (d is the screw diameter)

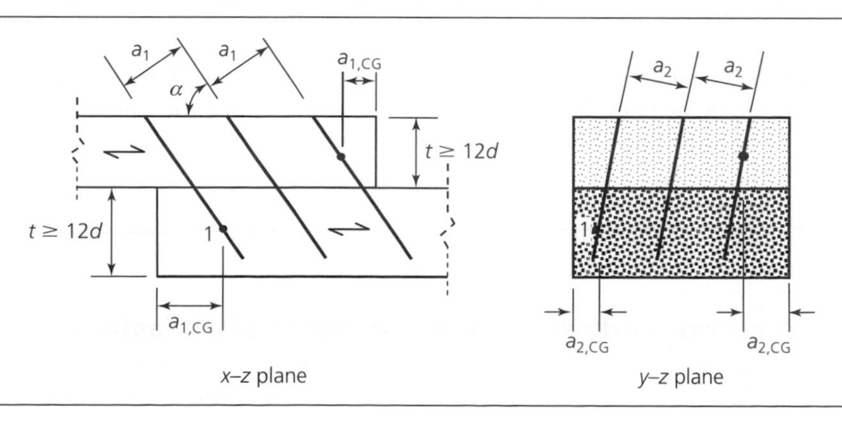

using the characteristic density of the timber in the connection and carried out in accordance with the requirements of EN 1382, $F_{ax,\alpha,Rk}$ is derived from

$$F_{ax,\alpha,Rk} = \frac{n_{ef} f_{ax,k} d l_{ef}}{1.2\cos^2\alpha + \sin^2\alpha} \quad \text{(D8.15)}$$

and when the value of $f_{ax,k}$ is able to be obtained from the screw manufacturer based on timber with a density ρ_a, $F_{ax,\alpha,Rk}$ is derived from

$$F_{ax,\alpha,Rk} = \frac{n_{ef} f_{ax,k} d l_{ef}}{1.2\cos^2\alpha + \sin^2\alpha}\left(\frac{\rho_k}{\rho_a}\right)^{0.8} \quad \text{(D8.16)}$$

The functions used in Equations D8.15 and D8.16 are as defined in EN 1995-1-1, and $\alpha \geq 30°$.

The characteristic pull-through resistance of screwed connections at an angle α to the grain with screws loaded axially, $F_{ax,\alpha,Rk}$, is calculated using *Equation 8.40b*, and the value used for the characteristic pull-through parameter, $f_{ax,head}$, can be derived from tests using the characteristic density of the timber in the connection, carried out in accordance with the requirements of EN 1383.

The characteristic tensile resistance of a screw, $f_{tens,k}$, referred to in *Equation 8.40c*, is derived from tests carried out in accordance with the requirements of EN 1383, using the weaker of the head tear off or the tensile strength of the shank.

For a connection with n screws, each subjected to a force component parallel to the shank, the effective number of screws in the connection, n_{ef}, will be

$$n_{ef} = n^{0.9} \quad \text{(D8.17)}$$

Clause 8.7.2

From the requirements of *clause 8.7.2*, the characteristic withdrawal capacity of the connection, $F_{ax,\alpha,Rk}$, with screws of diameter d, compliant with 6 mm $\leq d \leq$ 12 mm, loaded at an angle α to the grain, will be derived from the design value, $F_{ax,\alpha,Rd}$, where:

$$F_{ax,\alpha,Rd} = \min(\text{of the design value of } \textit{Equations 8.38, 8.40b, 8.40c}) \quad \text{for } 0.6 \leq d_1/d \leq 0.75 \quad \text{(D8.18)}$$

or

$$F_{ax,\alpha,Rd} = \min(\text{of the design value of } \textit{Equations 8.40a, 8.40b, 8.40c}) \quad \text{for } 0.75 < d_1/d \leq 0.9 \quad \text{(D8.19)}$$

Where $d < 6$ mm, Equation D8.18 will still apply, but *Equation 8.38* will be derived from the rules given in PD 6693-1. The characteristic value of the withdrawal capacity of the connection, $F_{ax,\alpha,Rk}$ will be obtained from

$$F_{ax,\alpha,Rk} = 1.25F_{ax,\alpha,Rd} \text{ (if screw fails in tension), otherwise} = F_{ax,\alpha,Rd}\frac{\gamma_M}{k_{mod}} \quad \text{(D8.20)}$$

where k_{mod} and γ_M are defined in Equation D8.1.

Clause 8.7.1
Clause 8.7.2

If the screws are fixed at right angles to the grain direction and loaded laterally, the characteristic lateral load-carrying capacity of the connection will be derived using the rules in *clause 8.7.1*, with the rope effect being calculated using the rules in *clause 8.7.2* to obtain the characteristic withdrawal capacity of a screw.

Clause 8.3.3

For screwed connections subjected to combined tension and lateral loading, the requirements of *clause 8.3.3* apply, as discussed in Section 8.3.3 in this guide.

8.8. Connections made with punched metal plate fasteners

Punched metal plate fasteners must comply with the requirements of EN 14545 (BSI, 2008b), and are used to connect two or more timber members that are adjacent to each other and have the

Figure 8.25. Typical punched metal plate connections

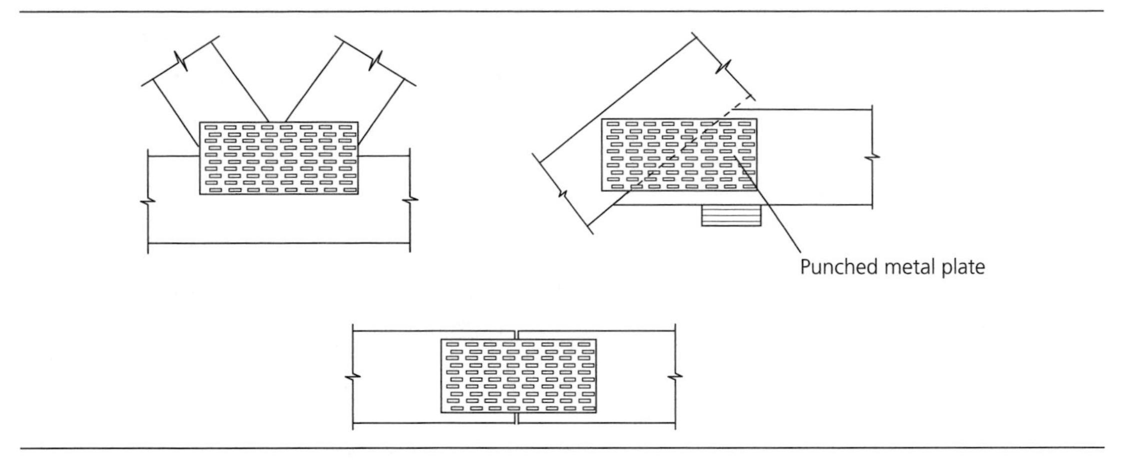

same thickness. They are most commonly used in the manufacture of trussed rafters and trussed floor structures in timber frame construction.

The metal plate incorporates indented areas that are pressed out of the plane of the plate into the timber to form the connection. Examples of this type of connection are shown in Figure 8.25.

The strength of this type of fastener is dependent on the metal plate and the configuration and geometry of the pressed indentations, and as these properties vary between metal plate manufacturers, strength properties are not given in the rules in *clause 8.8* and in any harmonised EN code. However, as a general guide, the lateral load able to be transferred over a surface area of 100 mm^2 of the metal plate is of the order of 500 N.

Clause 8.8

Information required for design is listed in *clause 8.8.3*, and the design rules used with these fasteners are given in the subsections of *clause 8.8*.

Clause 8.8.3
Clause 8.8

Reference should also be made to Appendix A, where proposed revisions to the content of *clauses 8.8.5.1(1)*, *8.8.5.1(2)* and *8.8.5.1(4)* as well as *8.8.5.2(1)* are indicated.

Clause 8.8.5.1(1)
Clause 8.8.5.1(3)
Clause 8.8.5.1(4)
Clause 8.8.5.2(1)

As connections formed with these fasteners are normally designed by the punched metal plate manufacturer, the design procedure is not discussed in this guide.

8.9. Split ring and shear plate connectors

These types of connector have been commonly used in Europe but not in the UK. The lateral load-carrying capacity of these connectors is much larger than can be achieved by bolted connections, and the connection stiffness is also much greater. They are used for connections in large-span trusses and, where shear plate connectors are used, structures have the advantage of being able to be preassembled at the fabrication facility, dismantled for transportation then reassembled at the site prior to erection. These connectors fit into preformed grooves in the timber members.

Ring connectors are either solid or formed with a split in their ring, and are only suitable for timber-to-timber connections, as shown in Figure 8.26. They are referenced as types A1 to A6 in EN 912, ranging in size from 60 mm to 260 mm in diameter, but the rules in *clause 8.9(1)* in EN 1995-1-1 do not allow connectors larger than 200 mm to be used. The connectors are held in place by bolts and washers that do not directly contribute to the strength of the connection.

Clause 8.9(1)

Shear plate connectors can be used for timber-to-timber joints, in pairs back-to-back, and also as a single unit for steel-to-timber connections, as shown in Figure 8.26. They are referenced as types B1 to B4 in EN 912, range in size from 65 mm to 190 mm in diameter and are held in place by bolts and washers. With these connectors, the bolt does not directly contribute to the connection strength, but is required to resist the shear stresses across the connection interface.

Figure 8.26. (a) Split ring and (b) shear plate connectors

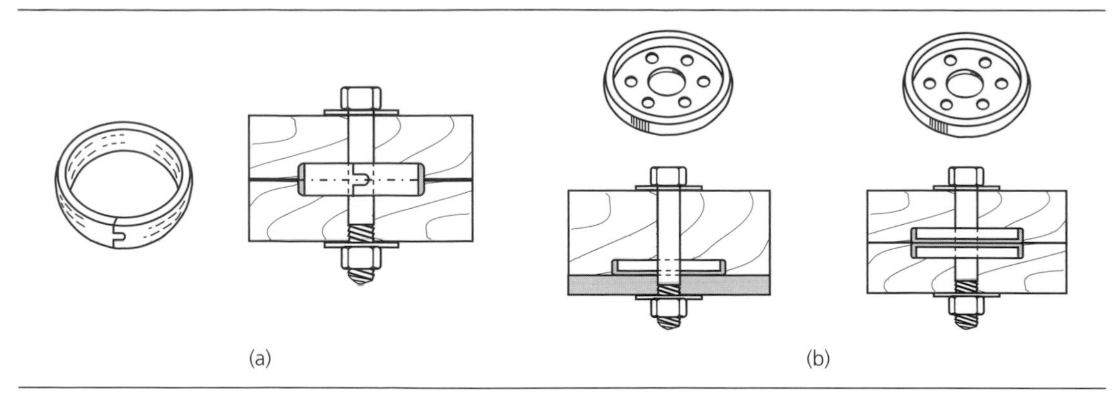

Clause 10.4.3(4)

The minimum diameter for bolts used with split ring and shear plate connectors is given in *clause 10.4.3(4)* and EN 13271 (BSI, 2002).

Load transfer in a ring connector is by embedment stresses in the timber and shear resistance of the ring, with the bolt holding the connection elements together. With a shear plate connector, the load transfer is again by embedment in the timber, and shear and bearing resistance within the shear plate and the bolt, in addition to holding the elements together, provides bearing resistance to the shear plate and shear resistance.

Clause 8.9(2)

The characteristic strength of these connectors loaded parallel to the grain is obtained from *Equations 8.61*, which model two failure modes. These are block shear failure (brittle failure) of the timber at the loaded end of the connection, and embedment failure. *Equation 8.61a* gives the characteristic block shear capacity, and *Equation 8.61b* gives the characteristic embedment capacity, and the lower value will be the characteristic load-carrying capacity per connector per shear plane parallel to the grain, $F_{v,0,Rk}$. As the shear strength of the bolt in a shear plate connector will always exceed $F_{v,0,Rk}$, it does not require to be checked. To prevent other forms of brittle failure occurring, minimum thickness criteria and also minimum spacings and end and edge distances must be complied with, and the requirements are given in *clause 8.9(2)* and *Table 8.7*, respectively. It is to be noted that the minimum loaded end spacing value ($a_{3,t}$) of $1.5d_c$ in *Table 8.7* is to be revised to read $2.0d_c$, as stated in Appendix A.

Up to maximum values, the connector strength will be influenced by the member thickness being used (factor k_1), the characteristic density of the timber (factor k_3) and, in the case of the block shear strength, the loaded end distance (factor k_2). For shear plate connectors in timber-to-steel connections a 10% increase in strength is permitted (factor k_4). The connector diameter, d_c, and the embedment depth, h, which must be in mm, are obtained from EN 912, and the value used for $a_{3,t}$ in *Equation 8.63* is the actual loaded end distance, and cannot be less than the minimum value given for $a_{3,t}$ in the revised *Table 8.7* referred to above. These connectors are generally unsuitable for use where the characteristic density of the timber is greater than approximately 610 kg/m^3.

To give a feel for the characteristic load-carrying capacity of these types of connector, the capacity of a type B2 shear plate and type A2 ring connector in C24 timber under optimum conditions is given in Table 8.5.

Clause 8.9(10)
Clause 8.9.(11)

Where connectors are staggered, spacing rules are given in *clauses 8.9(10)* and *8.9(11)*, permitting an elliptical arrangement in accordance with *Equation 8.69*, as shown on Figure 8.27.

Table 8.5. Characteristic capacities of type B2 and type A2 connectors

	Shear plate, type B2 (67 mm nominal diameter)	Split ring, type A2 (64 mm nominal diameter)
Characteristic load-carrying capacity	19.1 kN	21.4 kN

Figure 8.27. Reduced distances for connectors. (Based on EN 1995-1-1 (*Figure 8.13*). Reproduced with permission from EN 1995-1-1, © British Standards Institute, 2004)

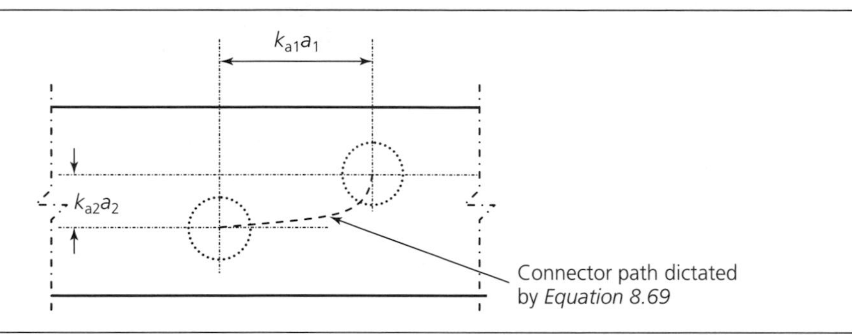

The spacing parallel to the grain, $k_{a1}a_1$, can further be reduced by the use of the factor $k_{s,red}$ but with an associated strength reduction, as stated in *clause 8.9(11)*. When $k_{a2}a_2 < 0.5k_{a1}a_1$, the connectors should be considered to be in one row parallel to the grain.

Clause 8.9(11)

Where the connector is loaded at an angle to the grain, the characteristic load-carrying capacity in that direction is reduced, and is obtained from *Equation 8.67*. For this condition, the force component perpendicular to the grain must also not exceed the strength referred to in *clause 8.1.4(3)*, as discussed in Section 8.1.4 in this guide, and the force component along the grain direction must be validated against the characteristic load-carrying capacity parallel to the grain and using the effective number of connectors.

Clause 8.1.4(3)

For loading parallel to the grain in a connection having n_{sp} shear planes, the effective number of connectors in a row is obtained from *Equation 8.71*, and the characteristic strength of the row will be

$$F_{v,0,Rk,con} = [2 + (1 - n/20)](n - 2)n_{sp}F_{v,0,Rk} \qquad \text{(D8.21)}$$

where n is the number of bolts in the row and $F_{v,0,Rk}$ is the characteristic load-carrying capacity per connector per shear plane parallel to the grain.

With regard to stiffness behaviour, the slip modulus K_{ser} per shear plane per fastener at the serviceability limit state is obtained from *clause 7.1*, and, as with metal dowel-type fasteners, it is suggested in this guide that the lateral stiffness of a connection is based on the number of connector bolts per shear plane being used, not the effective number per shear plane.

Clause 7.1

With these types of connector, no requirement is given in EN 1995-1-1 for clearance; however, particularly when using shear plate connectors, it is recommended in this guide that an allowance is made for initial slip due to tolerance/gap take up.

8.10. Toothed-plate connectors

As with ring and shear plate connectors, toothed-plate connectors are also commonly used in Europe but not in the UK. The lateral load-carrying capacity is smaller than can be obtained from ring and shear plate types, but they will have a higher strength and greater stiffness than obtained from bolted connections. They are used for connections in large-span trusses, and with single-sided types there is the advantage of being able to preassemble, dismantle and reassemble, as with shear plate connectors. These connectors are made from a metal plate with triangular teeth along the edges or spikes on the plate, and form the connection by being pressed into the timber. They can be single or double sided, as shown in Figure 8.28.

They are available in circular, square and other shapes, with sizes ranging from 38 mm to 165 mm, and are manufactured in accordance with the requirements of EN 912. The connectors are defined as types C1 to C11, and are categorised by the plate material, shape and whether single or double sided.

Figure 8.28. (a) Single- and (b) double-sided circular toothed-plate connectors

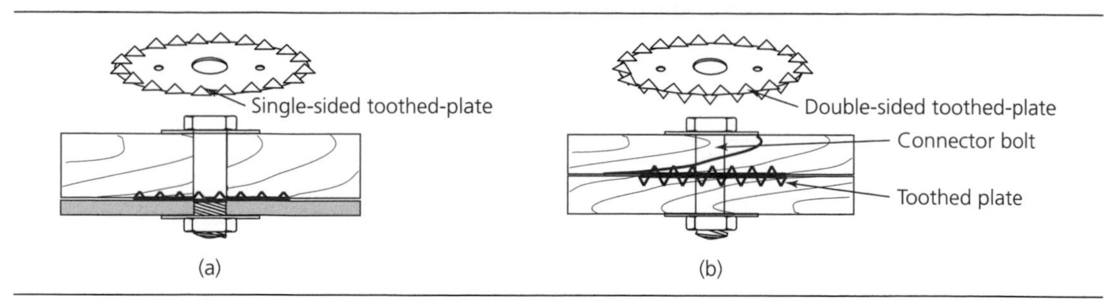

Single-sided toothed-plate connectors can be used to connect timber to steel or, in a back-to-back configuration, timber to timber. Double-sided toothed plates are suitable where non-demountable timber-to-timber connections are required. As stated, these connectors are pressed into the timber to form the connection, and because of the large force required to push the teeth fully into the timber, they are not suitable where the characteristic density of the timber is greater than about 500 kg/m^3. They are used with bolts, and the connection capacity is the sum of the connector and the bolt strength. The bolt size requirements are given in EN 13271.

In a double-sided toothed-plate joint, load transfer is by timber embedment stresses over the teeth area being transferred across the connection by the shear strength of the teeth. Because of embedment slip, the members will bear onto the bolt, which then contributes to the connection strength. With a single-sided toothed-plate the load transfer is slightly different but has the same result of transferring load to the bolt so that the joint strength is a combination of the strength of the toothed-plate connector and the bolt.

In accordance with the load transfer mechanism described above, the characteristic strength of a connection made using a toothed plate and a bolt, per shear plane, $F_{v,Rk,con}$, is the summation of the characteristic strength of the toothed-plate connector, $F_{v,Rk}$, and the characteristic strength of the laterally loaded bolt per shear plane, $F_{v,Rk,bolt}$, as follows:

$$F_{v,Rk,con} = F_{v,Rk} + F_{v,Rk,bolt} \qquad \text{(D8.22)}$$

Clause 8.5

The strength of the bolt is derived from the design rules in *clause 8.5*, as explained in Section 8.5 in this guide, and the characteristic strength of the toothed-plate connector, $F_{v,Rk,}$ is calculated from the relevant relation (*Equation 8.72*). With these connectors the design rules are structured to ensure ductile failure occurs, and this is achieved through the minimum thickness criteria and also the minimum spacings and end and edge distances given in the design rules in *clause 8.10(3)* and *Tables 8.8* and *8.9*, respectively.

Clause 8.10(3)

Up to maximum values, the connector strength is influenced by the member thickness being used (factor k_1), the loaded end distance (factor k_2) and the characteristic density of the member (factor k_3). The function d_c in *Equation 8.72*, which must be in mm, is dependent on the connector type being used, and is obtained from EN 912. It is also to be noted that the value for $a_{3,t}$ in *Equation 8.74* is the actual loaded end distance, and cannot be less than the $a_{3,t}$ value obtained from *Equation 8.75* for type C1 to C9 connectors or from *Equation 8.76* for types C10 and C11. These equations allow lower values to be used for $a_{3,t}$ than the 'minimum' values given in *Tables 8.8* and *8.9*. It is to be noted that the minimum spacing value of $2.0d_c$ in *Table 8.8* is incorrect and, as stated in Appendix A, is to be amended to read $1.5d_c$.

To give a feel for the characteristic load-carrying capacity of single- and double-sided connectors of similar size, the capacities of a type C1 double-sided connector and a similar-sized type C2 single-sided connector are listed in Table 8.6.

Clause 8.9

The stagger rules for circular connector types C1, C2, C6 and C7 are the same as stated in *clause 8.9* for split ring and shear plate connectors, but no guidance is given on the minimum lateral spacing within which they will function in a single-row configuration when loaded parallel to the grain.

Table 8.6. Characteristic capacities of type C1 and C2 connectors

	Single sided toothed-plate connector, type C2 (62 mm with 12 mm diameter steel bolt)	Double sided toothed-plate connector, type C1 (62 mm with 12 mm diameter steel bolt)
Characteristic load-carrying capacity	16.4 kN	16.4 kN

Where the connector is loaded at an angle to the grain, the characteristic load-carrying capacity of the connector remains unchanged, but the connection strength is reduced because of the reduced strength contribution from the bolt. Also, under this condition, for softwoods the force component perpendicular to the grain must not exceed the splitting capacity referred to in *clause 8.1.4(3)*, and, in accordance with *clause 8.1.2(5)*, the force component along the grain direction must be validated against the characteristic load-carrying capacity of the connection parallel to the grain.

Clause 8.1.4(3)
Clause 8.1.2(5)

When loading parallel to the grain in a connection having n_{sp} shear planes and n bolts per shear plane, the characteristic connection strength will be

$$F_{v,0,Rk,con} = n_{sp}(nF_{v,Rk} + n_{ef}F_{v,0,Rk,bolt}) \tag{D8.23}$$

where $F_{v,Rk}$ is as previously defined; $F_{v,0,Rk,bolt}$ is the characteristic strength of the bolt loaded along the grain direction; and n_{ef} is the effective number of bolts along the grain direction in the connection, referred to in Section 8.5.1.1 in this guide.

With regard to stiffness behaviour, the slip modulus K_{ser} per shear plane per fastener at the serviceability limit state is obtained from *clause 7.1* in EN 1995-1-1, and the comment in Section 8.9 in this guide regarding the occurrence of an initial slip with shear plate connectors is also relevant to single-sided toothed-plate connectors.

Clause 7.1

REFERENCES

Bejtka I and Blass HJ (2002) Joints with inclined screws. *35th Meeting of the Working Commission W18 – Timber Structures, International Council for Research and Innovation in Building and Construction*, Kyoto.

BSI (1999a) BS EN 1382: 1999. Timber structures – Test methods – Withdrawal capacity of timber fasteners. BSI, London.

BSI (1999b) BS EN 1383: 1999. Timber structures – Test methods – Pull-through resistance of timber fasteners. BSI, London.

BSI (2000) BS EN 10230-1: 2000. Steel wire nails – Part 1: Loose nails for general application. BSI, London.

BSI (2002) BS EN 13271: 2002. Timber fasteners – Characteristic load-carrying capacities and slip-moduli for connector joints. BSI, London.

BSI (2005) BS EN 1993-1-8: 2005. Eurocode 3: Design of steel structures – Part 1-8: Design of joints. BSI, London.

BSI (2006) BS EN 14358: 2006. Timber structures – Calculation of characteristic 5-percentile values and acceptance criteria for a sample. BSI, London.

BSI (2008a) BS EN 14592: 2008 + A1: 2012. Timber structures – Dowel-type fasteners – Requirements. BSI, London.

BSI (2008b) BS EN 14545: 2008. Timber structures – Connectors – Requirements. BSI, London.

BSI (2009a) BS EN 338: 2009. Structural timber. Strength classes. BSI, London.

BSI (2009b) BS EN 409: 2009. Timber structures – Test methods – Determination of the yield moment of dowel type fasteners. BSI, London.

Johansen KW (1949) *Theory of Timber Connections*. IABSE Publication No. 9. International Association for Bridge and Structural Engineers, Bern.

Designers' Guide to Eurocode 5: Design of Timber Buildings
ISBN 978-0-7277-3162-3

http://dx.doi.org/10.1680/dtb.31623.129

Chapter 9
Components and assemblies

This chapter is concerned with the requirements for the design of components and assemblies at ultimate limit states. It covers the material in *Section 9* of EN 1995-1-1, and addresses the following clauses:

- Components *Clause 9.1*
- Assemblies *Clause 9.2*

Components are the elements of a structure and design rules are given for the following types: glued thin-webbed beams, thin flanged beams and mechanically jointed beams. Also, requirements are given for mechanically jointed as well as glued columns.

Assemblies can be structures or substructures, and design rules are given for trusses, for roof, floor and wall diaphragms and for structural bracing.

9.1. Components

General

Within a framework of satisfying the requirements of sustainability, the design objective for any component is to try to maximise strength and stiffness at minimum cost, and glued thin-webbed and glued thin-flanged beams are composite sections that try to achieve this by optimising the section shape and by using materials for the web and flange component elements that will best provide the required strength and stiffness properties.

Because the stiffness properties of the web and flanges normally differ, to be able to apply simple elastic bending theory, either the radius of curvature approach or the equivalent section method, commonly referred to as the transformed-section method (Gere and Timoshenko, 1991), can be used. With the latter, the section is transformed into an equivalent section using only one of the materials from the untransformed section, and is the method adopted in this guide.

When subjected to stress at the ultimate limit states, because the creep behaviour of the materials used in these components will normally differ, instantaneous stresses will occur followed by final condition stresses. The creep behaviour is a function of the material deformation factor, k_{def} (referred to in Chapter 2 in this guide), and the higher the value of the factor the greater the creep effect. The instantaneous stress is generated immediately the load is applied, and with time the loading will cause creep, and the final condition stresses take this effect into account. For the instantaneous condition the mean value of stiffness properties will apply, and for the final condition, in accordance with the requirements of *clauses 2.2.2(1)P* and *2.3.2.2(2)* and as explained in Chapter 5 of this guide, final mean stiffness values must be used. The final mean values incorporate the quasi-permanent factor ψ_2, which is associated with the action causing the largest stress-to-strength ratio, and where the largest ratio is caused by a permanent action, ψ_2 is taken to equal 1. For most practical thin-webbed and thin-flanged beams, the difference between the instantaneous and the final stress is small, and, as required by *clause 2.2.2(1)P*, the stress analysis is undertaken at the final condition. Further explanation of the stiffness properties for this type of situation is given in Chapter 5 of this guide.

Clause 2.2.2(1)P
Clause 2.3.2.2(2)
Clause 2.2.2(1)P

9.1.1 Glued thin-webbed beams

In these sections, different materials are used for the web and the flanges and strength rules in *clause 9.1.1* give the validation requirements for each of these elements. In addition, stresses

Clause 9.1.1

Figure 9.1. Stresses in thin-webbed sections

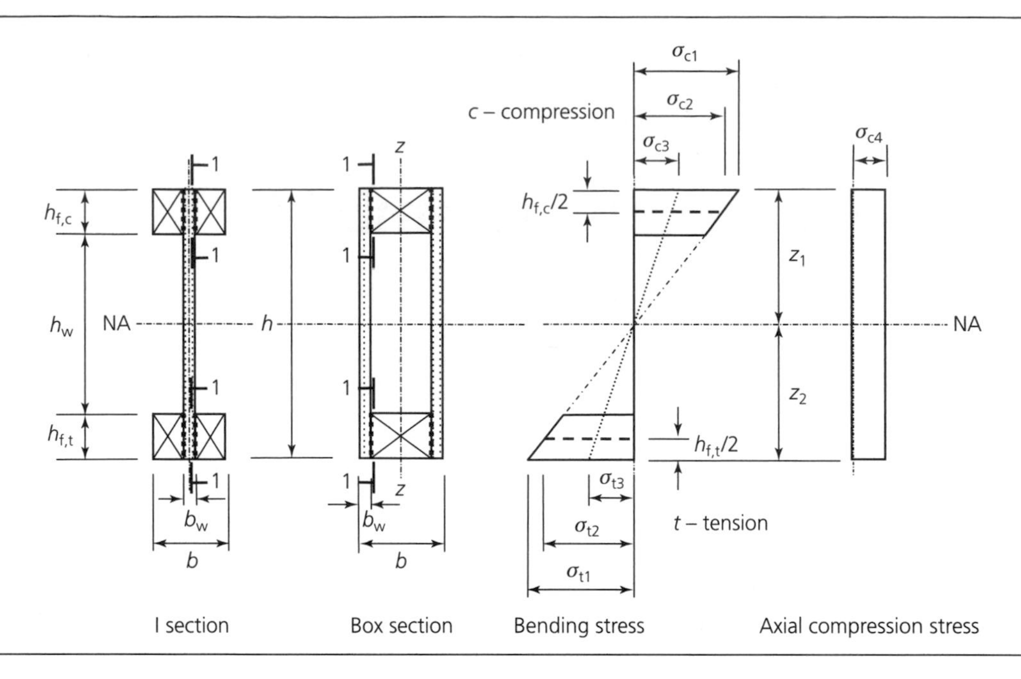

are transferred between the web and the flanges, and design rules are also given for the stresses at the glued connections between the elements.

The flanges are normally made from solid wood or wood-based materials, and the web is usually made from panel material. Typical thin-webbed sections used in practice are I and box beams, and the bending and compression stresses to be validated with these sections are shown on Figure 9.1. The figure also defines the terms and symbols used for elements of the beam as given in the design rules in *clause 9.1.1*.

Clause 9.1.1

For presentation purposes, the stress analysis in this guide assumes I and box sections are symmetrical in geometry about the z–z axis, and the top and bottom flanges are the same width and made from the same material, equating to the normal configuration for such sections. Also, the transformed section method referred to earlier in Section 9.1 is used, and section properties are based on the web material being transformed into flange material. Further, although not covered in the code, stresses arising from the addition of a nominal axial compression force on the section are given in this guide, assuming that the loading is applied along the centroidal axis; the member functions as a beam; compression instability of the composite section will not arise; and the extreme bending tension stress in the web and in the tension flange will always be in tension. For clarity, the sign convention shows compression stresses as positive and tensile stresses as negative, and for the condition where there is no axial compression, σ_{c4} will be 0.

For a section in which the mean value of the modulus of elasticity and the deformation factor for the web and for the flange materials are, respectively, $E_{w,mean}$, $k_{w,def}$ and $E_{f,mean}$, $k_{f,def}$, and the quasi-permanent factor ψ_2 is as defined in *clause 2.3.2.2(2)*, the respective final mean values will be

Clause 2.3.2.2

$$E_{w,fin} = E_{w,mean}/(1 + \psi_2 k_{w,def})$$

$$E_{f,fin} = E_{f,mean}/(1 + \psi_2 k_{f,def}) \quad \text{(D9.1)}$$

Based on these stiffness values, the transformed section geometrical properties will be

$$A_{ef,fin} = A_f + (E_{w,fin}/E_{f,fin})A_w \quad \text{(D9.2)}$$

$$I_{ef,fin} = I_f + (E_{w,fin}/E_{f,fin})I_w \quad \text{(D9.3)}$$

where $A_{ef,fin}$ and $I_{ef,fin}$ are the transformed area and the transformed second moment of area of the section, respectively, at the final condition and, using the symbols shown in Figure 9.1, A_w is the cross-sectional area of the untransformed web material ($b_w(h_w + h_{f,t} + h_{f,c})$ for an I section and $2b_w(h_w + h_{f,t} + h_{f,c})$ for a box section); A_f is the cross-sectional area of the flange material ($(b - b_w)(h_{f,t} + h_{f,c})$ for an I section and $(b - 2b_w)(h_{f,t} + h_{f,c})$ for a box section); I_w is the second moment of area of the untransformed web about the neutral axis of the transformed section; and I_f is the second moment of the area of both flanges about the neutral axis of the transformed section.

At the final condition the distance from the neutral axis to the extreme compression and extreme tension fibre will be $z_{1,fin}$ and $z_{2,fin}$, respectively, where

$$z_{1,fin} = \frac{1}{A_{ef,fin}}\left\{(b - a_1 b_w)\left[\frac{h_{f,c}^2}{2} + h_{f,t}\left(h - \frac{h_{f,t}}{2}\right)\right] + \frac{E_{w,fin}}{E_{f,fin}} A_w \frac{h}{2}\right\}$$

and

$$z_{2,fin} = h - z_{1,fin}$$

where a_1 is 1 for an I beam and 2 for a box beam.

Example 9.1

Calculate the transformed section properties of the I beam shown in Figure 9.2 at the final deformation condition when the ψ_2 factor associated with the action causing the largest stress-to-strength ratio is 0.3 and it functions in service class 2 conditions. The face grain of the plywood is parallel to the direction of the beam span.

Figure 9.2. I beam

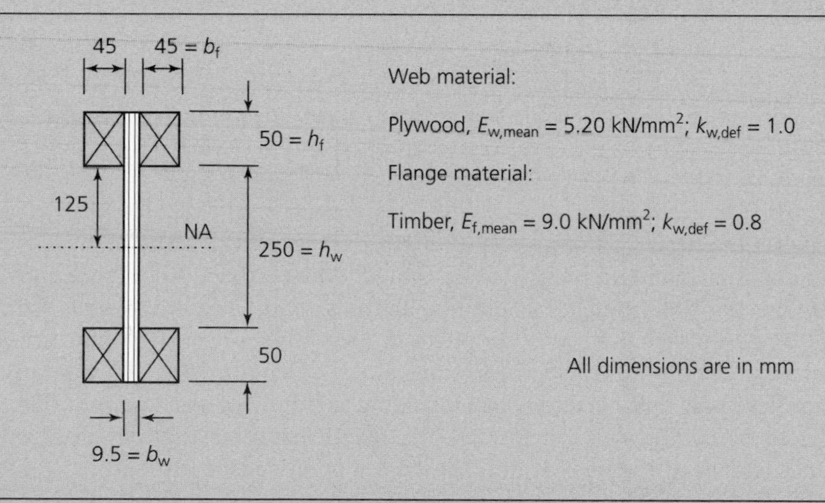

Web material:

Plywood, $E_{w,mean}$ = 5.20 kN/mm^2; $k_{w,def}$ = 1.0

Flange material:

Timber, $E_{f,mean}$ = 9.0 kN/mm^2; $k_{w,def}$ = 0.8

All dimensions are in mm

Convert to the flange material.

The neutral axis position will be the centroidal y–y axis position:

$$E_{w,fin} = E_{w,mean}/(1 + \psi_2 k_{w,def}) = 5.2/(1 + 0.3 \times 1.0) = 4.0 \text{ kN/mm}^2$$

$$E_{f,fin} = E_{f,mean}/(1 + \psi_2 k_{f,def}) = 9.0/(1 + 0.3 \times 0.8) = 7.258 \text{ kN/mm}^2$$

$$A_{ef,fin} = A_f + (E_{w,fin}/E_{f,fin})A_w = 2 \times 2 \times 45 \times 50 + (4.0/7.258) \times 9.5 \times (250 + 2 \times 50)$$

$$= 10\,832.46 \text{ mm}^2$$

$$I_{ef,fin} = I_f + (E_{w,fin}/E_{f,fin})I_w$$
$$= 4 \times 45 \times 50^3/12 + 2 \times 2 \times 45 \times 50 \times 150^2 + (4.0/7.258) \times 9.5 \times 350^3/12$$
$$= 2.2308 \times 10^8 \text{ mm}^4$$

When the section is subjected to a design bending moment, M_d, and a design axial force, N_d, the combined stress to be taken by the wood based flanges will be as follows.

The extreme fibre flange design compressive stress is

$$\sigma_{f,c,m,d} = \sigma_{c4} + \sigma_{c1} = N_d/A_{ef,fin} + M_d z_{1,fin}/I_{ef,fin} \quad \text{(D9.4)}$$

The extreme fibre flange design tension stress is

$$\sigma_{f,t,m,d} = \sigma_{c4} - \sigma_{t1} = N_d/A_{ef,fin} - M_d z_{2,fin}/I_{ef,fin} \quad \text{(D9.5)}$$

The design conditions to be satisfied for these stresses are given in *Equations 9.1* and *9.2*, and are

$$\sigma_{f,c,m,d} \le f_{m,d}$$

$$\sigma_{f,t,m,d} \le f_{m,d}$$

where $f_{m,d}$ is the design bending strength of the flange material and is obtained using the procedure described in Section 6.1.6 of this guide. If a lateral torsional instability analysis is undertaken, the design buckling strength derived from the analysis should be used rather than $f_{m,d}$.

The mean flange design compressive stress is

$$\sigma_{f,c,d} = \sigma_{c4} + \sigma_{c2} = N_d/A_{ef,fin} + M_d(z_{1,fin} - h_{f,c}/2)/I_{ef,fin} \quad \text{(D9.6)}$$

The design condition to be satisfied for this stress is given in *Equation 9.3*, and is

$$\sigma_{f,c,d} \le k_c f_{c,0,d} \quad \textit{(9.3)}$$

Clause 9.1.1(1)

where k_c converts the design compression strength parallel to the grain of the flange material, $f_{c,0,d}$, to the buckling strength, $k_c f_{c,0,d}$. As stated above, the design buckling strength of the beam can be derived from an instability analysis, and, if this is done, then $k_c = 1$. In *clause 9.1.1(1)* the approach used is to adopt the conservative assumption that the compression flange will act as a column restrained at positions along its length where it is laterally supported, and the design strength will be the design buckling strength of the column. The slenderness ratio is taken to be, $\sqrt{12}(\ell_c/b)$, where ℓ_c is the length between adjacent positions of lateral support; b is as shown in Figure 9.1; and the design compression strength, $f_{c,0,d}$, is calculated as described in Section 6.3.2 in this design guide. Where full lateral restraint to the beam is provided by the floor structure, k_c will be 1, and the full design compression strength of the flange, $f_{c,0,d}$, can be used.

The mean flange design tensile stress is

$$\sigma_{f,t,d} = \sigma_{c4} - \sigma_{t2} = N_d/A_{ef,fin} - M_d(z_{2,fin} - h_{f,t}/2)/I_{ef,fin} \quad \text{(D9.7)}$$

The design condition to be satisfied for this stress is given in *Equation 9.4*, and is

$$\sigma_{f,t,d} \le f_{t,0,d} \quad \textit{(9.4)}$$

where $f_{t,0,d}$ is the design tension strength of the flange material parallel to the grain, and is obtained from $f_{t,0,k}$ using the procedure described in Section 6.1.2 of this guide.

The combined stress in the web due to the bending moment and axial force will be as follows.

The extreme fibre web design compressive stress is

$$\sigma_{w,c,d} = \sigma_{c4} + \sigma_{c3} = (N_d/A_{ef,fin} + M_d z_{1,fin}/I_{ef,fin})(E_{w,fin}/E_{f,fin}) \quad \text{(D9.8a)}$$

The extreme fibre web design tension stress is

$$\sigma_{w,t,d} = \sigma_{c4} - \sigma_{c3} = (N_d/A_{ef,fin} - M_d z_{2,fin}/I_{ef,fin})(E_{w,fin}/E_{f,fin}) \quad \text{(D9.8b)}$$

The design conditions to be satisfied for these stresses are given in *Equations 9.6* and *9.7*, and are

$$\sigma_{w,c,d} \leq f_{c,w,d}$$

$$\sigma_{w,t,d} \leq f_{t,w,d}$$

where $f_{c,w,d}$ and $f_{t,w,d}$ are, respectively, the in-plane design compressive and tensile bending strengths of the web material, derived from the in-plane characteristic compressive bending strength, $f_{c,w,k}$, and tensile bending strength, $f_{t,w,k}$ as follows:

$$f_{c,w,d} = k_{mod} k_{sys} f_{c,w,k}/\gamma_M$$

$$f_{t,0,d} = k_{mod} k_{sys} f_{t,w,k}/\gamma_M$$

where γ_M is the material factor given in *Table NA.3* in the National Annex to EN 1995-1-1; k_{mod} is the modification factor referred to in *clause 3.1.3* and in Chapters 2 and 3 in this guide; and k_{sys} is the system strength factor referred to in *clause 6.6*.

Clause 3.1.3
Clause 6.6

If these characteristic strengths are not given, the characteristic compressive and tensile strengths can be used.

Where there are glued splice connections in the web, their strengths must be checked in accordance with the rules for glued connections given in PD 6693-1.

With thin-webbed beams, two types of shear failure can occur. The web can fail due to web buckling caused by the beam shear forces and/or it can fail due to shear. Web buckling can be checked from a buckling analysis or by using the simple conservative approach given in *Equation 9.8*, which requires that the clear distance between the flanges, h_w, is shown to be $\leq 70b_w$, where b_w is the width of the web (or of each web when a box beam is being used).

With regard to in-plane web shear, as $k_{cr} = 1$ for panel material the shear width able to be used will be b_w, and when $h_w \leq 35b_w$ the shear stress is taken to be uniform over a web depth of $(h_w + 0.5(h_{f,t} + h_{f,c}))$. Within the range $35b_w \leq h_w \leq 70b_w$, because of web buckling effects, the depth available for shear reduces linearly to a minimum value of $(0.5(h_w + 0.5(h_{f,t} + h_{f,c}))$ when $h_w = 70b_w$. Based on this criterion the shear force able to be taken by the web (or each web when a box section is being used) is obtained from *Equation 9.9*. The design panel shear strength, $f_{v,0,d}$, in *Equation 9.9* is derived from the characteristic panel shear strength of the web, $f_{v,0,k}$, from the procedure described in Section 6.1.7 of this guide.

The stress transfer between the web and the flanges is by horizontal shear, and the critical design condition will be the horizontal shear stress at positions 1–1 as shown in Figure 9.1. At these positions the design horizontal shear stress, $\tau_{mean,d}$, which is assumed to be uniform across the flange depth, will occur at the position of maximum design shear, V_d, and will be

$$\tau_{mean,d} = V_d S_{f,fin}/[I_{ef,fin}(nh_f)] \quad \text{(D9.9)}$$

where the functions are described above, and $S_{f,fin}$ is the first moment of area of the flange about the final condition neutral axis (e.g. for the compression flange of an I beam, $S_{f,fin} = h_{f,c}[(b - b_w)(z_{1,fin} - h_{f,c}/2])$; n is the number of glued connections between the web and the flange (=2 for an I beam or a box beam); and h_f is the flange height.

The strength of the glued connection will exceed the shear strength of the web or the flange material, and the design condition will relate to the material having the lowest shear strength. For this condition, the panel material will be sheared in rolling shear (refer to Section 6.1.7 in this guide), EN 1995-1-1 concluding this will always be lower than the shear strength of the wood-based flange. On this basis the design condition to be satisfied is given in *Equation (9.10)*, and is

$$\tau_{\mathrm{mean,d}} \le f_{\mathrm{v,90,d}} \text{ when } h_f \le 4b_{\mathrm{ef}} \quad \textbf{or} \quad \tau_{\mathrm{mean,d}} \le f_{\mathrm{v,90,d}}(4b_{\mathrm{ef}}/h_f)^{0.8} \quad \text{when } h_f > 4b_{\mathrm{ef}} \qquad (9.10)$$

where b_{ef} is $b_w/2$ for I beams and b_w for boxed beams, and $f_{\mathrm{v,90,d}}$ is the design planar (rolling) shear strength derived from the characteristic planar shear strength of the web material, $f_{\mathrm{v,90,k}}$, as follows:

$$f_{\mathrm{v,90,d}} = k_{\mathrm{mod}}k_{\mathrm{sys}}f_{\mathrm{v,90,k}}/\gamma_M$$

where γ_M is the material factor for the web material, given in *Table NA.3* in the National Annex to EN 1995-1-1 and the other symbols are as previously defined.

Example 9.2

The beam profile shown in Figure 9.2 is used in the construction of an office floor that functions in service class 2 conditions. Flanges are C18 timber, plywood web properties are given below and the beams are at 450 mm c/c (bs) with an effective span of 5.0 m (L) and a depth of 350 mm (H). The characteristic medium-duration floor loading supported is 2.5 kN/m^2, the characteristic permanent load from the floor structure is 1.0 kN/m^2 and there is no axial loading. Show that the beam complies with the ultimate limit state requirements of EN 1995-1-1. The compression flange of the beam is supported laterally by the floor decking structure.

Plywood properties: $f_{\mathrm{c,0,k}} = 12.7$ N/mm^2; $f_{\mathrm{t,0,k}} = 9.7$ N/mm^2; $f_{\mathrm{v,k}} = 3.5$ N/mm^2; (rolling shear strength) $f_{\mathrm{r,k}} = 0.89$ N/mm^2; $E_{\mathrm{w,mean}} = 5.2$ kN/mm^2; $G_{\mathrm{w,mean}} = 0.43$ kN/mm^2; $k_{\mathrm{w,def}} = 1.0$.

Design factors: (web material factor) $\gamma_{\mathrm{w,M}} = 1.2$; $\gamma_{\mathrm{f,M}} = 1.3$; $k_{\mathrm{mod,perm}} = 0.6$; $k_{\mathrm{mod,med}} = 0.8$; $k_{\mathrm{sys}} = 1.1$; (flange timber) $k_h = \min((150/h_f)^{0.2}, 1.3) = \min((150/50)^{0.2}, 1.3) = 1.246$; $k_c = 1$; $z_{1,\mathrm{fin}} = z_{2,\mathrm{fin}} = H/2 = 350/2 = 175$ mm.

Clause 2.3.2.2(2)

In accordance with *clause 2.3.2.2(2)*, the largest stress-to-strength ratio (r) will be associated with the larger of the permanent action/$k_{\mathrm{mod,perm}}$ and the variable action/$k_{\mathrm{mod,med}}$ values. For the beam, r_1 is proportional to

$$\gamma_G G_k/k_{\mathrm{mod,perm}} = 1.35 \times 1.0/0.6 = 2.25$$

and r_2 is proportional to

$$\gamma_G G_k + \gamma_Q Q_k)/k_{\mathrm{mod,med}} = (1.35 \times 1.0 + 1.5 \times 2.5)/0.8 = 6.38$$

that is, the largest stress in relation to strength will be associated with r_2 (i.e. due to the medium duration action) for which $\psi_2 = 0.3$, and the design loading condition will be the combined variable and permanent action:

$$F_d = (\gamma_G G_k + \gamma_Q Q_k)bs = (1.35 \times 1.0 + 1.5 \times 2.5) \times 0.45 = 2.295 \text{ kN/m}$$

$$V_d = F_d L/2 = 2.295 \times 2.5 = 5.74 \text{ kN}$$

Design moment on the beam:

$$M_d = F_d L^2/8 = 2.295 \times 5^2/8 = 7.17 \text{ kNm}$$

Design bending stress in the top and bottom timber flanges:

$$\sigma_{f,m,d} = M_d z_{1,fin}/I_{ef,fin} = 7.17 \times 175 \times 10^6/2.2308 \times 10^8 = 5.62\ \text{N/mm}^2$$

Design bending strength of the timber flange:

$$f_{m,d} = k_{mod,med} k_{sys} k_h f_{m,k}/\gamma_{f,M} = 0.8 \times 1.1 \times 1.246 \times 18/1.3 = 15.18\ \text{N/mm}^2$$

$$\sigma_{m,d}/f_{m,d} = 5.62/15.18 = 0.37 < 1$$

So, the flanges are OK in bending.

Design bending stress in the web:

$$\sigma_{w,d} = M_d z_{1,fin}(E_{w,fin}/E_{f,fin})/I_{ef,fin} = 7.17 \times 175 \times 10^6 \times (4.0/7.258)/2.2308 \times 10^8$$
$$= 3.1\ \text{N/mm}^2$$

Design bending strength of the web in compression:

$$f_{c,w,d} = k_{mod,med} k_{sys} f_{c,0,k}/\gamma_{w,M} = 0.8 \times 1.1 \times 12.7/1.2 = 9.31\ \text{N/mm}^2$$

$$\sigma_{w,d}/f_{c,w,d} = 3.1/9.31 = 0.33 < 1$$

So, the web is OK in compression.

Design bending strength of the web in tension:

$$f_{t,w,d} = k_{mod,med} k_{sys} f_{t,0,k}/\gamma_{w,M} = 0.8 \times 1.1 \times 9.7/1.2 = 7.11\ \text{N/mm}^2$$

$$\sigma_{w,d}/f_{t,w,d} = 3.1/7.11 = 0.44 < 1$$

So, the web is OK in tension.

Design mean axial stress in the flanges:

$$\sigma_{f,c,d} = \sigma_{f,t,d}$$

$$\sigma_{f,c,d} = M_d(z_{1,fin} - h_f/2)/I_{ef,fin} = 7.17 \times 10^6 \times (175 - 50/2)/2.2308 \times 10^8 = 4.82\ \text{N/mm}^2$$

Design compression strength of the flange:

$$k_c f_{c,0,d} = k_c k_{mod,med} k_{sys} f_{c,0,k}/\gamma_{f,M} = 1 \times 0.8 \times 1.1 \times 18/1.3 = 12.18\ \text{N/mm}^2$$

$$\sigma_{f,c,d}/k_c f_{c,0,d} = 4.82/12.18 = 0.4 < 1$$

So, the flange is OK in compression.

Design tension strength of flange:

$$f_{t,0,d} = k_{mod,med} k_h k_{sys} f_{t,0,k}/\gamma_{f,M} = 0.8 \times 1.246 \times 1.1 \times 11/1.3 = 9.28\ \text{N/mm}^2$$

$$\sigma_{f,t,d}/f_{t,0,d} = 4.82/9.28 = 0.52 < 1$$

So, the flange is OK in compression.

Buckling strength of the web (*Equation 9.8*):

$$h_w/b_w = 250/9.5 = 26.32 < 70$$

and is therefore OK.

Shear strength of the web (*Equation 9.9*):

$$f_{v,0,d} = k_{mod,med}k_{sys}f_{v,k}/\gamma_{w,M} = 0.8 \times 1.1 \times 3.5/1.2 = 2.57$$

and for

$$h_w/b_w = 250/9.5 = 26.32$$

the design shear force able to be taken by the web, $F_{v,w,Ed}$, will be

$$F_{v,w,Ed} = b_w h_w[1 + 0.5(2h_f)/h_w]f_{v,0,d} = 9.5 \times 250[1 + 50/250] \times 2.57/10^3$$

$$= 7.32\ \text{kN} > 5.74\ \text{kN}$$

and is therefore OK.

The shear strength of the glued connection along length 1–1 (see Figure 9.1) is calculated.

First moment of the area about the final condition neutral axis:

$$S_{f,fin} = 2b_f h_f(H/2 - h_f/2) = 2 \times 45 \times 50(350/2 - 50/2) = 675\,000\ \text{mm}^3$$

Total length of the glue line:

$$2h_f = 2 \times 50 = 100\ \text{mm}$$

Design shear stress along the glue line length:

$$\tau_{mean,d} = V_d S_{f,fin}/I_{ef,fin}2h_f = 5.74 \times 1000 \times 675\,000/(2.2308 \times 10^8 \times 100)$$

$$= 0.17\ \text{N/mm}^2$$

Design rolling shear strength of the web, with $b_{ef} = b_w/2 = 9.5/2 = 4.75$ mm:

$$f_{v,90,d} = k_{mod,med}k_{sys}f_{r,k}/\gamma_{w,M} = 0.8 \times 1.1 \times 0.89/1.2 = 0.65\ \text{N/mm}^2$$

As $h_f > 4b_{ef}$, from *Equation 9.10* the design planar shear strength is

$$f_{v,90,d}(4b_{ef}/h_f)^{0.8} = 0.65(4 \times 4.75/50)^{0.8} = 0.3\ \text{N/mm}^2$$

$$\tau_{mean,d}/f_{v,90,d} = 0.17/0.3 = 0.57 < 1$$

So, the web is OK in shear.

9.1.2 Glued thin-flanged beams

Clause 9.1.2

In these sections, different materials are used for the web and the flanges, and the strength rules in *clause 9.1.2* give the validation requirements for each of these elements. Stresses are also transferred between the web and the flanges across the glued connection, and design rules are given for the stresses at these locations.

Clause 9.1.2

The flanges are normally made from panel material, and the web from solid timber, but glued laminated timber or laminated veneer lumber (LVL) may also be used. The beams may have flanges on the top and bottom faces or only on the top face, and the example in Figure 9.3 shows the elements associated with a thin-flanged beam finished with panels on both faces. The figure also defines the terms and symbols used in the design rules in *clause 9.1.2* for elements of the beam, referred to as I and U beams.

Figure 9.3. Glued thin-flanged beam. (Based on EN 1995-1-1 (*Figure 9.2*). Reproduced with permission from EN 1995-1-1, © British Standards Institute, 2004)

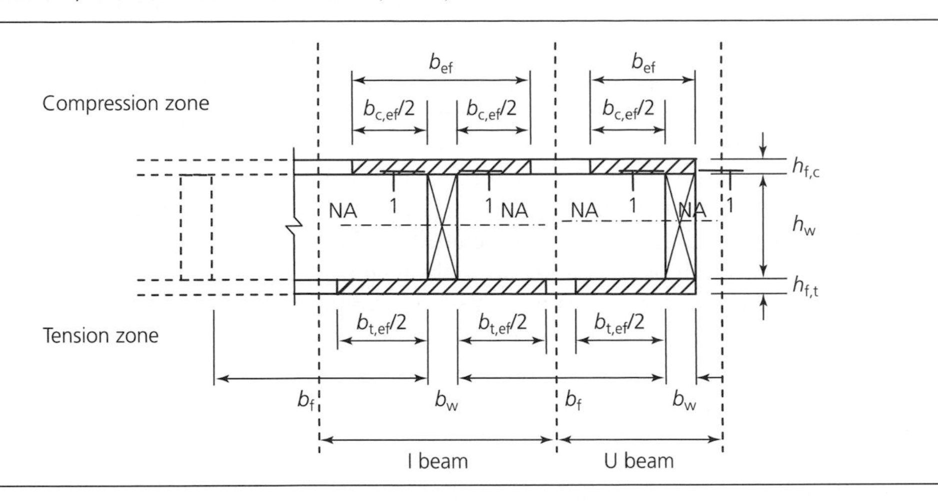

The top flange of these beams will normally function as a flooring member spanning over the beam webs at right angles to the direction of span of the thin-flanged beam. It may also have to function as a floor diaphragm, as discussed in Section 9.2.3.2 in this guide, and it will normally be these loading conditions that will determine the flange thickness. The bottom flange will also function this way, but as the vertical loading will generally be smaller than on the top flange, the flange thickness will normally be less. Consequently, with this type of composite section the flange depths must be validated to meet the above requirements, and the design rules in Chapters 6 and 7 in this guide will apply for that behaviour. The rules in *clause 9.1.2* only address the requirements of the thin-flanged beam functioning as a composite section.

Clause 9.1.2

The stress analysis in this guide assumes there is a linear strain variation over the beam depth, and the composite section can be considered to function as independent I and U beams as shown in Figure 9.3. However, where the k_{sys} factor for the webs can be applied, it can be argued it may function as a contiguous section. Flange widths contributing to the section strength are based on the 'effective width' criteria given in *clause 9.1.2(4)*. On this basis, for shear lag (which applies to both flanges), assuming that the flange material is stressed to its design strength, the maximum width that can be used will be as given in Table 9.1. For buckling instability, which will only apply to the compression flange, the effective width will be the width that can be stressed to the design compression strength without strength loss due to instability, and this is also defined in Table 9.1. In the table, ℓ is the effective span of the composite section, h_f is the flange thickness, and $b_{c,ef}$ and $b_{t,ef}$ are as shown in Figure 9.3. The effective width used for the compression flange will be the smaller of the shear lag and the plate buckling value.

Clause 9.1.2(4)

Table 9.1. Maximum value to be used for the effective flange width for shear lag and plate buckling effects

Panel material used for the flange	For shear lag: $b_{c,ef}$,[a] $b_{t,ef}$	For plate buckling: $b_{c,ef}$
Plywood with face grain (EN 636: BSI, 2008):		
Parallel to the web direction	0.1ℓ	$20h_f$
Perpendicular to the web direction	0.1ℓ	$25h_f$
OSB (EN 300: BSI, 2006)	0.15ℓ	$25h_F$
Particleboard (EN 312: BSI, 2010a) or Fibreboard (EN 622: BSI, 2003)	0.2ℓ	$30h_f$

Data from EN 1995-1-1
[a]The value of $b_{c,ef}$ for shear lag must not exceed the maximum value of $b_{c,ef}$ for plate buckling

In addition, unless a detailed buckling analysis is carried out, the clear distance between the web faces, b_f, must not exceed twice $b_{c,ef}$ derived from Table 9.1 and based on the plate buckling criterion.

Applying the transformed-section method referred to in Section 9.1 in this guide, the section properties in the analysis are derived from the web material transformed into flange material and, for simplicity, it has been assumed that the modulus of elasticity of the flanges will be the same. Where flange properties are different, an example of the application of the methodology is given in Example 9.3. As in Section 9.1.1, stresses arising from a nominal axial compression load in addition to bending induced stresses are included in this guide, assuming that the compression loading is applied through the centroidal axis; the member functions as a beam and compression instability of the composite section is not a design condition; and the extreme stresses in the web and flange on the 'tension' side of the section are in tension. The sign convention used in this guide assumes compression stresses are positive and tensile stresses negative.

Clause 2.3.2.2(2)

On this basis, if the mean value of the modulus of elasticity and the deformation factor for the web and for the flange materials are, respectively, $E_{w,mean}$, $k_{w,def}$ and $E_{f,mean}$, $k_{f,def}$, and the quasi-permanent factor ψ_2 is as defined in *clause 2.3.2.2(2)*, the final mean values, $E_{w,fin}$ and $E_{f,fin}$ will be obtained using Equation D9.1. If transformed into flange material, the transformed area, $A_{ef,fin}$, and the transformed second moment of the area of the section $I_{ef,fin}$ at the final condition will be obtained from Equations D9.2 and D9.3, where I_w and I_f are as previously defined, and, using the symbols in Figure 9.3:

A_w is the cross-sectional area of the untransformed web material

$= b_w h_w$

for an I beam and U beam.

A_f is the cross-sectional area of the flange material

$= h_{f,c}(b_{c,ef} + b_w) + h_{f,t}(b_{t,ef} + b_w)$

for the I beam

$= h_{f,c}(b_{c,ef}/2 + b_w) + h_{f,t}(b_{t,ef}/2 + b_w)$

for the U beam.

At the final condition, the distance from the neutral axis to mid-depth of the compression and tension flange is $z_{1,fin}$ and $z_{2,fin}$, respectively, and will be derived as follows:

$$z_{1,fin} = \frac{1}{A_{ef,fin}}\left[(b_{c,ef} + b_w)\left(\frac{h_{f,c}^2}{2}\right) + (b_{t,ef} + b_w)h_{f,t}\left(H - \frac{h_{f,t}}{2}\right) + \frac{E_{w,fin}}{E_{f,fin}}A_w\frac{H}{2}\right] - \frac{h_{f,c}}{2}$$

and

$$z_{2,fin} = H - \left(\frac{h_{f,c}}{2} + \frac{h_{f,t}}{2}\right) - z_{1,fin}$$

When the section is subjected to a design axial force, N_d, and a design bending moment, M_d, the axial and bending stresses will be as shown in Figure 9.4, and the design values will be as follows.

The mean flange design compressive stress is

$$\sigma_{f,c,d} = \sigma_{c3} + \sigma_{c1} = N_d/A_{ef,fin} + M_d\, z_{1,fin}/I_{ef,fin} \qquad \text{(D9.10)}$$

The mean flange design tension stress is

$$\sigma_{f,t,d} = \sigma_{c3} - \sigma_{t1} = N_d/A_{ef,fin} - M_d\, z_{2,fin}/I_{ef,fin} \qquad \text{(D9.11)}$$

Figure 9.4. Stress distribution in a glued thin flanged beam

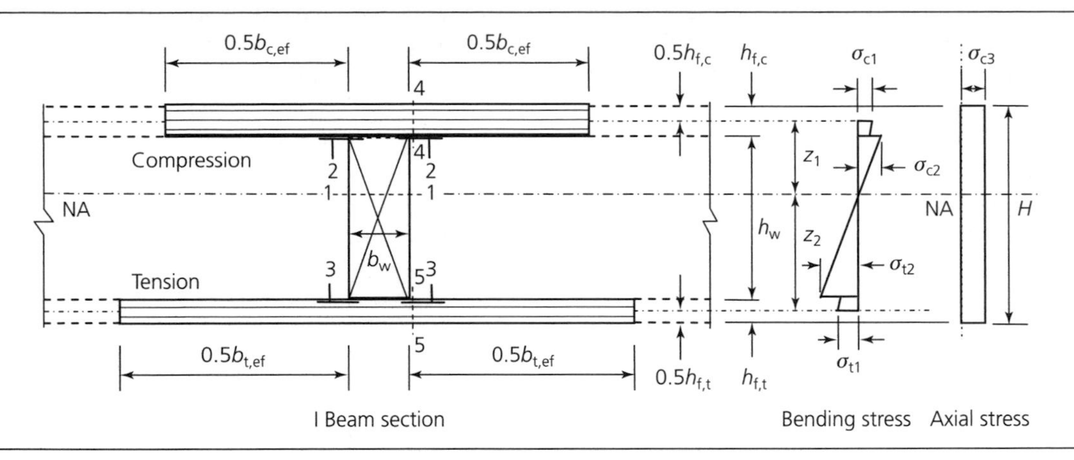

The design conditions to be satisfied for these stresses are given in *Equations 9.15* and *9.16*, and are

$$\sigma_{\mathrm{f,c,d}} \leq f_{\mathrm{f,c,d}}$$

$$\sigma_{\mathrm{f,t,d}} \leq f_{\mathrm{f,t,d}}$$

where $f_{\mathrm{f,c,d}}$ and $f_{\mathrm{f,t,d}}$ are the design compressive and tensile strengths of the flange material, respectively, and are derived from the characteristic compressive strength ($f_{\mathrm{c,0,k}}$ or $f_{\mathrm{c,90,k}}$) and characteristic tensile strength ($f_{\mathrm{t,0,k}}$ or $f_{\mathrm{t,90,k}}$), respectively as follows:

$$f_{\mathrm{f,c,d}} = k_{\mathrm{mod}} f_{\mathrm{c,(0/90),k}} / \gamma_{\mathrm{M}}$$

$$f_{\mathrm{f,t,d}} = k_{\mathrm{mod}} f_{\mathrm{t,(0/90),k}} / \gamma_{\mathrm{M}}$$

where γ_{M} is the material factor given in *Table NA.3* in the National Annex to EN 1995-1-1, and the other symbols are as previously defined.

Where the face grain of the flange material is loaded parallel to the grain, $f_{\mathrm{f,c/t,0,k}}$ will be used, and if loaded perpendicular to the grain, $f_{\mathrm{f,c/t,90,k}}$ will apply.

The combined stress in the web due to the bending moment and axial force will be as follows.

The extreme fibre web design compressive stress is

$$\sigma_{\mathrm{w,c,d}} = \sigma_{\mathrm{c3}} + \sigma_{\mathrm{c2}} = (N_{\mathrm{d}}/A_{\mathrm{ef,fin}} + M_{\mathrm{d}}(z_{1,\mathrm{fin}} - h_{\mathrm{f,c}}/2)/I_{\mathrm{ef,fin}})(E_{\mathrm{w,fin}}/E_{\mathrm{f,fin}}) \qquad \text{(D9.12)}$$

The extreme fibre web design tension stress is

$$\sigma_{\mathrm{w,t,d}} = \sigma_{\mathrm{c3}} - \sigma_{\mathrm{t2}} = (N_{\mathrm{d}}/A_{\mathrm{ef,fin}} - M_{\mathrm{d}}(z_{2,\mathrm{fin}} - h_{\mathrm{f,t}}/2)/I_{\mathrm{ef,fin}})(E_{\mathrm{w,fin}}/E_{\mathrm{f,fin}}) \qquad \text{(D9.13)}$$

and the design condition to be satisfied for these stresses is that the maximum value must be $\leq f_{\mathrm{m,d}}$, where $f_{\mathrm{m,d}}$ is the design bending strength of the flange material, obtained using the procedure described in Section 6.1.6 of this guide.

Clause 9.1.2

The web is also subjected to shear stresses, and *clause 9.1.2* only requires the shear stress at the glued joints between the web and the flanges (positions 2–2 and 3–3 in Figure 9.4) to be checked. However, depending on the geometry and materials used in the composite section, the stresses at 1–1, 4–4 and 5–5 in Figure 9.4 could also be critical, and the stress and strength requirements of these are also covered in this guide. When subjected to a design shear force, V_{d}, the shear stresses at the above positions will be as follows.

Clause 6.1.7

The shear stress in the web at the neutral axis position (1–1), $\tau 1_{v,fin,d}$, is uniform across the width of the web, and to take into account the effect of the k_{cr} factor referred to in *clause 6.1.7* it is proposed in this guide that it is applied to the shear strength rather than the beam width. On this basis the shear stress in the web will be

$$\tau 1_{v,fin,d} = V_d S1_{f,fin,NA}/I_{ef,fin} b_w \tag{D9.14}$$

where the functions are as previously defined and $S1_{f,fin,NA}$ is the first moment of the area of the transformed section above (or below) the neutral axis. An alternative is to take all of the shear force in the web, in which case

$$\tau 1_{v,fin,d} = 1.5 V_d / b_w h_w$$

Clause 6.1.7(2)

The verification for these alternatives will be $\tau 1_{v,fin,d} \leq k_{cr} f_{v,d}$, where $f_{v,d}$ is the design shear strength of the web material obtained using the procedure described in Section 6.1.7 of this guide, and the value of k_{cr} is as given in *clause 6.1.7(2)*.

The horizontal shear stress at the glue line positions 2–2 and 3–3, $\tau 2_{r,fin,d}$, and $\tau 3_{r,fin,d}$, respectively, are assumed to be uniform across b_w, with the crack factor not being considered to be relevant in this case (as failure is taken to be in the panel material), and will be

$$\tau 2_{r,fin,d} = V_d S2_{f,fin,NA}/I_{ef,fin} b_w$$

$$\tau 3_{r,fin,d} = V_d S3_{f,fin,NA}/I_{ef,fin} b_w \tag{D9.15}$$

where the functions are as previously defined, and $S2_{f,fin,NA}$ and $S3_{f,fin,NA}$, which relate to the transformed section, are the first moment of the area about the neutral axis of the flange of thickness $h_{f,c}$ and of thickness $h_{f,t}$, respectively. The design conditions to be satisfied for this stress situation are defined in *Equation 9.14*, and for each shear stress, $\tau 2,3_{v,fin,d}$:

$$\tau 2,3_{r,fin,d} \leq \begin{Bmatrix} f_{v,90,d} & \text{for} \quad b_w \leq 8h_f \\ f_{v,90,d}\left(\dfrac{8h_f}{b_w}\right)^{0.8} & \text{for} \quad b_w > 8h_f \end{Bmatrix} \tag{9.14}$$

where, for $\tau 2_{r,fin,d}$, $h_f = h_{f,c}$ and for $\tau 3_{r,fin,d}$, $h_f = h_{f,t}$. Also, $f_{v,90,d}$ is the design planar (rolling) shear strength of the flange material being considered, and is derived from its characteristic planar shear strength, $f_{v,90,k}$, as defined in Section 9.1.1 in this guide, but noting that the k_{sys} factor will be 1, as referred to in Section 6.6. Where U beams are being designed, *Equation 9.14* will still apply, but $4h_f$ is used rather than $8h_f$.

The panel shear stress at positions 4–4, $\tau 4_{v,fin,d}$, and 5–5, $\tau 5_{v,fin,d}$, are assumed to be uniform across the respective flange depths, and will be

$$\tau 4_{v,fin,d} = V_d S4_{f,fin,NA}/I_{ef,fin} n h_{f,c}$$

$$\tau 5_{v,fin,d} = V_d S5_{f,fin,NA}/I_{ef,fin} n h_{f,t} \tag{D9.16}$$

where the functions are as previously defined; $S4_{f,fin,NA}$ and $S5_{f,fin,NA}$ relate to the transformed section, and are the first moments of area of the flange of thickness $h_{f,c}$ and of thickness $h_{f,t}$, respectively, each excluding the area of flange over the width of the web; and n is the number of critical shear locations in the flange (i.e. 2 for an I beam and 1 for a U beam).

The design condition to be satisfied is that each flange shear stress (i.e. $\tau 4_{v,fin,d}$, and $\tau 5_{v,fin,d}$) must be $\leq f_{v,d}$, where $f_{v,d}$ is the design panel shear strength of the flange material. This strength is derived from the characteristic panel shear strength of the web, $f_{v,0,k}$, using the procedure described in Section 6.1.7 of this guide.

Example 9.3
Stressed skin panels are used for a flat roof structure spanning between supports 4.0 m (L) apart and functioning in service class 2 conditions. The panels are 211 mm deep (H), and the cross-section at a typical I beam position is as shown in Figure 9.4. The clear distance between adjacent webs is 525 mm (b_f), and the panel is glued between the flanges and the web. Including for the self-weight of the structure, the design loading from permanent and snow loading is 1.95 kN/m^2 (q_d), and the largest stress-to-strength ratio is due to the snow loading. Class C22 timber is used for the web, which is 40 mm wide (b_w) by 180 mm deep (h_w), and the top and bottom flanges are Canadian plywood, 18.5 mm ($h_{f,c}$) and 12.5 mm ($h_{f,t}$) thick, respectively, both with their face ply perpendicular to the direction of span of the stressed skin panel. The building is located at 300 m above sea level. Show that the I beam section of the panel will comply with the rules in EN 1995-1-1 at the ultimate limit states. (In the analysis the transformed section will be based on the top flange material.)

Properties of 18.5 mm plywood: $f_{c,90,k} = 6.5$ N/mm^2; $f_{v,k} = 3.5$ N/mm^2; (rolling shear strength) $f_{r,k} = f_{v,90,k} = 0.63$ N/mm^2; $E_{f,c,90,mean} = 2.67$ kN/mm^2.

Properties of 12.5 mm plywood: $f_{t,90,k} = 7.4$ N/mm^2; $f_{v,k} = 3.5$ N/mm^2; (rolling shear strength) $f_{r,k} = f_{v,90,k} = 0.64$ N/mm^2; $E_{f,t,90,mean} = 3.96$ kN/mm^2.

As the largest stress in relation to strength is due to the snow loading, from the National Annex to EN 1990, $\psi_2 = 0$ and the stiffness values will be based on mean properties.

Design factors: (flange material) $\gamma_{f,M} = 1.2$; $\gamma_{w,M} = 1.3$; $k_{mod,perm} = 0.6$; $k_{mod,short} = 0.9$; (web) $k_{w,sys} = 1.1$; (flange) $k_{f,sys} = 1.0$; (web timber) $k_h = 1$; and $k_c = 1$

Geometric properties of the I beam:

$$b_{c,ef} = \min(0.1L, 25h_{f,c}) = \min(0.1 \times 4000, 25 \times 18.5) = 400 \text{ mm}$$

$$b_f/2b_{c,ef} = 525/2 \times 462.5 = 0.57 < 1$$

So $b_{c,ef}$ is OK (*clause 9.1.2(5)*).

Clause 9.1.2(5)

$$b_{t,ef} = 0.1L = 400 \text{ mm}$$

and

$$b_{t,ef} < b_f$$

So $b_{t,ef}$ is OK.

Area of the top flange:

$$A_{f,c} = (b_{c,ef} + b_w)h_{f,c} = (400 + 40) \times 18.5 = 8140 \text{ mm}^2$$

Area of the bottom flange:

$$A_{f,t} = (E_{f,t,90,mean}/E_{f,c,90,mean})(b_{t,ef} + b_w)h_{f,t} = (3.96/2.67) \times (400 + 40) \times 12.5 = 8157.3 \text{ mm}^2$$

Area of web:

$$A_w = (E_{w,mean}/E_{f,c,90,mean})b_w h_w = (10/2.67) \times 40 \times 180 = 26\,966.3 \text{ mm}^2$$

Net area:

$$A_{ef,fin} = A_{f,c} + A_{f,t} + A_w = 8140 + 8157.3 + 26\,966.3 = 43\,263.6 \text{ mm}^2$$

Distances $z_{1,\mathrm{fin}}$ and $z_{2,\mathrm{fin}}$:

$$z_{1,\mathrm{fin}} = [A_{\mathrm{f,c}}(h_{\mathrm{f,c}}/2) + A_{\mathrm{f,t}}(H - h_{\mathrm{f,t}}/2) + A_{\mathrm{w}}(h_{\mathrm{w}}/2 + h_{\mathrm{f,c}})]/A_{\mathrm{ef,fin}} - h_{\mathrm{f,c}}/2$$

$$= [8140 \times 18.5/2 + 8157.3 \times (211 - 12.5/2) + 26\,966.3 \times (180/2 + 18.5)]$$

$$/43\,263.6 - 9.25$$

$$= 98.72 \text{ mm}$$

$$z_{2,\mathrm{fin}} = h_{\mathrm{w}} + (h_{\mathrm{f,c}} + h_{\mathrm{f,t}})/2 - z_{1,\mathrm{fin}} = 180 + (18.5 + 12.5)/2 - 98.72 = 96.78 \text{ mm}$$

Second moment of the area about the National Annex position:

$$I_{\mathrm{ef,fin}} = A_{\mathrm{f,c}}(h_{\mathrm{f,c}}^2/12 + z_{1,\mathrm{fin}}^2) + A_{\mathrm{f,t}}(h_{\mathrm{f,t}}^2/12 + z_{2,\mathrm{fin}}^2) + A_{\mathrm{w}}\{h_{\mathrm{w}}^2/12 + [z_{1,\mathrm{fin}} - (h_{\mathrm{f,c}} + h_{\mathrm{w}})/2]^2\}$$

$$= 8140 \times (18.5^2/12 + 98.72^2) + 8157.3 \times (12.5^2/12 + 96.78^2) + 26\,966.3$$

$$\times \{180^2/12 + [98.72 - (18.5 + 180)/2]^2\}$$

$$= 2.28889 \times 10^8 \text{ mm}^4$$

The design loading conditions are calculated.

Design moment on the beam:

$$M_{\mathrm{d}} = F_{\mathrm{d}}L^2/8 = 1.95 \times 4^2/8 = 3.9 \text{ kN m}$$

Design shear force:

$$V_{\mathrm{d}} = F_{\mathrm{d}}L/2 = 1.95 \times 4/2 = 3.9 \text{ kN}$$

Design mean flange compression stress in the top flange:

$$\sigma_{\mathrm{f,c,d}} = M_{\mathrm{d}}z_{1,\mathrm{fin}}/I_{\mathrm{ef,fin}} = 3.9 \times 10^6 \times 98.72/(2.28889 \times 10^8) = 1.68 \text{ N/mm}^2$$

Design compression strength of the top flange, $f_{\mathrm{c,0,d}}$:

$$f_{\mathrm{f,c,d}} = k_{\mathrm{mod,short}}f_{\mathrm{c,90,k}}/\gamma_{\mathrm{f,M}} = 0.9 \times 6.5/1.2 = 4.88 \text{ N/mm}^2$$

$$\sigma_{\mathrm{f,c,d}}/f_{\mathrm{f,c,d}} = 1.68/4.88 = 0.34 < 1$$

So, the flange is OK in compression.

Design mean flange tension stress in the bottom flange:

$$\sigma_{\mathrm{f,t,d}} = M_{\mathrm{d}}z_{2,\mathrm{fin}}(E_{\mathrm{f,t,90,mean}}/E_{\mathrm{f,c,90,mean}})/I_{\mathrm{ef,fin}}$$

$$= 3.9 \times 10^6 \times 96.78 \times (3.96/2.67)/(2.28889 \times 10^8) = 2.45 \text{ N/mm}^2$$

Design tension strength of the bottom flange, $f_{\mathrm{t,0,d}}$:

$$f_{\mathrm{f,c,d}} = k_{\mathrm{mod,short}}f_{\mathrm{t,90,k}}/\gamma_{\mathrm{f,M}} = 0.9 \times 7.4/1.2 = 5.55 \text{ N/mm}^2$$

$$\sigma_{\mathrm{f,t,d}}/f_{\mathrm{f,t,d}} = 2.45/5.55 = 0.44 < 1$$

So, the flange is OK in tension.

The maximum distance from the National Annex position to the extreme tension fibre in the web is

$$z_t = z_{2,fin} - h_{f,t}/2 = 96.78 - 12.5/2 = 90.53 \text{ mm}$$

Design bending stress in the web:

$$\sigma_{w,t,d} = M_d z_t (E_{w,mean}/E_{f,c,90,mean})/I_{ef,fin}$$

$$= 3.9 \times 10^6 \times 90.53 \times (10/2.67)/(2.28889 \times 10^8) = 5.78 \text{ N/mm}^2.$$

Design bending strength of the web:

$$f_{m,d} = k_{mod,short} k_h k_{sys} f_{m,k}/\gamma_{w,M} = 0.9 \times 1 \times 1.1 \times 22/1.3 = 16.75 \text{ N/mm}^2$$

$$\sigma_{w,t,d}/f_{m,d} = 5.78/16.75 = 0.35 < 1$$

So, the web is OK in bending.

The shear strength of the web is calculated.

First moment of area of the section above the National Annex position:

$$S1_{f,fin,NA} = A_{f,c} z_{1,fin} + b_w (E_{w,mean}/E_{f,c,90,mean})(z_{1,fin} - h_{f,c}/2)^2/2$$

$$= 8140 \times 98.72 + 40 \times (10/2.67)(98.72 - 18.5/2)^2/2 = 1.403 \times 10^6 \text{ mm}^3$$

Shear stress at the Neutral Axis position:

$$\tau 1_{v,fin,d} = V_d S1_{f,fin,NA}/I_{ef,fin} b_w = 3.9 \times 10^3 \times 1.403 \times 10^3/(2.28889 \times 10^8 \times 40)$$

$$= 0.6 \text{ N/mm}^2$$

Design shear strength of the web, with $k_{cr} = 0.67$:

$$f_{v,d} = k_{mod,short} f_{v,k}/\gamma_{w,M} = 0.9 \times 3.8/1.3 = 2.63 \text{ N/mm}^2$$

$$k_{cr} f_{v,d} = 0.67 \times 2.63 = 1.76 \text{ N/mm}^2$$

$$\tau 1_{v,fin,d}/f_{v,d} = 0.6/1.76 = 0.34 < 1$$

So, the web is OK in shear at the Neutral Axis position.

As the design shear stress condition at position 2–2 will have the largest stress-to-strength ratio, only that condition is given, and is derived as follows:

$$S2_{f,fin,NA} = A_{f,c} z_{1,fin} = 8140 \times 98.72 = 8.036 \times 10^5 \text{ mm}^3$$

$$\tau 3_{v,fin,d} = V_d S3_{f,fin,NA}/I_{ef,fin} b_w = 3.9 \times 10^3 \times 8.036 \times 10^5/(2.28889 \times 10^8 \times 40)$$

$$= 0.34 \text{ N/mm}^2$$

Design planar shear strength of the top plywood flange, as $b_w/h_{f,c} = 2.16 < 8$:

$$f_{r,d} = k_{mod,short} f_{r,k}/\gamma_{f,M} = 0.9 \times 0.63/1.2 = 0.47 \text{ N/mm}^2$$

$$\tau 3_{v,fin,d}/f_{r,d} = 0.34/0.47 = 0.72 < 1$$

So, the planar shear stress at the interface is OK.

The shear strength of the flange material at the face of the web is calculated.

As the largest shear stress condition at position 5–5 will have the largest stress-to-stress ratio, only that condition is given, and is derived as follows:

$$S5_{\text{f,fin,NA}} = [(E_{\text{f,t,90,mean}}/E_{\text{f,c,90,mean}})(b_{\text{t,ef}}) - b_{\text{w}}]h_{\text{f,t}}(z_{2,\text{fin}})$$

$$= ((3.96/2.67)440 - 40) \times 12.5 \times (96.78) = 7.4107 \times 10^5\ \text{mm}^3$$

$$\tau5_{\text{v,fin,d}} = V_{\text{d}}S5_{\text{f,fin,NA}}/I_{\text{ef,fin}}2h_{\text{f,t}} = 3.9 \times 10^3 \times 7.4107 \times 10^5/(2.28889 \times 10^8 \times 2 \times 12.5)$$

$$= 0.51\ \text{N/mm}^2$$

Design panel shear strength of the bottom plywood flange:

$$f_{\text{v,d}} = k_{\text{mod,short}}f_{\text{v,k}}/\gamma_{\text{f,M}} = 0.9 \times 3.5/1.2 = 2.63\ \text{N/mm}^2$$

$$\tau3_{\text{v,fin,d}}/f_{\text{r,d}} = 0.51/2.63 = 0.19 < 1$$

So, the panel shear stress at the interface is OK.

9.1.3 Mechanically jointed and glued columns

The types of section that come into this category of structure or structural element are discussed in Chapter 11 of this guide.

9.2. Assemblies

9.2.1 Trusses

Clause 9.2.1

The design rules in *clause 9.2.1* cover all types of truss, including trussed rafters used in domestic construction. 'Trussed rafter' is the term used in the UK for a triangulated structure that is assembled in one plane from same-thickness material with joints formed using punched metal plate fasteners.

Clause 9.2.1(3)

Where trusses are predominantly loaded at the node positions and member instability is not a factor, the sum of the combined bending and compression stress ratios given in *Equations 6.19* and *6.20* must not exceed 0.9. When deriving the slenderness ratio for in-plane instability of a compression member, the effective length is taken to be the member length between adjacent points of contraflexure (i.e. points of zero bending moment) derived from the structural analysis. With fully triangulated trusses where the conditions defined in *clause 9.2.1(3)* apply, the effective length of a compression member should be taken to be the bay length, which is defined in Chapter 5 of this guide.

Clause 9.2.1(4)

Where the simplified analysis referred to in Chapter 5 is used for fully triangulated trusses formed with metal gusset plate fasteners, the effective lengths to be used for the verification of member design are given in *clause 9.2.1(4)*. It is also to be noted that the calculated axial forces derived from this type of analysis have to be increased by 10% when verifying the strength of connections and members in compression.

For trusses loaded at the nodes and analysed using a simplified analysis, the design stress-to-strength ratios in compression and in tension must not exceed 0.7, and only 70% of the associated connection capacity can be used.

Out-of-plane instability of compression members must also be checked. Where members are supported laterally by adequate bracing members, the effective length can be taken to be the distance between adjacent bracing support positions, and where diaphragm structures give continuous lateral support, out-of-plane instability will not arise.

Trusses must be designed for handling and erection forces, and guidance is given in *clause 9.2.1(8)* on the force and load duration the joints should be checked against for this condition.

Clause 9.2.1(8)

When checking member strength, the effect of loss of area at connections due to the type of fastener being used has to be taken into account, and this is referred to in Section 5.2 of this guide.

9.2.2 Trusses with punched metal plate fasteners

The requirements of *clause 9.2.1* will also apply to trusses formed with punched metal plate fasteners, but these trusses must also comply with the requirements of EN 14250 (BSI, 2010b), which, among other things, limits the maximum truss length to 35 m. The method of analysis used with these trusses shall comply with the requirements of *clause 5.4.1*, and is referred to in Section 5.4.1 of this guide.

Clause 9.2.1

Clause 5.4.1

The plate dimensioning rules given in *clause 9.2.2* for minimum overlap on members and at chord splices must be complied with, and where a member supports a concentrated force less than 1.5 kN acting at right angles to the direction of the member span, and under the design loading the axial stress-to-strength ratio in the member is less than 0.4, the member need only be verified against bending, as required by *Equation 9.19*.

Clause 9.2.2

9.2.3 Roof and floor diaphragms

9.2.3.1 General

Where roof or floor structures constructed from wood-based panel materials are fixed by mechanical fasteners and detailed to be able to transfer in plane shear forces, they can be used as diaphragm structures, transferring lateral forces to the supporting structure. The supporting structure can be shear walls, as discussed in *clause 9.2.4*, braced framed structures or indeed any structure that will provide adequate lateral support.

Clause 9.2.4

Diaphragm structures are particularly common in timber-framed buildings, and a typical example of a horizontal floor diaphragm is shown in Figure 9.5.

Figure 9.5. Typical floor diaphragm with alternative panel layouts

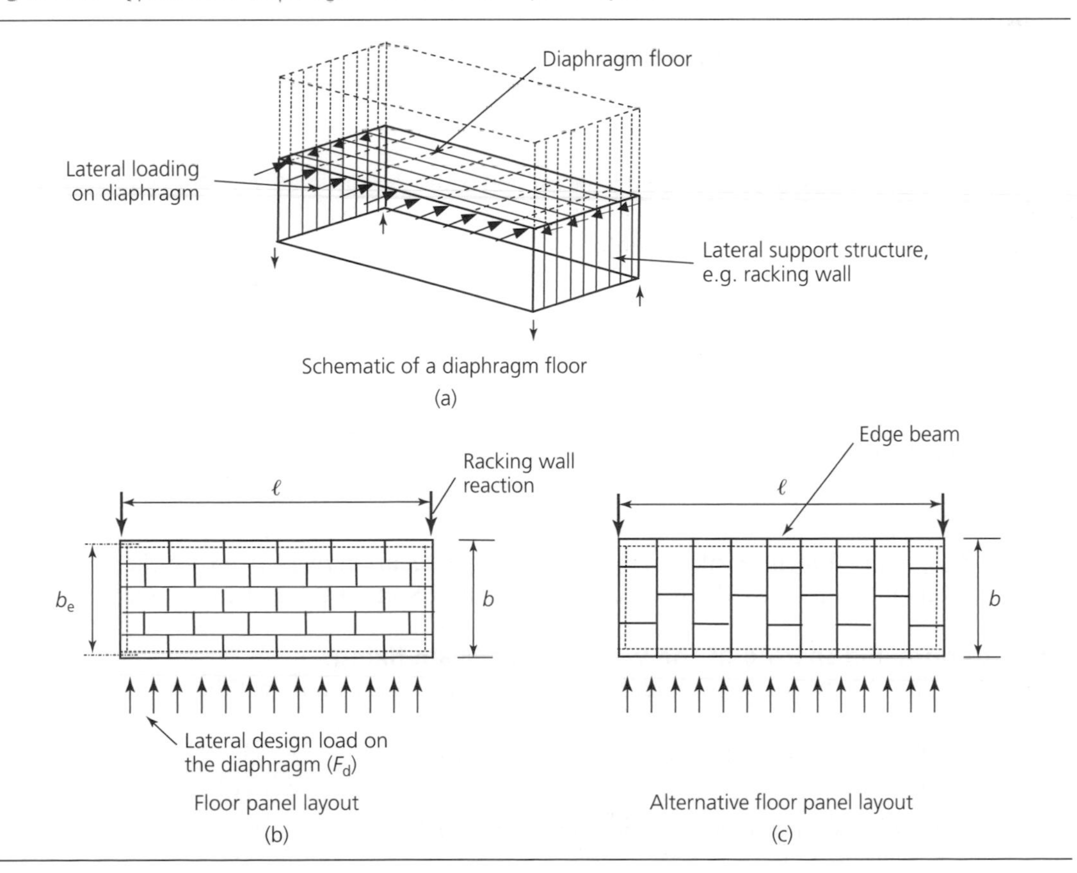

Where gypsum plasterboard ceilings are used in the construction, they can also be designed to function as structural diaphragms, and the design rules can be adjusted to suit such structures. Design rules for the use of plasterboard ceilings as diaphragm structures in domestic construction are given in PD 6693-1.

Wood-based panels will generally be supported by floor beams or the roof structure, and the strength of the fixings used to connect the panels to the support structure may be taken to be 20% greater than the values derived from the design rules in *Section 8*. This is permitted, as the design condition involves a large number of fasteners in a common configuration being subjected to the same load, and in such circumstances the probability of all of the fasteners failing at the characteristic strength is considered to be outwith the design basis. For this situation, to achieve the reliability target set for the ultimate limit state design condition, the mean value strength (taken to be 1.2 times the fastener characteristic strength) may be used.

9.2.3.2 Simplified analysis of roof and floor diaphragms

The simplified method assumes that the panel sheathing will act as the web of a horizontal deep beam spanning between end shear walls. The flanges of the beam, which may be the head binder plate of timber fame walls, ring beams, trimmer joists, etc., are, unless a more detailed analysis is undertaken, designed to take the tension/compression force arising from the maximum bending moment generated by the design load (F_d in Figure 9.5) in addition to any other loading they support. Within the web area, the panels transfer the shear forces, and the most effective layout for shear transfer will be with the staggered arrangement as shown in Figure 9.5(b), However, for the condition where the structure must function as a diaphragm designed to resist wind from any direction, the Figure 9.5(c) configuration will also occur.

Clause 10.8.1

The simplified method can only be used where the diaphragm is subjected to uniformly distributed loading; the span ℓ as shown in Figure 9.5 is between $2b$ and $6b$ (where b is the diaphragm width); the critical ultimate limit state design condition will be failure of the fasteners and not the panels; and the panels are fixed in accordance with the detailing rules given in *clause 10.8.1*.

Design requirements for the diaphragm under the lateral load shown in Figure 9.5(b)

(i) Axial strength of the edge beams

(Assuming that each beam is vertically supported along its length and instability effects can be ignored.)

The maximum axial stress in each edge beam under the design load, F_d, (N) will be

$$\sigma_{t,d} = \frac{F_d \ell}{8 b_e A}$$

where b_e (mm) is the distance between the centroidal axes of the edge beams and A (mm^2) is the net cross-sectional area of each edge beam.

On the basis that the tensile strength will be less than the compression strength, the verification requirement will be:

$$\sigma_{t,d} \leq f_{t,0,d}$$

where $f_{t,0,d}$ is the design tensile strength of the edge beam, derived from its characteristic tensile strength, $f_{t,0,k}$, using the procedure described in Section 6.1.2 of this guide.

(ii) Strength and spacing of the web to edge beam fixings

The shear stress between the web and the edge beam, τe_d, (N/mm^2) will be

$$\tau e_d = \frac{F_d}{2 b_e t}$$

where t (mm) is the thickness of the web panel material.

Using fasteners with a design strength, $Fc_{v,Rd}$, derived using the design rules in *Section 8*, and increasing by the 1.2 factor, the required fastener spacing, sp_e (mm), will be

$$sp_e \leq \frac{1.2Fc_{v,Rd}}{t\tau e_d}$$

(iii) Shear strength of the web panels at the racking wall

The shear stress in the web panels, $\tau_{v,d}$, (N/mm^2) will be

$$\tau_{v,d} = \frac{F_d}{2bt}$$

where b (mm) is the overall width of the diaphragm. For this condition, the verification requirement will be

$$\tau_{v,d} \leq f_{v,d}$$

where $f_{v,d}$ is the design shear strength of the web panel material derived from its characteristic panel shear strength, $f_{v,k}$, using the procedure described in Section 6.1.7 in this guide.

(iv) Strength and spacing of the fixings to the shear wall structure and to the supporting joists within the web area

The force/unit length, Fm_d (N/mm), to be taken in lateral shear by the diaphragm fixings along each shear wall will be

$$Fm_d = \frac{F_d}{2b}$$

Using fasteners with a design strength $Fs_{v,Rd}$, (N) derived using the design rules in *Section 8* and increasing by a factor of 1.2, the required spacing, sp_s (mm), will be

$$sp_s = \frac{1.2Fs_{v,Rd}}{Fm_d}$$

Notes

1 The diaphragm edge beams must be continuous elements and if not, they have to be connected together to function as a continuous member.
2 Where connections are required to make edge beams continuous, they must be designed to transmit a maximum tensile/compression design force $= F_d\ell/8b_e$, and where fasteners are used, because of the local nature of the connection, the lateral design strength of the fastener derived from the rules in *Section 8* **cannot** be increased by the 1.2 statistical factor used for the web fastener design.

The shear force across the width of the web is taken to be uniform, as is the shear force along the connection between the edge beam and the edge panels of the web.

Although the shear force in the web reduces to zero at mid-span, standard detailing is used for connections between the panels and their support members except along discontinuous panel edges, where increased spacing is permitted. At these positions, the fixing spacing can be increased by a factor of 1.5, but the spacing must not exceed 150 mm.

The fixings between the web and the compression edge beam will prevent local buckling of the beam, enabling the full compression strength to be used so that the design condition will be the edge beam tensile strength. Where cut-outs occur in the diaphragm (e.g. stairwell areas), additional structure capable of transferring the compression and tension forces in these areas and across the reduced web structure will probably be required.

9.2.4 Wall diaphragms

9.2.4.1 General

Walls in timber construction are commonly formed from vertical timber members (referred to as studs), fixed at regular intervals and faced on one or both sides with sheathing material. The sheathing material may be plasterboard or a combination of plasterboard and wood-based panel products (e.g. OSB), and such walls are capable of exhibiting considerable in-plane resistance to horizontal and vertical imposed actions. The wall structure, when held laterally by the building structure, will form a wall diaphragm, and providing it is adequately restrained to prevent overturning and sliding and serviceability limits will not be exceeded, it can be designed to function as a racking wall. 'Racking strength' is the term used to define the in-plane withstand capability of the wall to the effect of in-plane horizontal and vertical wall loading, where the horizontal load is generally wind loading.

The racking resistance of a wall diaphragm can be determined by testing in accordance with EN 594 (BSI, 2011) or by calculation, and EN 1995-1-1 gives two simplified calculation methods, Method A and Method B. Method A applies where wall diaphragms are tied down at their ends, to prevent any uplift of the diaphragm. This will apply where the wall ends are anchored (by metal straps or equivalent) throughout their height or where vertical permanent action(s) at these positions are sufficient to prevent uplift. Method B applies where the diaphragm resists uplift by the sheathing to timber fasteners and the fasteners/connection between the base of the diaphragm and the supporting substructure, and this form of construction is the most commonly used method for domestic buildings in the UK. Method B has been derived from a conversion of the permissible strength-based racking resistance procedure given in BS 5268 Part 6 to a limit state design approach, but in this process some factors have been misinterpreted, and also the use of plasterboard as a sheathing material, permitted in BS 5268, is not allowed, as only wood-based panel products compliant with *clause 3.5* can be used. Further, Method B underestimates the racking strength of walls with openings as it fails to take into account the resistance of the timber structure framing the opening areas in the diaphragm, which is also the case for racking design in accordance with Method A.

Clause 3.5

To overcome the deficiencies associated with Method B the 'simplified method of wall diaphragms' was developed for use in the UK, and the procedure is included in PD 6693-1. The method draws on the plastic lower-bound method of analysis developed by Kallsner and Girhammar (2004), and incorporates modifications relating to the sheathing fastener strength to take account of fastener spacing. Also, in line with the method in BS 5268, it incorporates a modification factor to make some allowance for the strength contribution obtained from opening areas in walls. PD 6693-1 also includes a procedure to take into account the racking strength of plasterboard-clad timber frame walls using the above approach.

The National Annex to EN 1995-1-1 requires the simplified method in PD 6693-1 to be used in the UK for situations where the diaphragm resists uplift by the sheathing to timber fasteners and the fasteners/connection between the base of the diaphragm and the supporting substructure, and Method A to be used where wall ends are anchored or prevented from uplift by vertical loading. Further information on the simplified method and guidance on how it should be applied in the design of racking walls is given in Chapter 13.

9.2.5 Bracing

Bracing rules are given for members subjected to direct compression forces or to compression forces arising from bending actions. Rules are given for the design of lateral bracing to single members in *clause 9.2.5.2*, and, for determining the loading for which the bracing for beam or truss systems should be designed as well as the limiting deflection permitted for the bracing system, in *clause 9.2.5.3*.

Clause 9.2.5.2
Clause 9.2.5.3

9.2.5.1 General

Where it is necessary to prevent failure by instability, to increase stability strength, or to prevent failure by excessive deflection, some form of bracing will be necessary.

Clause 9.2.5.2
Clause 9.2.5.3
Clause 10.2

When designing bracing, the effect of deviation from straightness of members has to be taken into account, and the design rules in *clauses 9.2.5.2* and *9.2.5.3* are deemed to have incorporated the effect of the limiting out of straightness values given in *clause 10.2*.

When dealing with the bracing requirements of trussed rafters, although the general principles for the design of bracing members given in *clause 9.2.5* will apply, specific requirements are given in PD 6693-1.

Clause 9.2.5

It should be noted that, depending on the structural layout being used, bracing members will have to be designed to function as compression as well as tension members, and consideration should be given to the bracing system to be adopted to enable the design strength of the bracing members to be optimised. Also, bracing members may be required to transmit external loading (e.g. wind loading) in addition to bracing forces, and the bracing members as well as bracing systems have to be designed to withstand the worst loading combination that can arise.

9.2.5.2 Single members in compression

Clause 9.2.5.2

In *clause 9.2.5.2*, design rules are given for the stiffness and strength requirements of bracing members used to provide lateral support to a **single** member in compression.

The stiffness requirement of each bracing member is defined in *Equation 9.34*, being derived from classical elastic buckling theory. As the equation does not include for the effect of initial out of alignment of the member, theoretically it reduces the stiffness value required, however, this is offset by other conservative factors that are not taken into account in the design rules. The modification factor, k_s, in *Equation 9.34* is a non-linear function dependent on the ratio of the member span, ℓ, to the bay length, a, shown in *Figure 9.9*, and theoretically ranges from 2 to a maximum value of 4. The requirement of the National Annex to EN 1995-1-1 is that a value of 4 should always be used, which is also the recommended value in EN 1995-1-1.

With single-beam members, the moment causing instability is based on the **maximum** design moment in the member, M_d, minus the stability moment able to be resisted by the beam, $k_{crit}M_d$. The instability force, N_d, caused by this net moment, $(1 - k_{crit})M_d$, is conservatively obtained by dividing this moment by the beam depth, as defined in *Equation 9.36*.

The stiffness value, C, obtained from *Equation (9.34)*, is the stiffness that must be provided to the compression member at each bracing position, and for the general case shown in Figure 9.6 where a bracing member is connected to a single compression member with a connection having a lateral stiffness, k_1, and to its rigid end support by a connection having a lateral stiffness, k_2, the axial stiffness to be provided by the bracing member, C_b, must be

$$C_b \geq \frac{1}{\left(\frac{1}{C} - \frac{1}{k_1} - \frac{1}{k_2}\right)} \qquad \text{(D9.17)}$$

For the above condition, the minimum cross-sectional area of the bracing member will be $C_bL/E_{0,mean}$, where $E_{0,mean}$ is the mean modulus of elasticity parallel to the grain of the bracing member and L is the bracing member length.

Each bracing member has to be designed to resist the stabilising force given in *Equation 9.35*, and the value used for the $k_{f,1}$ and $k_{f,2}$ modification factors in that equation differ because the factors are dependent on the permitted deviation from straightness of the compression member material.

Figure 9.6. Combined stiffness

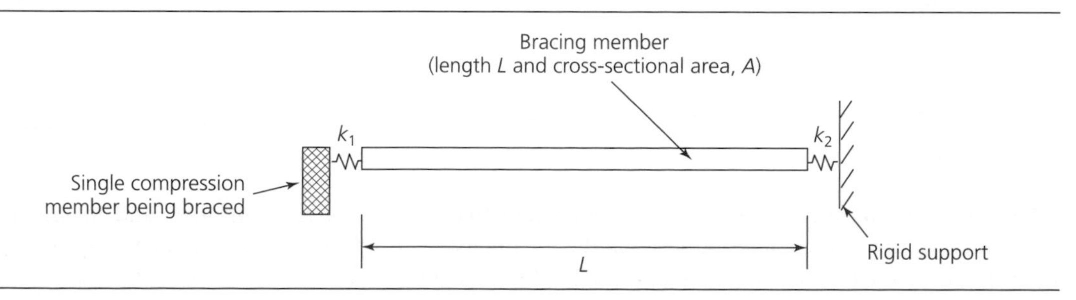

For a solid timber compression member, the code recommends $k_{f,1} = 50$, and for glued laminated timber or LVL, $k_{f,2} = 80$; however the National Annex to EN 1995-1-1 requires that the respective values to be used are 60 and 100.

Example 9.4
A 9 m-long (ℓ) glued laminated column carrying a design load of 120 kN (N_d) is braced at the third points within its height. The stiffness of the connection between the bracing member and the column at these positions and the bracing member and its rigid support at the other end is 0.5 × the bracing member axial stiffness (k_b). What is the value of design stabilising force to be provided by each bracing member and the minimum bracing member axial stiffness?

Factors:

$$k_s = 4$$

$$k_{f,2} = 100$$

Bay length of the column:

$$a = \ell/3 = 9/3 = 3 \text{ m}$$

Stiffness of the bracing member and its end connections (*Equation 9.34*):

$$C = k_s N_d / a = 4 \times 120/3 = 160 \text{ kN/m}$$

Axial stiffness of bracing member: as $1/C = 1/k_b + 2/k_b + 2/k_b = 5/k_b$, so

$$k_b = 5C = 5 \times 160 = 800 \text{ kN/m}$$

Design stabilising force:

$$F_d = N_d / k_{f,2} = 120/100 = 1.2 \text{ kN}$$

9.2.5.3 Bracing of beam or truss systems

This clause addresses the situation where there are several parallel members that require to be braced at intermediate positions along their length and where lateral resistance is provided to the bracing members by a bracing system. A typical example involving the bracing of roof beams is shown in Figure 9.7.

In Figure 9.7 there are n compression members laterally supported by bracing members, each a distance a apart, and the bracing members are laterally supported by a bracing system, which in this example is shown as a truss system at one end of the braced structure. The bracing system can be positioned anywhere within or attached to the braced structure.

Clause 9.2.5.3

The design rules in *clause 9.2.5.3* are derived from classical elastic theory, and incorporate approximations. The internal stability load per unit length imposed on the bracing system from the bracing members, q_d, is taken to be uniformly distributed along the length of the bracing system, as shown in Figure 9.8, and is obtained from *Equation 9.37*:

$$q_d = k_\ell \frac{nN_d}{k_{f,3} l} \qquad (9.37)$$

where n is the number of compression members that require lateral support, and all must be connected to the bracing members; N_d is the mean design compression force in each compression member, derived as described in Section 9.2.5.2 in this guide; l is the overall span of the stabilising

Figure 9.7. Braced compression system

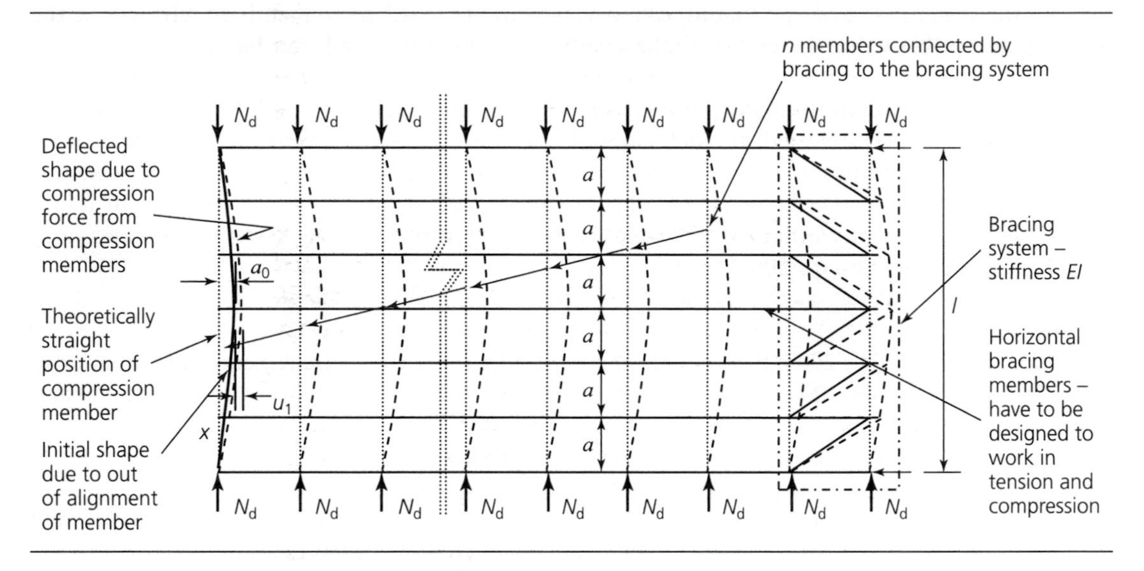

system (in m); and $k_{f,3}$ is a modification factor, which, in accordance with the requirements of the National Annex to EN 1995-1-1 is 50 when $a \leq 600$ mm and 40 when $a > 600$ mm.

In *Equation 9.37*, factor k_ℓ covers for cases where the span of the compression member is greater than 15 m, and is derived from *Equation 9.38*. The code assumes that for this condition there will be contract tolerance limits that will require the deviation from straightness of the members to be less than the limits given in *clause 10.2(1)*. On this understanding, the value of k_ℓ will be less than 1, and will reduce the value of the stability load, q_d, to be taken on the bracing system. It is essential that the designer ensures that the contract includes relevant tolerance requirements that will achieve this.

Clause 10.2(1)

The maximum axial force in each bracing member will be $q_d a$, and the strength of these members must be validated against this force. There is no requirement in the design rules to check the bracing stiffness for these conditions.

Figure 9.8 shows the bracing system in Figure 9.7 subjected to the internal stability load arising from the n members being braced, q_d, plus any external applied loading, w_d, and for the most unfavourable condition they will be acting in the same direction. There will also be the

Figure 9.8. Bracing system

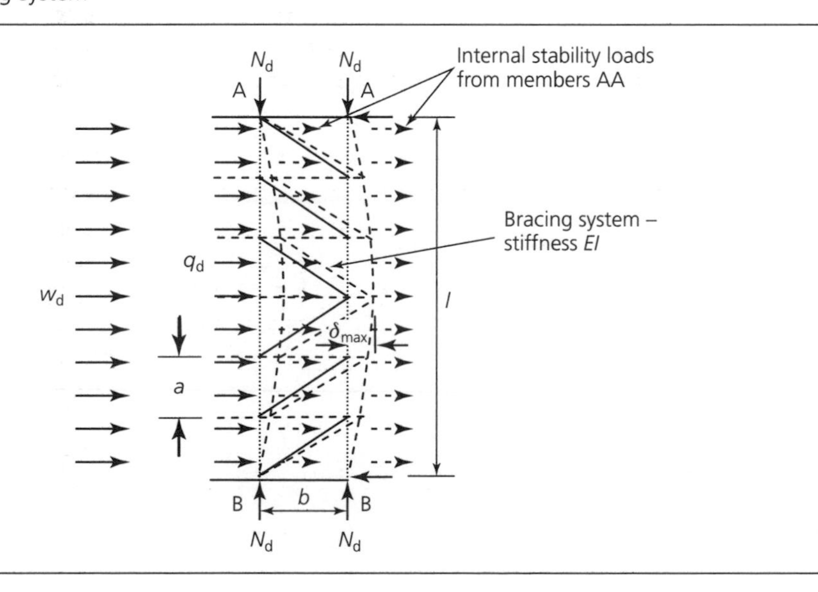

internal stability loads on each of the bracing system members, AB, which will add to the lateral load on the system. The design loading ($q_d + w_d +$ stability loads from members AB) will be the most unfavourable value derived from the combination loading condition being used.

One of the approximations incorporated into the design rules is that the deflection of the compression members caused by deformation of the bracing system will be small, and, because of this assumption, its effect has been omitted in *Equation 9.37*. To ensure that this requirement is not infringed, the maximum horizontal deflection of the bracing system under the action of q_d plus any other external load that can be applied to the system (e.g. wind, w_d), say, δ_{max}, must not exceed $l/500$.

REFERENCES

BSI (2003) BS EN 622-1: 2003. Fibreboard – Specifications – General requirements. BSI, London.

BSI (2006) BS EN 300: 2006. Oriented Strand Boards (OSB) – Definitions, classification and specifications. BSI, London.

BSI (2008) BS EN 636: 2008. Plywood – Specifications. BSI, London.

BSI (2010a) BS EN 312: 2010. Particleboards – Specifications. BSI, London.

BSI (2010b) BS EN 14250: 2010. Timber structures – Product requirements for prefabricated structural members assembled with punched metal plate fasteners. BSI, London.

BSI (2011) BS EN 594: 2011. Timber structures – Test methods – Racking strength and stiffness of timber frame wall panels. BSI, London.

Gere JM and Timoshenko SP (1991) *Mechanics of Materials*, 3rd SI edn. Chapman and Hall, London.

Kallsner B and Girhammar UA (2004) Influence of the framing joints on plastic capacity of partially anchored wood-framed shear walls. *Working Commission W18 – Timber Structures, Meeting 37*, Edinburgh.

Designers' Guide to Eurocode 5: Design of Timber Buildings
ISBN 978-0-7277-3162-3

http://dx.doi.org/10.1680/dtb.31623.153

Chapter 10
Structural detailing and control

This short section of EN 1995-1-1 relates to issues of detailing and quality control under the following clause headings:

- General *Clause 10.1*
- Materials *Clause 10.2*
- Glued Joints *Clause 10.3*
- Connections with mechanical fasteners *Clause 10.4*
- Transportation and erection *Clause 10.5*
- Control *Clause 10.6*
- Special rules for diaphragm structures *Clause 10.7*
- Special rules for trusses with punched metal plate fasteners *Clause 10.8*

10.1. General

This clause draws attention to the requirements of the section as a whole, which relate to construction details or procedures necessary to validate the assumption of the design rules given in the previous sections, and which should be incorporated into a project specification. In this regard, the National Building Specification (NBS, 2012) is widely used for creating project specifications. It contains clauses in open format that cover materials and workmanship, and allows project-specific clauses to be inserted. The reference 'NBS' when used below indicates that the specification contains relevant entries.

Some clauses are straightforward in their meaning, and no further comment is offered.

10.2. Materials

Clause 10.2(1)
Clause 6.2.2

Clause 10.2(1) gives the rules for deviation from straightness of compression members or frame members of solid or glulam timber that are referred to in *clause 6.2.2* (the design of members in compression). They are more stringent than the straightness rules in most grading standards, and should therefore be included in the project specification. As an example, this clause only allows a deviation of $L/300$ in a compression member of solid timber, whereas the grading standard for structural timber, EN 14081 (BSI, 2006), allows a bow of 10 mm in 2 m (i.e. $L/200$) for timber in C18 and above.

Clause 10.2(2)
Clause 10.2(3)

Clauses 10.2(2) and *10.2(3)* relate to the risk of wetting of timber components during the construction process. If this happens to members of solid timber, there will eventually be some shrinkage and risk of fissuring as the members dry out in service, and, for some wood-based components, a risk of more serious damage. These risks should be evaluated, and, where they are acceptable, care taken to ensure adequate ventilation in the built conditions. It should be pointed out that large glulams, even if protected from direct rain, will nevertheless have spent a period of time in a service class 2 environment during construction, and small surface checks (which are not generally strength-reducing) are likely to occur as the moisture content reduces.

10.3. Glued joints

Clauses 10.3(1)–10.3(3)

Clauses 10.3(1) to *10.3(3)* emphasise the importance of quality control in the fabrication of glued components. The various factors that are critical to the strength of a glued joint, given in Section 3.6 of this guide, must be part of the control plan, not least because they can only be checked at the time that the fabrication takes place (NBS).

Rules for the design of glued connections are not given in the code, but guidance on the design procedure to be used in the UK for overlapping glued connections is given in PD 6693-1.

10.4. Connections with mechanical fasteners

10.4.1 Nails

Clause 10.4.2(3)

Clause 10.4.2(3) defines the diameter of predrilled holes for nails in hardwood. A high degree of site control would have to be in place to rely on this detail.

10.4.2 Bolts and washers

Clause 10.4.3(1)
Clause 7.1

Clause 10.4.3(1) specifies the maximum allowable differences between diameters of the bolt and the bolt hole. These are the clearances referred to in *clause 7.1* (*Table 7.1,* footnote a), and which should be added into the calculation of joint slip.

Clause 10.4.3(2)

Clause 10.4.3(2) defines the washer size for bolted connections. When bearing on timber they are of course larger than the washers used in conventional steelwork. If there was a preference for smaller washers, the Johansen equations could be reviewed to see if the full rope effect was needed, or the bolts could be designed as dowels. Up to 20% more bolts may be needed, but nominal 'steel' washers (normally twice the bolt diameter) could then be used.

Clause 10.4.3(3)

Clause 10.4.3(3) refers to the tightening of lag screws and bolts, following the enclosure of the structure. Since most buildings are heated in occupation, resulting in a service class 1 environment, it may be necessary to make a final check on bolt tightness after handover.

10.4.3 Screws

To rely on the accuracy needed for predrilled screw holes, a high degree of site control would have to be in place.

10.5. Transportation and erection

The careful handling of members until they are finally built into the work is primarily the responsibility of the contractor. Nevertheless, it is up to the designer to anticipate any critical temporary condition that might reasonably arise as part of the erection process, and either draw the contractor's attention to it or ask for a procedural statement for review.

10.6. Control

This clause is essentially a reminder of the typical elements of a contractor's quality control plan, in which the designer might take a more detailed interest, depending on the criticality of the item. Of more significance are maintenance items for the future service life of the structure, which have to be communicated to the building owner.

10.7. Special rules for diaphragm structures

Clause 9.2.3
Clause 9.2.4

The details given in this section apply to the forms of construction described in *clauses 9.2.3* and *9.2.4*, with these as part of a standard range of details (see also Chapter 13 in this guide).

10.8. Special rules for trusses with punched metal plate fasteners

Trusses with punched metal plate fasteners, or trussed rafters, are made by specialist manufacturers and delivered by them to site. Individually, the trusses could have a length/thickness ratio of 150 or more, and so they are relatively flexible. This clause is aimed at keeping the internal and out-of-plumb distortions of the trusses after erection within acceptable limits, when they are then fixed by bracing members.

REFERENCES

BSI (2006) BS EN 14081-1: 2005 + A1: 2011. Timber structures. Strength graded structural timber with rectangular cross section. General requirements. BSI, London.

NBS (2012) NBS Building. NBS, London.

Designers' Guide to Eurocode 5: Design of Timber Buildings
ISBN 978-0-7277-3162-3

http://dx.doi.org/10.1680/dtb.31623.155

Chapter 11
Informative annexes

11.1. Informative annexes

This chapter is concerned with the requirements of informative *Annexes A*, *B* and *C* at the end of the code. It covers the following topics:

- Block shear and plug shear failure at multiple dowel-type steel-to-timber connections — *Annex A*
- Mechanically jointed beams — *Annex B*
- Built-up columns — *Annex C*

These topics are given for information, and the National Annex to EN 1995-1-1 states that the content of *Annexes B* and *C* can be used for design in the UK.

With regard to *Annex A*, the requirement is that in the UK this should only be used for connections where there are 10 or more dowel-type fasteners of diameter $\leq$6 mm or five or more dowel-type fasteners of diameter $>$6 mm and where the fasteners are in line parallel to the grain.

Clause 2.2.2

Clause 2.3.2.2(2)
Clause 2.3.2.2(3)
Clause 2.3.2.2(4)
Clause 2.4.1(2)P

Where stiffness properties are given in the relevant annexes, the values used are mean values. However, for ultimate limit state analyses the requirements of *Clause 2.2.2* must be followed, and where a first-order linear analysis is used and the distribution of internal forces is affected by the stiffness distribution within the structure, final mean values adjusted to the load component causing the largest stress in relation to strength, as defined in *clause 2.3.2.2(2)* (including the requirements of *clauses 2.3.2.2(3)* and *2.3.2.2(4)*) shall be used rather than mean values. Also, for a second-order linear elastic analysis, design values derived from the rules in *clause 2.4.1(2)P* will apply. For presentation and comparison reasons, mean values have been retained in this chapter; however, where the content of any annex is being used in an ultimate limit state analysis the stiffness value used shall be that required by the rules in EN 1995-1-1.

11.2. Annex A (informative): block shear and plug failure at multiple dowel-type steel-to-timber connections

The design guidance in *Annex A* applies to steel-to-timber connections formed with multiple dowel-type fasteners where there is a force component in the connection acting parallel to the grain and in a loaded end configuration, as defined in *Figure 8.7*. Under this type of loading condition, in addition to the possibility of ductile failure, as covered in *Section 8*, there is also the risk of a brittle failure, and this annex gives the brittle failure modes and associated characteristic strength equations that have to be validated.

Two types of brittle failure can arise:

(*a*) block type failure – where the timber within the connection fails in shear and tension around the perimeter of the fastener group, and the connection fails as a block, as shown in Figure 11.1(a)
(*b*) plug type failure – where there is flexural yielding of the fasteners in combination with shear and tension failure in the timber, and the connection fails as a plug as shown in Figure 11.1(b).

Clause NA.3.1

From *clause NA.3.1* in the National Annex to EN 1995-1-1, for UK designs these types of brittle failure only have to be checked when the fastener diameter is $\leq$6 mm and there are 10 or more

Figure 11.1. Steel-to-timber connection: block shear and plug shear failures

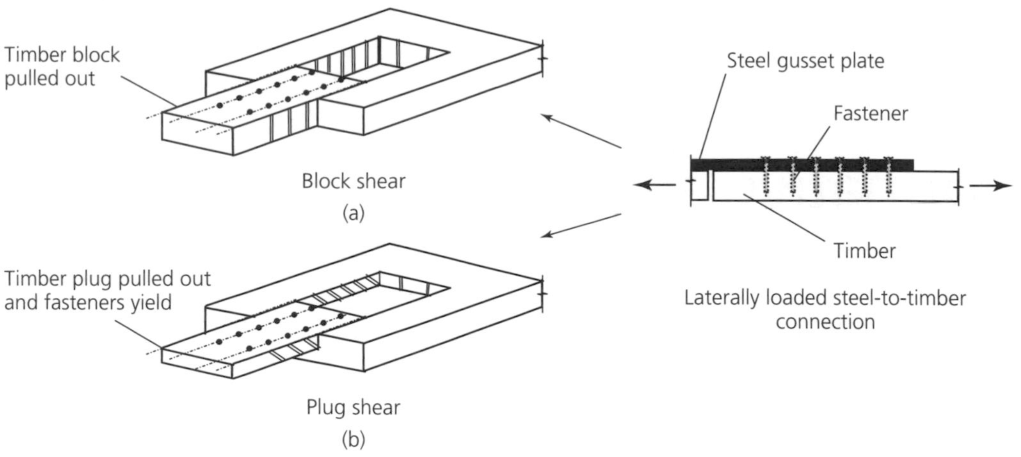

fasteners in a line parallel to the grain or when the fastener diameter exceeds 6 mm and there are five or more fasteners in a line parallel to the grain.

The brittle failure modes will be initiated by either a tension or a shear failure, and as these strengths function independently and are brittle mode failures, they will not interact with each other. The characteristic block shear or plug shear capacity of each failure mode, $F_{bs,Rk}$, will be obtained from *Equation A.1*, and will be the greater of the tension strength of the timber at the tension end of the connection $(1.5A_{net,t}f_{t,0,k})$ and the shear strength over the failed shear surface of the timber $(0.7A_{net,v}f_{v,k})$.

Where the ductile failure mode of the connection is type (c), (f), (j/l), (k) or (m) (as shown in *Figure 8.3*), the brittle failure assessment is to be based on a block shear failure condition, and the net shear area in the parallel to grain direction, $A_{net,v}$, will be $L_{net,v}t_1$, as defined in *Equation A.3*. For this condition, t_1 will be the thickness of the timber member or, if the fastener is in single shear, the pointside penetration depth. If the ductile failure is due to any of the other failure modes referred to in *Figure 8.3*, the plug shear failure condition will apply, and $A_{net,v}$ will be

$$\frac{L_{net,v}}{2}(L_{net,t} + 2t_{ef})$$

as also defined in *Equation A.3*.

It is to be noted that for failure modes (d) or (g) the relation given for *Equation A.7* for t_{ef} is incorrect, and, as stated in Appendix A, should read

$$t_{ef} = t_1\left[\sqrt{2 + \frac{4M_{y,Rk}}{f_{h,k}dt_1^2}} - 1\right]$$

Example 11.1: characteristic load-carrying capacity of a steel-to-timber connection

Clause 8.2.3

A tension connection between two C16 timber members, 150 mm wide by 60 mm thick (t), is formed using a 5 mm-thick (t_1) mild steel flitch plate and 10 mm diameter (d) grade 5.6 bolts in two rows, as shown on Figure 11.2. The predrilled bolt holes in the timber are 11 mm in diameter. From the application of the design rules in *clause 8.2.3*, the characteristic load-carrying capacity is evaluated, and the failure mode shown to be type (f), as defined in *Figure 8.3*. Based on the rules in *Annex A*, what is the characteristic load-carrying capacity of fracture of the connection?

Figure 11.2. Tension connection between two C16 timber members

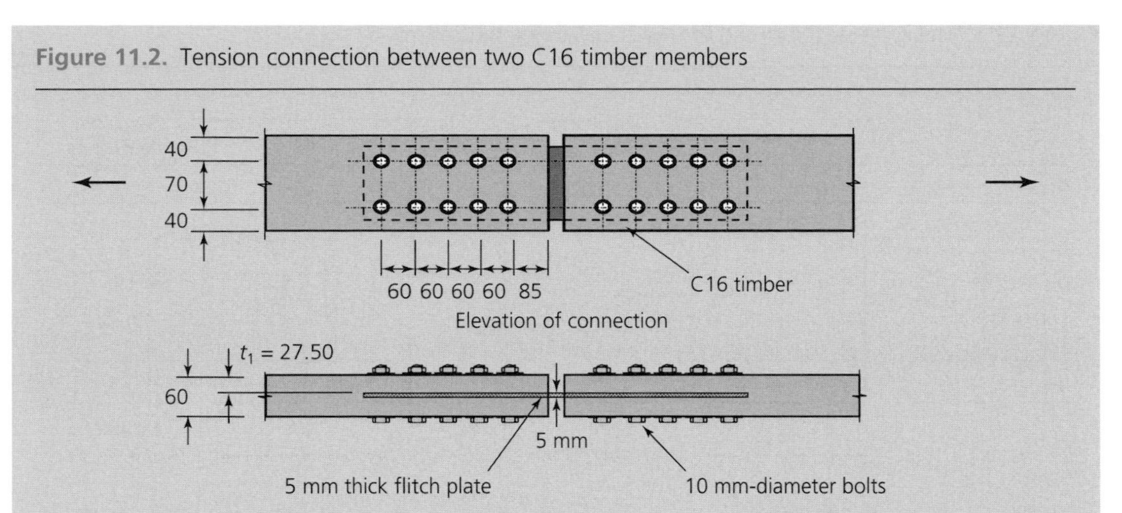

Properties of timber: $f_{t,0,k} = 10$ N/mm^2; $f_{v,k} = 3.2$ N/mm^2; $a_1 = 60$ mm; $a_2 = 70$ mm; $a_{3,t} = 85$ mm; $t_1 = 27.5$ mm.

$$L_{net,v} = 2\{4[a_1 - (d + 1 \text{ mm})] + [a_{3,t} - 0.5(d + 1 \text{ mm})]\}$$

$$= 2(4(60 - (10 + 1)) + (85 - 0.5(10 + 1))) = 551 \text{ mm}$$

$$L_{net,t} = a_2 - (d + 1 \text{ mm}) = 70 - (10 + 1) = 59 \text{ mm}$$

$$A_{net,t} = 2L_{net,t}t_1 = 2 \times 59 \times 27.5 = 3245 \text{ mm}^2$$

As the ductile failure mode is (f):

$$A_{net,v} = 2L_{net,v}t_1 = 2 \times 551 \times 27.5 = 30\,305 \text{ mm}^2$$

As the ductile failure mode is (f), block shear failure will occur, and the characteristic load-carrying capacity of fracture, $F_{bs,Rk}$ will be the greater of:

$F_{bs,Rk}$ (due to tension strength of block) $= 1.5A_{net,t}f_{t,0,k} = 1.5 \times 3245 \times 10 = 48.68$ kN

$F_{bs,Rk}$ (due to shear strength of block around perimeter)

$$= 0.7A_{net,v}f_{v,k} = 0.7 \times 30\,305 \times 3.2 = 67.88 \text{ kN}$$

Characteristic load-carrying capacity of fracture:

$$F_{bs,Rk} = 67.88 \text{ kN}$$

(A check of the pure tensile failure of the timber at the joint root shows that the tensile strength = 70.4 kN ($>F_{bs,Rk}$).)

11.3. Annex B (informative): mechanically jointed beams

This annex covers the design of beams of uniform cross-section that are formed from wood or wood-based elements connected together using mechanical fasteners such as nails, screws, bolts, dowels, etc. Slip will occur at the element interfaces, making the section semi-rigid and unable to be designed in accordance with the rules given in *Section 9* for glued components.

11.3.1 Simplified analysis

The analysis used for these sections is based on a first-order elastic analysis applying simple bending theory to each element of the beam, and the force exerted by the fasteners per unit

Figure 11.3. The cross-sections used for mechanically jointed beams

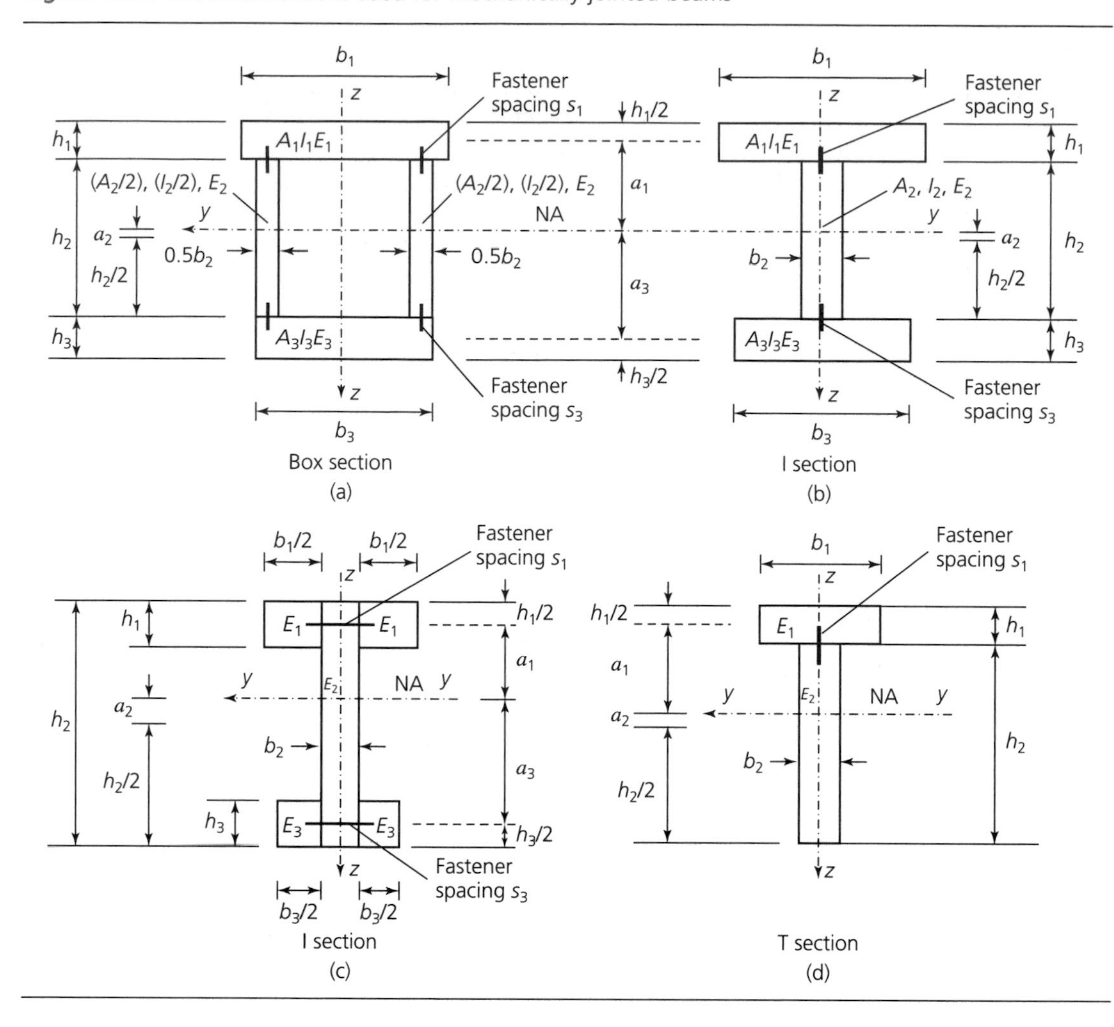

length at each interface is taken to be constant and continuous along the beam length. The stiffness of the fasteners at the interfaces is assumed to be linear, and the value used for a serviceability limit state analysis will be the slip modulus, K_{ser}, obtained from *Section 7*, and, for an ultimate limit state analysis, will be the value associated with the analysis requirements of *clause 2.3.2.2*. The approach ignores the effect of shear deformations.

Clause 2.3.2.2

11.3.2 Cross-sections

The cross-sections considered in the annex to which the simplified method has been applied are shown in Figure 11.3; however, the theory can be applied to other profiles.

11.3.3 Assumptions

The design procedure requires that simple bending theory will apply, stresses due to shear deformation effects are ignored and the following assumptions are met:

- The beams must be simply supported with a span ℓ. For a continuous beam, the expressions given can be applied to beam sections taking the beam span $\ell = 0.8$ times the relevant beam span. Where there is a cantilever on the beam, it can be designed as a beam with ℓ equal to twice the cantilever length; however, for the stress configuration shown in *Figure B.1*, compression stresses will change to tension stresses and vice versa.
- The wood or wood-based elements are either full length or, if they are jointed along their length, the connection must be a full-strength glued joint.
- The beam elements are connected to each other along their interfaces by mechanical fasteners with a slip modulus K. The value used for K will be the appropriate value for the type of analysis being undertaken.

- In order to comply with the assumption that the fasteners will exert a constant force per unit length along the length of the beam, the fastener spacing, s, should be constant. However, it is allowed to vary uniformly according to the shear force distribution along the beam, and will be assumed to be constant providing the maximum spacing, $s_{max} \leq 4s_{min}$, when s_{min} is the minimum spacing being used.
- The stiffness and stress analysis solutions given in *clauses B.2* to *B.5* are only valid for loading applied in the z **direction** that generates a sinusoidal or parabolic shaped bending moment and produces a shear force that will effectively vary linearly along the beam length (e.g. a beam subjected to uniformly distributed loading conditions).

Clauses B.2–B.5

11.3.4 Spacings

Where the flange of the beam consists of two elements connected to a web or a web consists of two elements, the spacing, s_i, between the fasteners is determined by the sum of the fasteners per unit length in the two joining planes. In other words, the fastener spacing used in *Equations B.5* and *B.10* for s_i in section (A) and section (C) in Figure 11.3 will be $s_1/2$ and $s_3/2$, as shown.

11.3.5 Deflections arising from bending moments

To calculate the beam deflection due to bending energy with these types of section, the effective bending stiffness, $(EI)_{ef}$, referred to in *clause B.2* and in Section 11.3.6 in this guide should be used. Shear deformation can be determined by calculating the shear deformation of the web, assuming all shear to be supported by the web.

Clause B.2

11.3.6 Values for the effective bending stiffness, $(EI)_{ef,y}$, a_1, a_2, γ_1 and γ_3 for the sections shown in *Figure B.1*

Applying the theory used in *Annex B*, the effective bending stiffness about the y–y axis and the associated γ_1 and γ_3 in terms of a_1, a_2, a_3, s_1 and s_3 for each of the cross-sections shown in *Figure B.1* is given in Table 11.1.

Distances s_1 and s_2 are the fastener spacings defined in *Figure B.1*, and for sections A and C the values to be used in the relevant equations are $s_1/2$ and $s_2/2$ respectively.

11.3.7 Normal stresses

The stresses given in *Equations B.7* and *B.8* are, respectively, the axial stress induced across the depth of element i and the bending stress at the extreme faces of element i, both induced by the moment M at the section being considered. The axial stress is caused by the axial strain in element i arising from the relative slip between the elements when subjected to bending. The bending stresses arise from the stresses induced by the proportion of the bending moment being taken by the element.

The axial and bending stresses are combined in each element to form the resultant design bending stress distribution across the section, and at no position should the combined stress exceed the design bending strength of the material. The design bending strength is obtained as described in Chapter 6 of this guide.

11.3.8 Maximum shear stress

The greatest shear stress will occur at the ends of the beam where the shear force is at its maximum, V, and at these positions the bending and axial stresses will be zero. As shown in *Figure B.1* the maximum shear stress across the beam depth will occur at the neutral axis position, and shall not exceed the design shear strength of the member, as described in Chapter 6 in this guide.

Also, as stated in Appendix A, h_2^2 should be replaced by h^2 (where h is as shown in the stress diagram in *Figure B.1*) in *Equation B.9*.

11.3.9 Fastener load

The load in a fastener will be obtained from *Equation B.10*, noting that for sections (A) and (C) in Figure 11.3 the spacing to be used will be $s/2$.

Table 11.1. Values for the effective bending stiffness of sections shown in Figure 11.3 (as detailed in *Figure B.1*) about the *y–y* axis

Profile in Figure 11.3	Effective bending stiffness about the *y–y* axis, $(EI)_{ef,y}$[a] (properties and sizes as shown in *Figure B.1*)
Built-up sections (A) and (B)	$(EI)_{ef,y} = E_1\frac{A_1h_1^2}{12} + E_2\frac{A_2h_2^2}{12} + E_3\frac{A_3h_3^2}{12} + \gamma_1E_1(A_1)a_1^2 + E_2(A_2)a_2^2 + \gamma_3E_3(A_3)a_3^2$ where for section (A) $\gamma_1 = 1/\left[1 + \pi^2E_1(A_1)\left(\frac{s_1}{2K_1}\right)\frac{1}{l^2}\right]$ $\gamma_3 = 1/\left[1 + \pi^2E_3(A_3)\left(\frac{s_3}{2K_3}\right)\frac{1}{l^2}\right]$ and for section (B) $\gamma_1 = 1/\left[1 + \pi^2E_1(A_1)\left(\frac{s_1}{K_1}\right)\frac{1}{l^2}\right]$ $\gamma_3 = 1/\left[1 + \pi^2E_3(A_3)\left(\frac{s_3}{K_3}\right)\frac{1}{l^2}\right]$ $a_2 = [\gamma_1E_1A_1(h_1 + h_2) - \gamma_3E_3A_3(h_2 + h_3)]/2(\gamma_1E_1A_1 + \gamma_3E_3A_3 + E_2A_2)$ If a_2 is positive, $a_1 = \frac{(h_1 + h_2)}{2} - a_2$ and if a_2 is negative, $a_1 = \frac{(h_1 + h_2)}{2} + a_2$ and $a_3 = \frac{(h_1 + h_3)}{2} + h_2 - a_1$
Built-up section (C)	$(EI)_{ef,y} = E_1\frac{A_1h^2}{12} + E_2\frac{A_2h_2^2}{12} + E_2\frac{A_3h_3^2}{12} + \gamma_1E_1(A_1)a_1^2 + E_2(A_2)a_2^2 + \gamma_3E_3(A_3)a_3^2$ where: $\gamma_1 = 1/\left[1 + \pi^2\frac{E_1}{2}(A_1)\left(\frac{s_1}{K_1}\right)\frac{1}{l^2}\right]$ and $\gamma_3 = 1/\left[1 + \pi^2\frac{E_1}{2}(A_3)\left(\frac{s_3}{K_3}\right)\frac{1}{l^2}\right]$ and a_1, a_2 and a_3 are as defined for sections (A) and (B).
Built-up section (D)	$(EI)_{ef,y} = E_1\frac{A_1^2}{12} + E_2\frac{A_2^2}{12} + \gamma_1E_1(A_1)a_1^2 + E_2(A_2)a_2^2$ where $\gamma_1 = 1/\left[1 + \pi^2E_1(A_1)\left(\frac{s_1}{K_1}\right)\frac{1}{l^2}\right]$ and $a_2 = \gamma_1E_1A_1(h_1 + h_2)/2(\gamma_1E_1A_1 + E_2A_2)$ $a_1 = \frac{(h_1 + h_2)}{2} - a_2$

Data from EN 1995-1-1

[a]Where the element interfaces are glued, $\gamma = 1$ and areas $A_1 = b_1h_1$, $A_2 = b_2h_2$ and $A_3 = b_3h_3$

Example 11.2: design of a mechanically jointed beam

The typical T section of a floor structure comprising a 20.5 mm-thick (h_1) Canadian Douglas fir plywood flange and 75 mm (b_2) by 300 mm (h_2) C24 timber webs at 400 mm c/c (b) is shown in Figure 11.4, and functions in service class 2 conditions. The effective span of the floor is 4.5 m (L), and the flange is connected to the web by 3.4 mm-diameter (d) smooth round nails at 50 mm c/c (s) along the beam length. The design bending moment and shear force on the beam are 9.60 kN m (M_d) and 8.54 kN (V_d), respectively, both arising from combined permanent and medium-duration variable actions, and the value of ψ_2 for the loading causing the largest stress in relation to strength is 0.3. Show that the bending stresses and shear strength of the section are acceptable, and calculate the maximum lateral force in the fixing nails. The face grain of the plywood is perpendicular to the direction of the floor span, and predrilling is not used for the nails.

Figure 11.4. T section of a floor structure

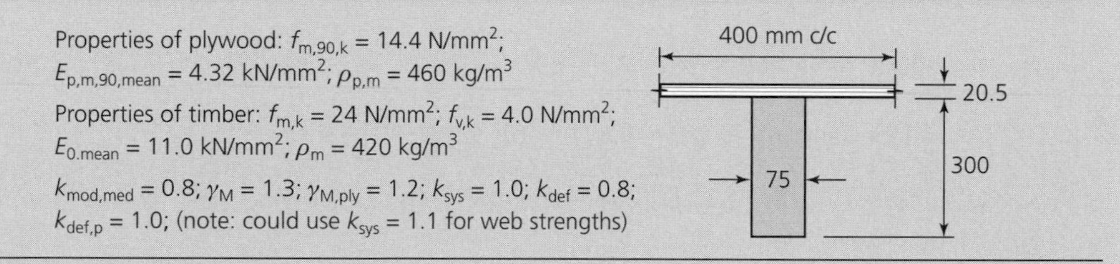

Stiffness properties – plywood and the final mean value:

$$E_{P,mean,fin} = E_{p,m,90,mean}/(1 + \psi_2 k_{def,p}) = 4.32 \times 10^3/(1 + 0.3 \times 1.0) = 3.323 \times 10^3 \text{ N/mm}^2$$

$$E_{0,mean,fin} = E_{0,m,90,mean}/(1 + \psi_2 k_{def}) = 11.0 \times 10^3/(1 + 0.3 \times 0.8) = 8.871 \times 10^3 \text{ N/mm}^2$$

Slip modulus of a nail:

$$K_{ser} = [(\rho_m \rho_{p,m})^{0.5}]^{1.5} d^{0.8}/30 = [(420 \times 460)^{0.5}]^{1.5} \times 3.4^{0.8}/30 = 817.65 \text{ N/mm}$$

Final mean value:

$$K_{ser,fin} = K_{ser}/[1 + \psi_2 2(k_{def} k_{def,p})^{0.5}] = 817.65/(1 + 0.3 \times 2(0.8 \times 1.0)^{0.5}) = 532.1 \text{ N/mm}$$

Effective width of a compression flange (rules in *clause 9.1.2*):

Clause 9.1.2

$$b_{c,ef} = \min[0.1L, 25h_1, (b - b_2)] = \min[0.1 \times 4500, 25 \times 20.5, (400 - 75)] = 325 \text{ mm}$$

The buckling length of the compression flange must not exceed twice the plate buckling flange width:

$$(b - b_2)/2(25h_1) = (400 - 75)/2 \times 25 \times 20.5 = 0.32$$

which is OK.

Geometric properties:

$$A_1 = bh_1 = 400 \times 20.5 = 8200 \text{ mm}^2;$$

$$A_2 = b_2 h_2 = 75 \times 300 = 22\,500 \text{ mm}^2$$

$$I_1 = bh_1^3/12 = 400 \times 20.5^3/12 = 287\,171 \text{ mm}^4$$

$$I_2 = b_2 h_2^3/12 = 75 \times 300^3/12 = 168\,750 \times 10^3 \text{ mm}^4$$

$$\gamma_1 = [1 + \pi^2 E_{P,mean,fin} A_1 s/(K_{ser,fin} L^2)]^{-1}$$

$$= [1 + \pi^2 \times 3.323 \times 10^3 \times 8200 \times 50/(532.1 \times 4500^2)]^{-1} = 0.4448$$

$$\gamma_2 = 1$$

$$a_2 = \gamma_1 E_{P,mean,fin} A_1 (h_1 + h_2)/2(\gamma_1 E_{P,mean,fin} A_1 + E_{0,mean,fin} A_2)$$

$$= 0.4448 \times 3.323 \times 10^3 \times 8200 \times (20.5 + 300)/(2 \times (0.4448 \times 3.323 \times 10^3$$

$$\times 8200 + 8.871 \times 10^3 \times 22\,500) = 9.175 \text{ mm}$$

$$a_1 = (h_1 + h_2)/2 - a_2 = (20.5 + 300)/2 \times 9.175 = 151.08 \text{ mm}$$

$$(EI)_{ef} = E_{P,mean,fin} b_1 h_1^3/12 + E_{0,mean,fin} b_2 h_2^3/12 + \gamma_1 E_{P,mean,fin} A_1 a_1^2 + E_{0,mean,fin} A_2 a_2^2)$$

$$= 3.323 \times 10^3 \times 400 \times 20.5^3/12 + 8.871 \times 10^3 \times 75 \times 300^3/12 + 0.4448 \times 3.323$$

$$\times 10^3 \times 8200 \times 151.08^2 + 8.871 \times 10^3 \times 22\,500 \times 9.175^2$$

$$= 0.95427 \times 10^6 + 1496.98 \times 10^6 + 276.66 \times 10^6 + 16.802 \times 10^6$$

$$= 1791.39 \times 10^9 \text{ kN/mm}^2$$

Bending stresses in the section are calculated.

Maximum compression bending stress in the plywood:

$$(\sigma_1 + \sigma_{m,1}) = M_d E_{P,mean,fin} (\gamma_1 a_1 + 0.5 h_1)/(EI)_{ef}$$

$$= 9.6 \times 10^6 \times 3.323 \times 10^3 \times (0.4448 \times 151.08 + 0.5 \times 20.5)/1791.39 \times 10^9$$

$$= 1.38 \text{ N/mm}^2$$

Maximum bending stress in timber, $(\sigma_2 + \sigma_{m,2})$:

$$(\sigma_2 + \sigma_{m,2}) = M_d E_{0,mean,fin} (a_2 + 0.5 h_2)/(EI)_{ef}$$

$$= 9.6 \times 10^6 \times 3.323 \times 10^3 \times (9.175 + 0.5 \times 300)/1791.39 \times 10^9$$

$$= 7.57 \text{ N/mm}^2$$

Design bending strength of the plywood:

$$f_{m,90,d} = k_{mod,med} k_{sys} f_{m,90,k}/\gamma_{M,ply} = 0.8 \times 1.0 \times 14.4/1.2 = 9.6 \text{ N/mm}^2$$

$$(\sigma_1 + \sigma_{m,1})/f_{m,90,d} = 1.38/9.6 = 0.14 < 1$$

which is OK.

Design bending strength of the timber:

$$f_{m,d} = k_{mod,med} k_{sys} f_{m,k}/\gamma_M = 0.8 \times 1.0 \times 24/1.3 = 14.77 \text{ N/mm}^2$$

$$(\sigma_2 + \sigma_{m,2})/(f_{m,90,d}) = 7.57/14.77 = 0.51 < 1$$

which is OK.

Maximum design shear stress in the beam web:

$$\tau_{v,d} = 0.5E_{0,mean,fin}h^2V_d/(EI)_{ef}$$

$$= 0.5 \times 8.871 \times 10^3 \times 150^2 \times 8.54 \times 10^3/1791.39 \times 10^9 = 0.48 \text{ N/mm}^2$$

Design shear strength:

$$f_{v,d} = k_{mod,med}k_{sys}f_{v,k}/\gamma_M = 0.8 \times 1.0 \times 4/1.3 = 2.46 \text{ N/mm}^2$$

$$\tau_{v,d}/f_{v,d} = 0.48/2.46 = 0.2 < 1$$

which is OK.

Maximum lateral force in the 3.4 mm diameter nail:

$$F_{v,Rd} = \gamma_1 E_{P,mean,fin}A_1a_1sV_d/(EI)_{ef}$$

$$= 0.4448 \times 3.323 \times 10^3 \times 8200 \times 151.08 \times 50 \times 8.54 \times 10^3/1791.39 \times 10^9$$

$$= 436.27 \text{ N}$$

(will be shown to be less than the design lateral strength of the nail, based on the rules in Chapter 8).

11.4. Annex C (informative): built-up columns

This annex covers the design of built-up columns formed from wood and wood-based products with the elements connected by glue or mechanical fasteners. The general requirements are covered in *clauses C.1.1* to *C.2.3*.

Clauses C.1.1–C.2.3

Spaced columns formed with packs or gusset plates that are nailed, glued or formed with connectors are covered in *clauses C.3.1* to *C.3.3*, and rules for latticed columns with glued or nailed joints are covered in *clauses C.4.1* to *C.4.3*.

Clauses C.3.1–C.3.3
Clauses C.4.1–C.4.3

11.4.1 Assumptions

The assumptions made for built-up columns are:

- the columns are pin jointed at their ends so that the effective column length is the actual column length, ℓ
- the elements of the built up column are all full length.

The only force on the column is the axial design loading, F_{cd}, acting along the centroidal axis of the column. Where there are small moments on the column due to self-weight or an equivalent, the effect can be taken into account in addition to the axial loading, as referred to in *clause C.2.3*.

Clause C.2.3

11.4.2 Load-carrying capacity

For spaced and latticed columns under axial loading and deflecting in the y direction (as shown in *Figures C.1* and *C.3*, respectively) there will be no stiffness contribution between the elements. For this condition the load-carrying capacity will be the sum of the strengths of the individual elements of the column. The slenderness ratio of each element will be based on the effective length ℓ of the column buckling in that direction, and its strength will be derived in accordance with the rules in *clause 6.3.2*.

Clause 6.3.2

For spaced and latticed columns under axial loading and deflecting in the z direction (also shown in *Figures C.1* and *C.3*, respectively) there will be a stiffness contribution between the column elements. The load-carrying capacity will be dependent on the effective slenderness ratio, λ_{ef}, of the column deflecting in this direction, which is determined in accordance with the

Clause C.3.2
Clause C.4.2
Clause 6.3.2

requirements of *clause C.3.2* for spaced columns and *clause C.4.2* for latticed columns. The associated instability factor, k_c, will be derived from the rules in *clause 6.3.2*.

When built-up columns are being used, the design procedures are not clearly defined in the annex.

To determine the stress in each element in a built-up column, because the axial load is applied along the centroidal axis the strain across the section is taken to be uniform, and it is proposed in this guide that the stress in each column element i, will be

$$\sigma_{c,0,d,i} = \frac{E_i F_{c,d}}{\sum_{i=1}^{n} E_i A_i} \qquad \text{(D11.1)}$$

where E_i is the mean value of the modulus of elasticity parallel to the grain of element i; $F_{c,d}$ is the axial design load on the column; and A_i is the cross-sectional area of element i. Where all of the elements have the same E value, Equation D11.1 will reduce to *Equation C.2*.

If the column section is asymmetrical about its y–y axis due to profile shape or to material properties, as is the case for sections (A) to (D) in Figure 11.3, and the column is deflecting in the y direction, it is proposed in this guide that the load-carrying capacity is obtained from the sum of the strengths of the individual elements of the section. When, however, the section is symmetrical both in shape and material properties, the stiffening effect between the elements can be taken into account, and the effective slenderness ratio, λ_{ef}, as explained in Section 11.4.3 of this guide, will be used to determine the strength. The same argument will also apply to built-up sections deflecting in the z direction, and for the sections shown in Figure 11.3, as they are all symmetrical in material and profile about the z–z axis, the column strength is determined using the effective slenderness ratio for deflection in that direction. Where the interfaces between the elements are connected by glue, when calculating the effective bending stiffness referred to in Section 11.4.3, function γ should be taken as equal to 1.

Clause 6.3.2

The strength of each element will be a function of the relative slenderness ratio of the built-up column, $\lambda_{rel,i}$. When there is no stiffening effect between the elements, the strength of each column element will be the sum of the strengths of the elements derived from the rules in *clause 6.3.2*. Where stiffening applies, λ_{rel} will be a function of the effective slenderness ratio, λ_{ef}, defined in Section 11.4.3, and the relative slenderness ratio for each element i in the column for deflection in the z direction, $\lambda_{rel,i,y}$, and in the y direction, $\lambda_{rel,i,z}$, will be

$$\lambda_{rel,i,y} = \frac{\lambda_{ef,y}}{\pi}\sqrt{\frac{f_{c,0,k,i}}{E_{0.05,i}}} \qquad \lambda_{rel,i,z} = \frac{\lambda_{ef,z}}{\pi}\sqrt{\frac{f_{c,0,k,i}}{E_{0.05,i}}} \qquad \text{(D11.2)}$$

where, for each element i of the column, $f_{c,0,k,i}$ is the characteristic axial compression strength parallel to the grain and $E_{0.05,i}$ is the 5 percentile modulus of elasticity parallel to the grain.

Clause 6.3.2

Adopting the β values given in *Equation 6.29* for the relevant materials being used in the column, the instability factors $k_{c,y,i}$ and $k_{c,z,i}$ are derived by following the rules in *clause 6.3.2*, enabling the design buckling strength, $k_c f_{c,0,d,i}$, for each direction to be obtained. The value of the design axial compression strength parallel to the grain of each material, $f_{c,0,d,i}$, will be obtained from $f_{c,0,k,i}$, as defined in Section 6.1.4 in this guide.

The validation requirement will be to show that for each material the design axial stress will not exceed its design buckling strength.

11.4.3 Mechanically jointed columns – effective slenderness ratio

With built-up columns of length ℓ in which stiffening between elements occurs, the effective slenderness ratio, λ_{ef}, is taken to be

$$\lambda_{ef} = \ell\sqrt{\frac{E_{mean}A_{tot}}{(EI)_{ef}}} \qquad \text{(D11.3)}$$

Table 11.2. Values for the effective bending stiffness of sections shown in Figure 11.3 (as detailed in *Figure B.1*) about the z–z axis

Profile in Figure 11.3	Effective bending stiffness about the z–z axis, $(EI)_{ef,y}$ [a] (properties and sizes as shown in *Figure B.1*)
Built-up section (A) where $E_1 = E_3$, $b_1 = (b_2 + b_3)$ and $h_1 = h_3$	$(EI)_{ef,z} = 2E_1\frac{h_1 b_1^3}{12} + 2E_2\frac{h_2(b_2/2)^3}{12} + 2\gamma_1 E_2\left(h_2\frac{b_2}{2}\right)\left(\frac{b_1}{2}-\frac{b_2}{4}\right)^2$ where $\gamma_1 = 1/\left[1 + \pi^2 E_2\left(h_2\frac{b_2}{2}\right)\left(\frac{s_1}{2K_1}\right)\left(\frac{1}{l^2}\right)\right]$
Built-up section (B) where $E_1 = E_3$, $b_1 = b_3$ and $h_1 = h_3$	$(EI)_{ef,z} = 2E_1\frac{h_1 b_1^3}{12} + E_2\left(\frac{h_2 b_2^3}{12}\right)$
Built-up section (C) where $E_1 = E_3$, $b_1 = b_3$ and $h_1 = h_3$	$(EI)_{ef,z} = 4E_1\frac{h_1(0.5b_1)^3}{12} + E_2\frac{h_2 b_2^3}{12} + 4\gamma_1 E_1(h_1 0.5b_1)\left(\frac{b_2}{2}+\frac{b_1}{4}\right)^2$ where $\gamma_1 = 1/\left[1 + \pi^2 E_1(h_1 b_1)\left(\frac{s_1}{2K_1}\right)\frac{1}{l^2}\right]$

Data from EN 1995-1-1
[a]Where the element interfaces are glued, $\gamma = 1$

where E_{mean} is the mean value of the modulus of elasticity parallel to the grain of all of the column elements (with no weighting to take account of the area of each material being used); A_{tot} is the total cross-sectional area of the built-up column; and $(EI)_{ef}$ is the effective bending modulus of the column derived in accordance with *Annex B*, referred to in Section 11.3.

The effective slenderness ratio will differ depending on the direction in which the column will deflect under the effect of the axial load. For built-up columns formed from the sections shown in Figure 11.3, where the deflection is in the z direction the relative slenderness ratio will be $\lambda_{ef,y}$, and for this condition the effective bending modulus will be $(EI)_{ef,y}$, as given in Table 11.1 for the relevant section being used.

Where the deflection is in the y direction, as stated in Section 11.4.2 the effective slenderness ratio will only apply when the built-up column section is symmetrical both in shape and material properties about the z–z axis. In Table 11.2 the sections referred to are symmetrical and made from the same materials, and the effective stiffness ratio, $\lambda_{ef,z}$, will be a function of the effective bending modulus $(EI)_{ef,z}$, as given in the table for the relevant section being used.

11.4.4 Mechanically jointed columns – load on fasteners

Because of the out-of-plane movement of the built-up column, the fasteners connecting the column elements will be subjected to lateral loading, and the fastener load is derived from the requirements of *clause B.5*. The value to be used for the design shear force, V_d, in *Equation B.10*, is obtained from *Equation C.5*, and will be dependent on the relative slenderness ratio associated with the bending axis.

Clause B.5

11.4.5 Spaced columns with packs or gussets

The permitted configurations for spaced columns are shown in *Figure C.1*, and either spacer packs or gusset plates can be used to connect the shafts. The columns are assumed to be pin jointed and effectively held in position at their ends from movement in the y and z directions.

About the z–z axis of the spaced column, the column shafts, which are b deep and h thick, behave as individual timber sections of effective length ℓ, and the slenderness ratio of each shaft about this axis will be

$$\lambda_1 = \sqrt{12}\frac{\ell}{b}$$

About the y–y axis, the shafts are stiffened by the packs/gussets, and the effective slenderness ratio for movement about this axis, λ_{ef}, is taken to be a function of the slenderness ratio based on the semi-composite behaviour of the latticed column about this axis, λ, and the slenderness ratio of an individual shaft about its weak axis, λ_1, taking account of the degree of fixity offered by the packs/gussets. λ_{ef} is obtained from *Equation C.10*, and to ensure it will be the critical design condition about the y–y axis, the value used for λ_1 in *Equation C.10* has to be at least 30.

Clause 6.3.2

The design strength of the spaced column will be $k_c f_{c,0,d}$, where k_c is derived in accordance with the rules in *clause 6.3.2*, as explained in Section 6.3.2 of this guide, using the larger of λ_z and λ_{ef}. The design compression strength parallel to the grain, $f_{c,0,d}$, is derived from the characteristic compression strength parallel to the grain of the shaft material, as described in Section 6.1.4 of this guide.

For a spaced column with n members subjected to an axial load $F_{c,d}$, the compressive stress on the column, $\sigma_{c,0,d}$, will be $F_{c,d}/nbh$, and the validation requirement will be to show that $\sigma_{c,0,d} \leq k_c f_{c,0,d}$.

When the spaced column deflects in the z direction, the packs/gussets will be subjected to shear forces as shown in *Figure C.2*, and the value of T_d shown in *Figure C.2* will be obtained from *Equation C.13*. It is a function of V_d, which is obtained from *Equation C.5*, and the value used for the effective slenderness, λ_{ef}, in that equation is obtained from *Equation C.10*. In addition to checking that the shear strength of the packs/gussets will not be exceeded under the action of the shear force, the connection at the shafts must also be checked for the combined effect of the shear force and the bending moment due to this force.

11.4.6 Lattice columns with glued or nailed joints

In lattice columns, two shafts are braced together by bracing members forming an N or V truss arrangement as shown in *Figure C.3*. The bracing used on each side must have the same configuration, but can be offset along the column by a distance $\ell_1/2$, where ℓ_1 is the bay length used for the bracing. As with spaced columns, these columns are also assumed to be pin jointed and effectively held in position from movement in the y and z directions at the ends.

About the z–z axis of the latticed column, the column flanges, which are b deep and h_1 thick as shown in Figure 11.5, behave as individual timber sections of effective length ℓ, and the slenderness ratio of each flange about this axis will be

$$\lambda_{zz} = \sqrt{12}\frac{\ell}{b}$$

Clause C.4.1(2)

About the y–y axis, there is a risk of local failure of a flange about its weak axis as well as failure when functioning as a composite section. For local failure, it is a requirement that the slenderness ratio of each flange about its weak axis when buckling between node positions is not greater than 60, and one of the assumptions made in *clause C.4.1(2)* is that the flange will not buckle locally.

Figure 11.5. Lattice columns – member sizes

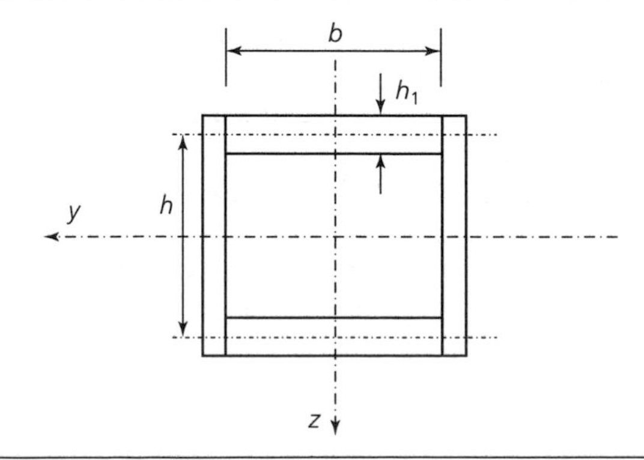

Although it is not a stated requirement in *clause C.4*, it is suggested in this guide that the actual slenderness ratio of a flange about its weak axis,

Clause C.4

$$\lambda_z = \sqrt{12}\frac{\ell_1}{h_1}$$

(where ℓ_1 is as shown on *Figure C.3*) is included as part of the strength assessment. For buckling as a composite section about the y–y axis, the effective slenderness ratio, λ_{ef}, is taken to be the maximum value obtained from *Equation C.14*, and, as for spaced columns, it is also a function of the degree of fixity offered by the type of fixing used for the bracing as well as the bracing shape.

The design strength of the latticed column will be $k_c f_{c,0,d}$, where k_c is derived in accordance with the rules in *clause 6.3.2*, as explained in Section 6.3.2 of this guide, using the largest of λ_z, λ_{zz} and λ_{ef}. k_c is then derived as described in Section 11.4.5 for spaced columns.

Clause 6.3.2

For a latticed column subject to an axial load $F_{c,d}$, the compressive stress on the column, $\sigma_{c,0,d}$, will be

$$\sigma_{c,0,d} = \frac{F_{c,d}}{2bh_1}$$

and the validation requirement will be to show that, $\sigma_{c,0,d} \le k_c f_{c,0,d}$.

When the latticed column deflects in the z direction, there will be a shear force V_d acting across the column. It is derived from *Equation C.5*, and the value used for the effective slenderness, λ_{ef}, in *Equation C.5* is obtained from *Equation C.14*. The forces in the connections at the shafts and in the lattice bracing will be derived from this force.

Example 11.3: design of a spaced column

A spaced column, 5.6 m long (L), is fabricated from 3 (n) shafts of equal cross-section, 97 mm (h) by 220 mm (b), as shown in Figure 11.6. The spaced column is pin jointed at each end, and held laterally in position at these locations. It is fabricated from C24 timber, and functions under service class 2 conditions and two 6 mm-thick mild steel gusset plates are fastened to the timber shafts using 6 mm-diameter nails and formed without predrilling. Check that the spaced column can support a combined design load of 28.8 kN ($F_{c,d}$), based on a combination of permanent and short-duration variable action (snow) in accordance with the design rules in EN 1995-1-1, and calculate the design forces that the gusset plates fasteners and steel gusset plates must be designed to withstand.

Timber properties and general issues: $f_{c,0,k} = 21$ N/mm^2; $\rho_k = 350$ kg/m^3; $E_{0.05} = 7.4$ kN/mm^2; $k_{mod} = 0.9$; $k_{sys} = 1$; $\gamma_M = 1.3$; $\beta = 0.2$. The section complies with the detailing requirements in *clause C.3.1(2)*, and, from *Table C.1*, $\eta = 4.5$.

Clause C.3.1(2)

The geometric properties are calculated.

Clear distance between shafts:

$$a = 100 \text{ mm}$$

$$A_{tot} = 3hb = 3 \times 97 \times 220 = 64\,020 \text{ mm}^2$$

Effective length about the z–z axis:

$$L_Z = L = 5600 \text{ mm}$$

Effective length of a shaft between the gusset plates:

$$L_{shaft} = 1300 \text{ mm}$$

Figure 11.6. Spaced column

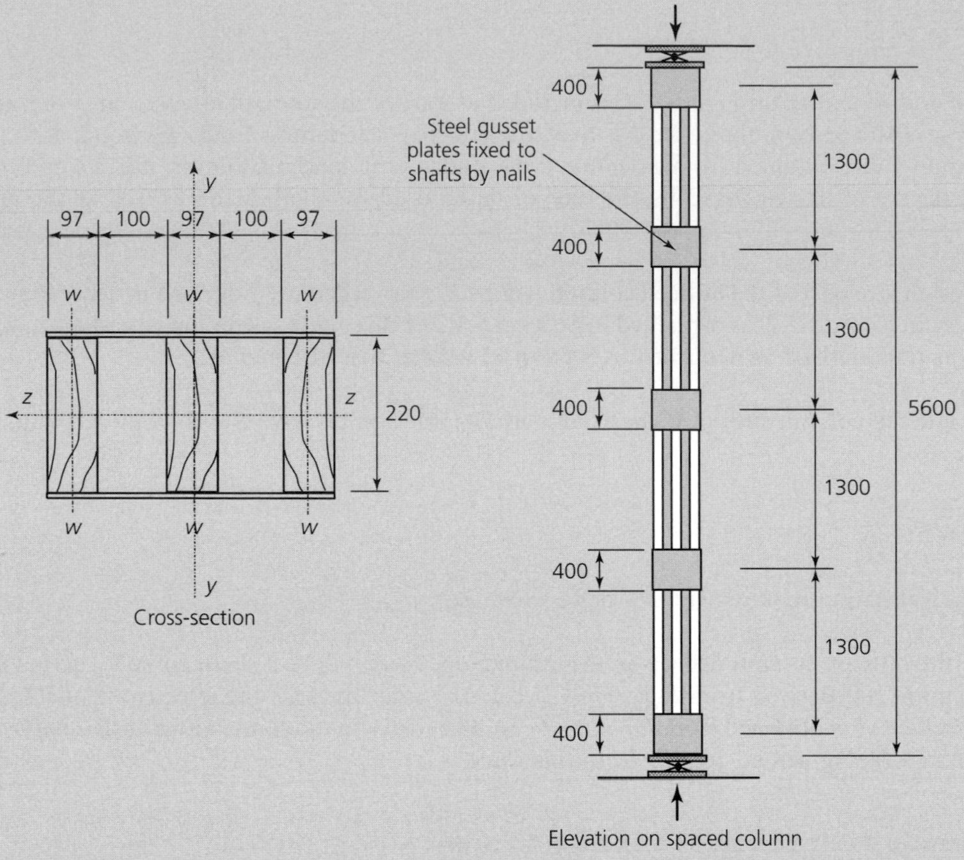

Second moment of the area of a spaced column about the z–z axis:

$$I_z = 3hb^3/12 = 3 \times 97 \times 220^3/12 = 25\,821.4 \times 10^4\ \text{mm}^4$$

$$i_z = (I_Z/A_{tot})^{0.5} = (25\,821.4 \times 10^4/64\,020)^{0.5} = 63.51\ \text{mm}$$

$$\lambda_z = L_z/i_z = 5600/63.51 = 88.18$$

Second moment of the area of a shaft about the w–w axis:

$$I_w = bh^3/12 = 220 \times 97^3/12 = 16\,732.34 \times 10^3\ \text{mm}^4$$

$$i_w = (I_w/bh)^{0.5} = (16\,732.34 \times 10^3/220 \times 97)^{0.5} = 28.00\ \text{mm}$$

$$\lambda_w = L_{shaft}/i_w = 1300/28.00 = 46.43$$

This exceeds 30, so is OK.

Second moment of the area of a spaced column about the y–y axis:

$$I_{tot} = b[(3h + 2a)^3 - (h + 2a)^3 + h^3]/12$$

$$= 220 \times [(3 \times 97 + 2 \times 100)^3 - (97 + 2 \times 100)^3 + 97^3]/12 = 1\,706\,565.14 \times 10^3\ \text{mm}^4$$

$$\lambda = L(A_{tot}/I_{tot})^{0.5} = 5600 \times [64\,020/(1\,706\,565.14 \times 10^3)]^{0.5} = 34.30$$

$$\lambda_{ef} = (\lambda^2 + \eta n \lambda_w^2/2)^{0.5} = (34.30^2 + 4.5 \times 3 \times 46.43^2/2)^{0.5} = 125.4$$

Maximum slenderness ratio of the spaced column:

$$\lambda_{max} = \max(\lambda_w, \lambda_{ef}, \lambda_z) = \max(46.43, 125.4, 88.18) = 125.4 = \lambda_{ef}$$

so buckling occurs about the y–y axis.

Design compression stress:

$$\sigma_{c,0,d} = F_{c,d}/A_{tot} = 28.8 \times 10^3/64\,020 = 0.45\ \text{N/mm}^2$$

Instability factor about the y–y axis:

$$\lambda_{rel,y} = \lambda_{ef}(f_{c,0,k}/E_{0.05})^{0.5}/\pi = 125.4 \times (21/7400)^{0.5}/\pi = 2.126$$

$$k_y = 0.5[1 + \beta(\lambda_{rel,y} - 0.3) + \lambda_{rel,y}^2] = 0.5[1 + 0.2(2.126 - 0.3) + 2.126^2] = 2.943$$

$$k_{cy} = 1/[k_y + (k_y^2 - \lambda_{rel,y}^2)^{0.5}] = 1/[2.943 + (2.943^2 - 2.126^2)^{0.5}] = 0.201$$

Design compression strength:

$$f_{c,0,d} = k_{mod}k_{sys}f_{c,0,k}/\gamma_M = 0.9 \times 1.0 \times 21/1.3 = 14.54\ \text{N/mm}^2$$

Design buckling strength of the spaced column:

$$k_{c,y}f_{c,0,d} = 0.201 \times 14.54 = 2.92$$

$$\sigma_{c,0,d}/k_{c,y}f_{c,0,d} = 0.45/2.92 = 0.15 < 1$$

and therefore OK.

The design shear force and moment (in the gusset plate and the nailed connection), $0.5T_d$ and Mg_d respectively, are calculated.

From *Equation C.5*, for $\lambda_{ef} = 125.4$:

$$V_d = F_{c,d}/60k_{c,y} = 28.8 \times 10^3/(60 \times 0.201) = 2388.1\ \text{N}$$

From *Equation C.13*, the values for $0.5T_d$ (design connection shear force at each shaft) and Mg_d (the design moment to be resisted by the gusset plates and their connections at each shaft) are

$$0.5T_d = 0.5V_dL_{shaft}/(a + h) = 0.5 \times 2388.1 \times 1300/(100 + 97) = 7879.52\ \text{N}$$

$$Mg_d = 0.5T_d(a + h)2/3 = 7879.52 \times (100 + 97) \times 2/3 = 1034.84 \times 10^3\ \text{N mm}$$

(The shear force and moment in each gusset plate will be half the above values.)

Designers' Guide to Eurocode 5: Design of Timber Buildings
ISBN 978-0-7277-3162-3

http://dx.doi.org/10.1680/dtb.31623.171

Chapter 12
Timber in fire (EN 1995-1-2)

12.1. Introduction

The structural fire design of timber is covered by EN 1995-1-2 (BSI, 2004), and strictly is outside the scope of this book. However, it has been considered useful to give a brief outline of the design approach, more particularly in relation to the period of fire resistance of unprotected load-bearing timber structures. The fire parts of various structural codes, in accordance with the Eurocode system, essentially supplement their main material code, and deal only with specific aspects of passive fire protection. They do not cover, for instance, the installation of sprinkler systems, or conditions in relation to the occupancy of the building.

Clause 2.4.1(6)

While EN 1995-1-2 mainly presents design rules based on calculations, a route (*clause 2.4.1(6)*) is left open for compliance based on testing. This alternative is often used to establish, for example, the compartmentation resistance of a partition.

EN 1995-1-2 also has a National Annex (BSI, 2006a), which does not recommend the use of some of the code annexes.

12.2. The basis of design

The code defines two basic functions of a structure under fire conditions:

- **mechanical resistance**, whereby the structure maintains its load-bearing function (criterion R) during the relevant fire exposure
- **fire compartmentation**, when the elements forming the boundaries of the fire compartment maintain their separating function (criterion E) or insulating function (criterion I) during the relevant fire exposure.

This chapter, as noted above, deals only with the load-bearing function. Two approaches to design are given in the code:

- **nominal fire exposure**, in which elements maintain their load-bearing function for a specified period of time – the relevant period for structural elements (e.g. $\frac{1}{2}$ hour or 1 hour) is generally set by national regulations
- **parametic fire exposure**, in which the development of the fire, including the decay phase, is considered.

This chapter deals only with the design of exposed timber elements for **nominal fire exposure**.

12.3. The behaviour of timber in a fire

The process of burning is a series of rapid chemical reactions between the timber (the fuel) and oxygen (in the air). The process produces heat (and light, as visible flames). For combustion to occur, fuel, air and heat must all be present: the process terminates on the removal of any one.

When timber is heated, gasses are released. For gasses to ignite, there is normally a source of ignition, which would need to have a temperature of around 250–300°C. If there is no ignition source, but the temperature increases, then spontaneous ignition may occur, as happens when a fire radiates heat onto a remote piece of timber. Once ignited, the burning gasses will now heat up adjacent timber, and the process continues. The heat transfer from the flame to unburnt material is mainly by radiation from the flames, and convection from the burning vapours. Changing the orientation of a burning match from the horizontal to the vertical demonstrates the different modes of heat transfer. Because timber is a good insulator, conduction of heat back into the unburnt material plays a minor role.

As the gasses burn off, the residue, charcoal, is largely pure carbon. The charcoal intumesces in its formation, expanding in volume and creating microscopic voids. As such, it is an excellent insulant, and the timber a short distance behind the charring layer is virtually undamaged. The layer has the effect of controlling the rate at which combustion occurs, and there is a linear relationship between charring depth and the time for a constant fire temperature, which is known as the charring rate.

The charring rate is, for obvious reasons, related to the timber density and the fire temperature. Fire curves, that is, the development of fire temperature with time for different fuels, are given in EN 1363 (BSI, 1999, 2012). The standard fire curve, which is the curve usually used for building fires and upon which the code charring rates are based, is that for a cellulose fuel source. However, if an open timber building frame was likely to store large quantities of, for example, petrochemicals, then the appropriate fire curve for hydrocarbons, with a higher charring rate, should be considered.

12.4. Actions in fire

Fire is an accidental condition, and the design values of the action combinations are given in Eurocode 0:

$$\sum_{i>1} G_{k,j} + (A_d) + \psi_{1,1} + Q_{k,1} + \sum_{i>1} \psi_{2,i} Q_{k,i} \qquad \text{(EN 1990: 6.11(b))}$$

The value (A_d) is to be omitted, because it does not apply to the situation after the commencement of a fire. In accidental situations, all values of the partial factors for actions and materials are 1.0, and are therefore also omitted.

Clause 2.4.2(3)

Clause NA(2.4)

Clause 2.4.2

Since calculations will probably already have been made to determine the action combinations for normal temperature conditions, *clause 2.4.2(3)* gives an expression for a reduction factor η_{fi} that can be applied to the results of that analysis to give reduced values for the fire situation. *Clause NA(2.4)* does not permit the values to be less than 0.4. As a simplification, and with qualifications, the value (Note 2 to *clause 2.4.2*) of η_{fi} may be taken as 0.6.

12.5. The design values of material properties in a fire

Design values for strength in the fire condition are calculated from the following expression:

$$f_{d,fi} = k_{mod,fi} \frac{f_{20}}{\gamma_{M,fi}} \qquad (2.1)$$

where $f_{d,fi}$ is the design strength in fire, f_{20} is the 20% fractile of a strength property at normal temperatures, $k_{mod,fi}$ is the modification factor for fire and $\gamma_{M,fi}$ is the partial safety factor for timber in a fire.

The value of $k_{mod,fi}$ (which replaces the modification factor for normal temperature design, k_{mod}) is taken as 1.0 in the reduced cross-section method described below. As already noted, the recommended partial safety factor for material properties in fire is 1.0. Thus, the design strength properties of materials in fire are simply based on the 20% fractile. *Table 2.1* of the code gives values of k_{fi}, a factor that upgrades characteristic values f_{05} to f_{20} values. For solid timber, for instance, the value is 1.25.

12.6. Charring rates and reduction in element cross-section

For unprotected timber members under standard fire exposure, the charring rate is taken as constant with time. For a larger member, or where charring is occurring on a single face, it is known as the one-dimensional charring rate, β_0. The more usual case, where charring is occurring on several or all faces, the notional charring rate, β_n, allows for the additional loss at corner roundings and fissures. The values of β_0 and β_n are given in *Table 3.1*. For solid softwood timber, for instance, the value of β_n is 0.8 mm/min.

Figure 12.1. Charring depths. (Based on EN 1995-1-2 (*Figure 3.2*). Reproduced with permission from EN 1995-1-1, © British Standards Institution, 2004)

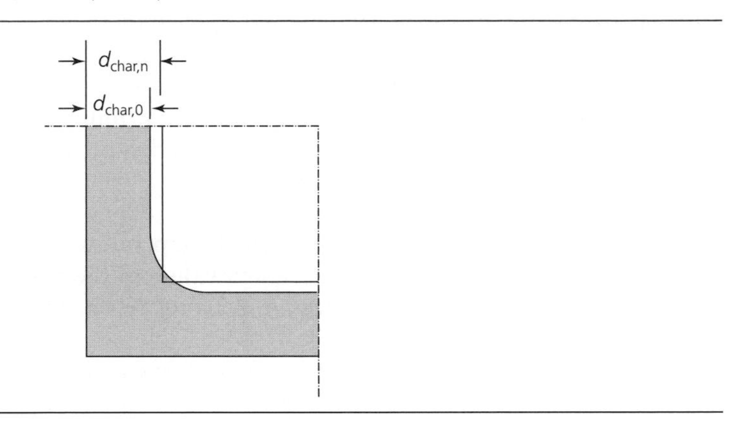

Figure 12.2. Definitions of residual cross-section and effective cross-section. (Based on EN 1995-1-2 (*Figure 4.1*). Reproduced with permission from EN 1995-1-1, © British Standards Institution, 2004)

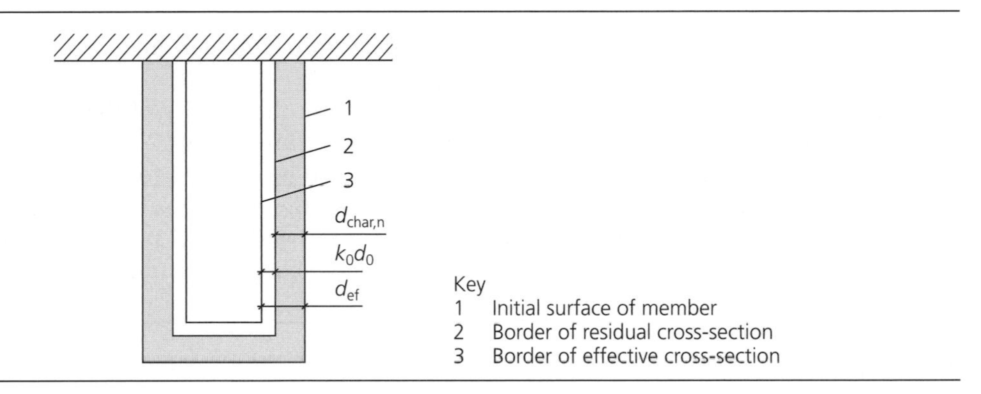

The charring depth, $d_{char,n}$, after exposure time t, is given as

$$d_{char,n} = \beta_n t \qquad (3.2)$$

with a similar expression for $d_{char,0}$ (Figure 12.1).

Clause 3.4.3 elaborates these rules for surfaces that are initially protected from fire exposure.

Clause 3.4.3

The effective charring depth d_{eff} in the reduced cross-section method described below (Figure 12.2) is then determined from

$$d_{eff} = d_{char,n} + k_0 d_0$$

where $d_{char,n}$ is given in *clause 3.4.2(2)*, $d_0 = 7$ mm and k_0 is 1, for values of $t \geq 20$ minutes (*Table 4.1*).

Clause 3.4.2(2)

This additional layer just under the char line is assumed to have zero strength. Inside this layer, the timber is taken to be undamaged by the fire.

The charring rates given in *Table 4.1* for glued laminated members require the integrity of the adhesive to be maintained for the relevant period. Adhesives of phenol-formaldehyde and aminoplastic type 1 adhesives complying with EN 301 (BSI, 2006b) satisfy this requirement, but some other adhesives soften at a temperature considerably below the charring temperature of the wood.

12.7. The design procedure for the calculation of the strength of an exposed timber element in a fire

This procedure is based on the assumption that the element has been shown to have adequate strength in normal conditions. Two methods of calculation are given in EN 1995-1-2: the

Clause 4.2.3
Clause 4.2.2

reduced properties method (*clause 4.2.3*) and the **reduced cross-section method** (*clause 4.2.2*). The latter method is recommended in the code and in the UK National Annex, and a calculation sequence is described below. The method applies to an unjointed element in the form of, for example, a beam or a column.

1. Establish the period of fire resistance required (t minutes) from national regulations.
2. (Section 12.4 in this guide.) Determine the design action on the element in the fire condition by:
   - the use of the EN 1990 equations or
   - the reduction factor applied to normal temperature values.
3. (Section 12.5.) Determine the fire design values of the relevant material properties.
4. (Section 12.6.) Calculate the effective charring depth, and hence the residual cross-section of the element.
5. Calculate the design stresses on the reduced cross-section.
6. Calculate the relevant design strengths, and check that they are equal to or greater than the design stresses.

In the analysis of the reduced cross-section:

Clause 4.3.1

Clause 4.3.2

- Compression perpendicular to the grain, and shear in rectangular and circular sections may be disregarded (*clause 4.3.1*).
- Notched beams should have a residual cross-section in the vicinity of the notch of at least 60% of the cross-section required for normal-temperature design.
- For an element that relies on bracing, this should also be checked for the relevant period of fire resistance (*clause 4.3.2*). If the residual area of the bracing is 60% of the initial value required for normal temperature design, and it is fixed with metal fasteners, it may be assumed not to fail. Otherwise, the element should be checked for the unbraced condition.

12.8. Connections

Most connections in designed timber structures are made with metal components such as nails, screws and bolts, sometimes used in conjunction with plates as splices or flitches. However, steel behaves very differently to timber in fire. It is a very good conductor of heat, and at temperatures of 600°C or so (a temperature achieved after about 10 minutes on the standard fire curve) it has lost about 50% of its strength, although it does not actually melt until reaching around 1100°C. It follows that steel connections between timber elements, even if the steel is only partly exposed, are very vulnerable in a fire.

Clause 6.2.1.1(2)

Rules for the period of fire resistance of typical fasteners are given in *Section 6*, which covers the standard range of, for example, nails, bolts and screws. Even though only the heads of these fasteners would be exposed in the fire, they are classed as 'unprotected'. Rules for the period of fire resistance of such **unprotected connections**, where spacings and edge and end distances comply with Part 1-1 of the code (Section 8), are given in *Table 6.1*. It will be seen that they are only able to achieve a resistance period of between 15 and 20 minutes. For fasteners with non-projecting heads, *clause 6.2.1.1(2)* gives a rule for increasing the period of fire resistance (up to a maximum of 30 minutes) by increasing the thickness of the timber members, and end and edge spacings.

Resistance times in excess of 30 minutes generally require **protected connections.** Protection is normally achieved by:

- recessing the fastener head, and filling the recess with a glued-in timber plug, or
- covering a group of fasteners (recessed as necessary) with a wood, wood-based or gypsum panel.

Clause 6.2.1.2

Rules for these forms of protection are given in *clause 6.2.1.2*. The rules include a method of fixing the panel to prevent its premature failure.

Clause 6.2.1.3

For connections that include **internal steel plates** as flitches, rules are given in *clause 6.2.1.3* for thickness and edge protection, which can achieve a fire resistance time of up to 60 minutes.

Separate consideration would have to be given to the fastener protection as above. **External steel plates**, acting as splices, are covered in *clause 6.3*. Where these are unprotected, a design reference is made to EN 1993-1-2 (BSI, 2005), the fire part of the steel code. A rule is given for the thickness of covering panels necessary to provide protection.

Clause 6.3

Section 7 gives detailing rules that relate to the various construction forms covered by this part of the code.

Example 12.1: the fire resistance of an exposed timber beam

The beam is in GL28 softwood glulam, spans 4.5 m and carries a fire-protected floor (Figure 12.3). The loaded width between beams is 2.5 m.

Figure 12.3. Exposed timber beam

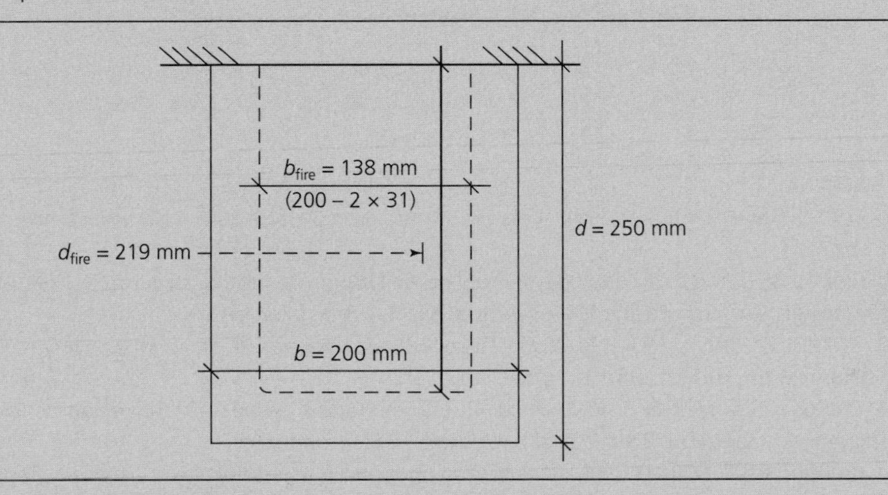

Loading: self-weight = 1.75 kN/m^2 (permanent); superimposed = 2.5 kN/m^2 (medium term).

The required time of fire exposure is 30 minutes.

Analysis at normal temperatures

$G_k = 2.5 \times 4.5 \times 1.75 = 19.7$ kN

$Q_k = 2.5 \times 4.5 \times 2.5 = 28.1$ kN

$F_d = \gamma_G G_k + \gamma_Q Q_k$

$= 1.35 \times 1.97 + 1.5 \times 28.1 = 68.7$ kN

as a uniformly distributed load.

$M_d = F_d \times 4.5/8 = 68.7 \times 4.5/8$

$= 38.6$ kN m

$\sigma_{m,d} = M_d/Z = 38.6 \times 10^6/2.08 \times 10^6$

$= \mathbf{18.5\ N/mm^2}$

$f_{m,k} = 28$ N/mm^2, $k_{mod} = 0.8$, $k_n = 1.09$, $\gamma_m = 1.25$

$f_{m,d} = 28 \times 0.8 \times 1.09/1.25 = \mathbf{19.5\ N/mm^2}$

Thus OK.

Normal section modulus

$Z_{norm} = 200 \times 250^2/6 = \mathbf{2.08 \times 10^6\ mm^3}$

Section reduction for *t* = 30 minutes

Softwood glulam: $\beta_n = 0.8$

$d_{char,u} = 0.8 \times 30 = 24$ mm

$k_0 = 1$, $d_0 = 7$

Thus

$k_0 d_0 = 7$

$d_{elf} = d_{char,n} + K_0 d_0 = 24 + 7 = \mathbf{31\ mm}$

Analysis in fire condition

$M_{d,fire} = \eta_{fi} M_{d,norm} = 0.6 \times 38.6$

$= 23.2$ kN m

$\sigma_{m,d,fire}$

$= M_{d,fire}/Z_{fire}$

$= 23.2 \times 10^6/1.1 \times 10^6 = 21.0$ N/mm^2

$k_{mod,fire} = 1.1 \gamma_{m,fire} = 1$

$f_{m,d,fire} = 28 \times 1.1 \times 1 \times 0.9 =$ **33.5 N/mm^2**

Thus OK.

Reduced section modulus

$b_{fire} = 200 - 2 \times 31 = 138$ mm

$d_{fire} = 250 - 31 = 219$ mm

$Z_{fire} = 138 \times 219^2/6 = 1.1 \times 10^6$ mm^3

REFERENCES

BSI (1999) BS EN 1363-2: 1999. Fire resistance tests. Alternative and additional procedures. BSI, London.

BSI (2004) BS EN 1995-1-2: 2004. Eurocode 5: Design of timber structures – General – Structural fire design. (Incorporating Corrigendum No. 1.) BSI, London.

BSI (2005) BS EN 1993-1-2: 2005. Eurocode 3: Design of steel structures – General rules – Structural fire design. BSI, London.

BSI (2006a) NA to EN 1995-1-2: 2004. UK National Annex to Eurocode 5: Design of timber structures – General – Structural fire design. BSI, London.

BSI (2006b) BS EN 301: 2006. Adhesives, phenolic and aminoplastic, for loading bearing timber structures. Classification and performance requirements. BSI, London.

BSI (2012) BS EN 1363-1: 2012. Fire resistance tests – General requirements. BSI, London.

Designers' Guide to Eurocode 5: Design of Timber Buildings
ISBN 978-0-7277-3162-3

http://dx.doi.org/10.1680/dtb.31623.177

Chapter 13
The design of multi-storey timber frame wall diaphragms

13.1. Introduction

The aim of this chapter is to outline an approach to the design of multi-storey timber frame wall diaphragms, and in particular to illustrate the application of a design method for their analysis given in PD 6993-1. The form of construction, known as platform frame, is described in detail in *Timber Frame Construction* (*TFC*) (Lancashire and Taylor, 2011), using framed wall panels stabilised with a sheet material and floors of joists with a board covering (Figure 13.1). The form has been used in the UK and elsewhere in Europe for the last 50 years, for dwellings up to four storeys in height that were built by specialist companies and, in the UK, were largely brick-clad.

Over this period there has been a significant rise in the standard of overall building performance requirements, and in 1991 the Building Regulations in England were amended to allow this form of construction to be extended to seven storeys in height. Although this chapter is concerned primarily with the frame walls, the following factors will influence the structural form:

- **Robustness:** the principle of 'robustness' is given in EN 1990, clause 2.1(4)P – 'A structure shall be designed and executed in such a way that it will not be damaged by accidental events to an extent disproportionate to the original cause'. For timber frames, such robustness is generally achieved by providing effective connections at floor and roof levels. Guidance is given in EN 1991-1-7 and the *Designers' Guide to Eurocode 1: Actions on Buildings: EN 1991-1-1 and -1-3 to -1-7* (Gulvanessian *et al*., 2008).
- **Thermal performance:** the last 30 years have seen a fourfold increase in the insulation requirements for buildings generally. Although not directly a structural issue, the required thickness of insulation may determine the stud widths for the external walls.
- **Acoustic performance:** increasing requirements for sound insulation in a multi-storey block are sometimes satisfied by the introduction of 'floating floors', or cavity walls with breaks in the floor construction at dwelling boundaries. The latter may restrict the capacity of the floor to distribute wind forces between the vertical structural elements.
- **Moisture movement of the frame:** the timber in each floor zone (Figure 13.1(d)) is inevitably loaded perpendicular to the grain, which results in small vertical deflections as the moisture content of this timber responds to the commissioning of the heating system. (The height of the wall panels themselves, where the studs are loaded parallel to the grain, is relatively unaffected by variations in moisture content.) Guidance on the degree of movement is given in *TFC*, Figure 9.1. Movement may be reduced by, for example, using engineered wood products for the floor construction.
- **Form of cladding:** buildings up to four storeys in height are often clad in brick, as a single outer leaf 100 mm thick, transferring its weight directly to the foundations, and stabilised by cavity ties to the timber frame forming the inner leaf of the wall (Figure 13.2). The brickwork provides a modest degree of wind shielding to the structure, and a calculation method, applicable to buildings up to three storeys high, is given in PD 6993-1. For taller structures, the differential movement between the brick (subject to thermal and moisture movements) and the frame (subject to the movements described above) become difficult to accommodate at openings such as windows, which necessarily bridge between the two leaves (Figure 13.3). Taller buildings are thus generally clad in relatively light-weight materials, such as tiles or renders, fixed to the frame, but able to accommodate small

Figure 13.1. (a) General arrangement; (b) wall panel; (c) window panel; (d) floor construction (left), corner detail (right)

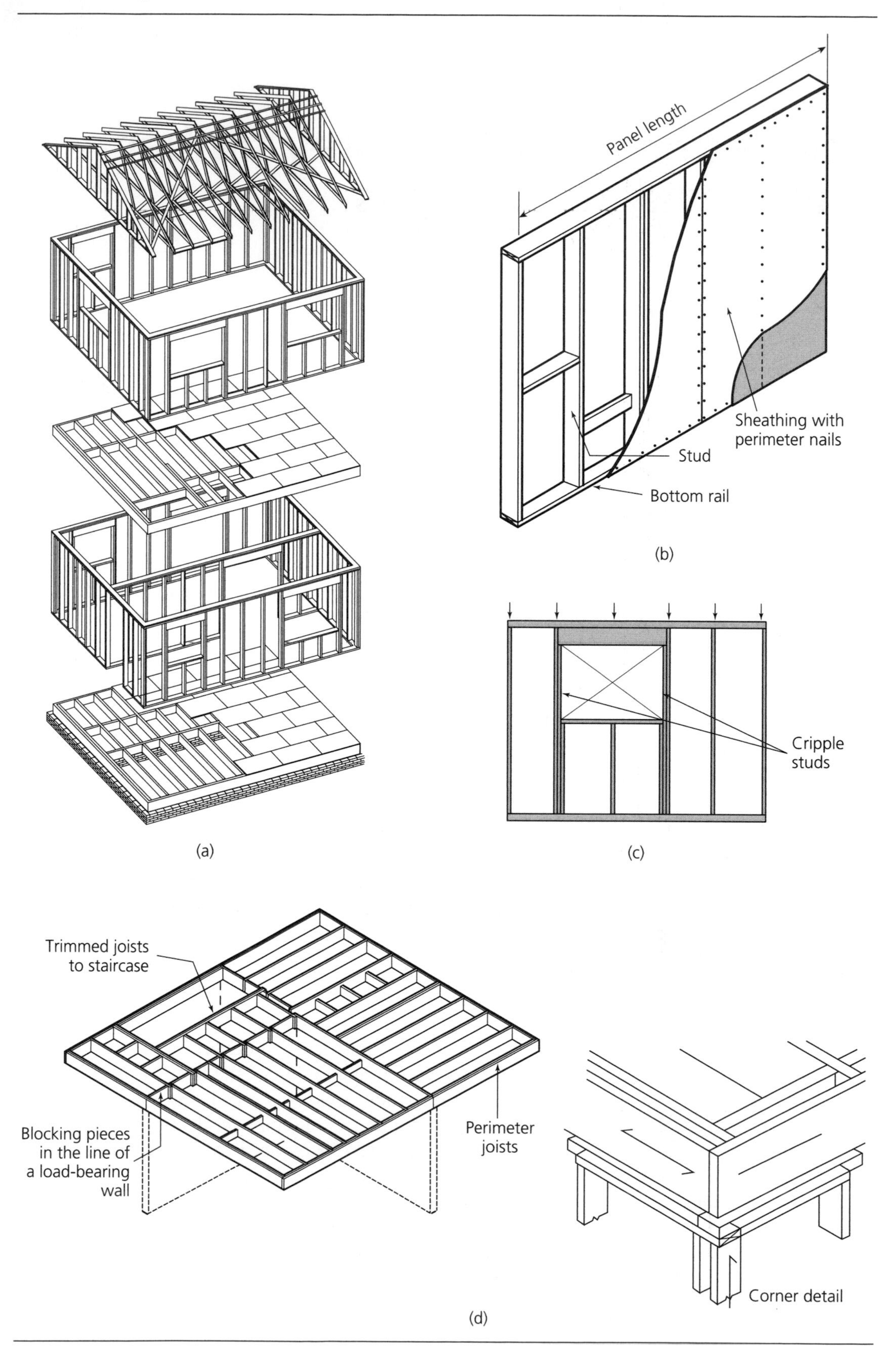

Figure 13.2. Ties connecting an external leaf of brickwork to the timber frame

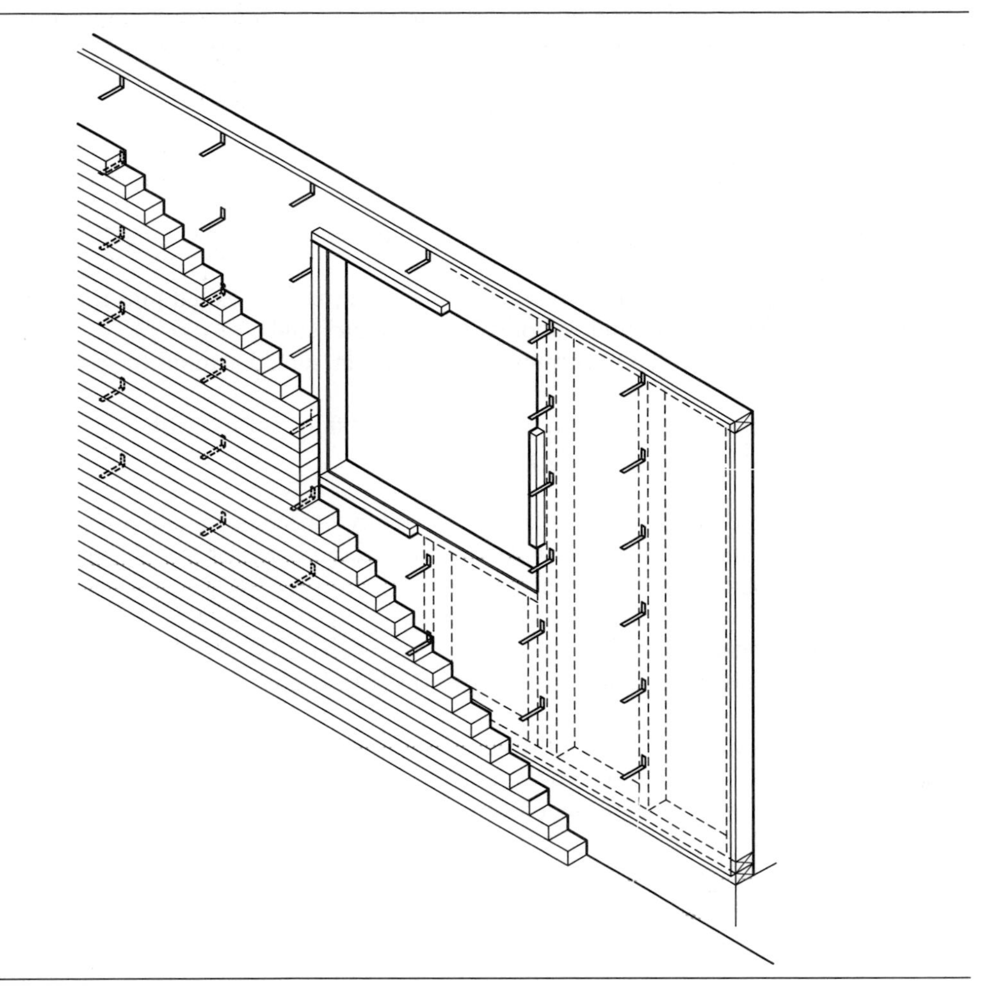

movements at each floor level. They afford no wind shielding to the building, but their self-weight can contribute to the permanent load on the structure.

- **Vertical circulation:** most tall buildings are fitted with lifts, but in a timber structure the mechanism might have difficulty with the variation in floor levels due to the moisture movements noted above. The shafts can be built with end-butted sheets of, for example,

Figure 13.3. Window frame, bridging the cavity between the timber frame and an external leaf of brickwork

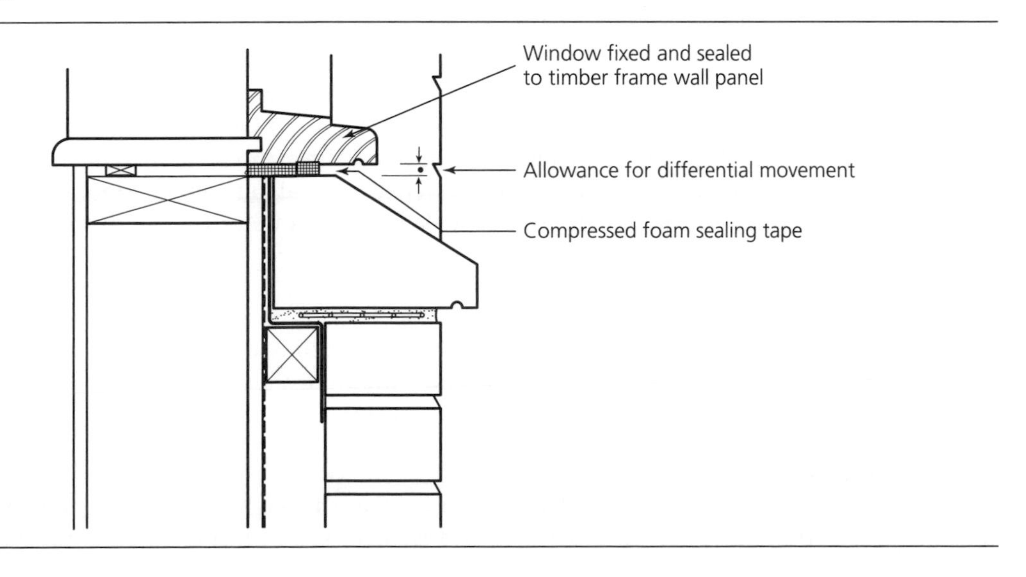

laminated veneer lumber, effectively eliminating the floor cross-grain, but some bridging detail would be needed at the lift door thresholds to accommodate the moisture movements of the main frame.

13.2. The standard form of construction

13.2.1 Wall panels

A typical wall panel is shown in Figure 13.1(b). It consists of:

- vertical studs: minimum thickness 38 mm, minimum depth 72 mm, maximum spacing 610 mm
- head plate (delete – binder) (above)) and sole plate (below), nailed into the ends of the studs (for an opening) a lintel, supported by 'cripple' studs
- sheathing: in sheets with a maximum width of 1.2 m, nailed around the perimeter and to intermediate studs
- fire resistance: provided by cladding the walls (and floor soffits) with a material that has a period of fire resistance, generally plasterboard.

The panel frame takes the vertical loads, but is virtually a mechanism under horizontal load. For this condition, racking stiffness is provided by the sheathing, nailed to the frame. A head binder fixed to the head plate links the panels together.

13.2.2 Floor panels

A typical floor arrangement is shown in Figure 13.1(d). It consists of:

- joists: spanning over or between the walls, of solid timber or proprietary form
- trimmers: to, for example, staircases
- blocking pieces in the line of the load-bearing walls
- perimeter joists.

In order to standardise the height of the wall panels, the floor joists have generally a constant thickness. The blocking pieces and perimeter joists carry the wall loads through the floor zone, and the perimeter joists may in addition form part of a system to prevent disproportionate collapse.

13.2.3 Foundations

It is assumed that foundations that are adequate to take the vertical, horizontal and possible uplift loads are provided, but they are not dealt with in more detail in this chapter.

13.3. Layout of the structural walls

13.3.1 General arrangement

The designer generally has to identify structural (i.e. load-bearing) walls from within the overall wall layout of the building plan. To carry the vertical loads, the walls should be distributed in a reasonably even pattern, to optimise the use of floor joists of a uniform depth. To provide wind resistance, the walls should have a total length in the two orthogonal directions capable of providing adequate overturning and racking resistance to wind, ideally without a disproportionate difference in length of the individual walls (Figure 13.4). The walls in this design method are assumed to be in a consistent plan position throughout the height. Load transfer systems at, for example, first-floor level to a different wall alignment at ground floor would have to be the subject of a special analysis. The definitions of descriptive terms is given in PD 6993-1, Figure 1.

13.3.2 Openings

A doorway, because of its height, is taken as a divider between two separate wall diaphragms. In the design method given below, the insertion of a window within certain dimensional limits is allowed, with a corresponding reduction in panel strength.

13.3.3 Vertical load

The vertical loads on an individual wall are the sum of the self-weight of the panels above, and the incoming loads from the floor. In the simple arrangement shown in Figure 13.4, the floor joists

Figure 13.4. Simple multi-storey block layout

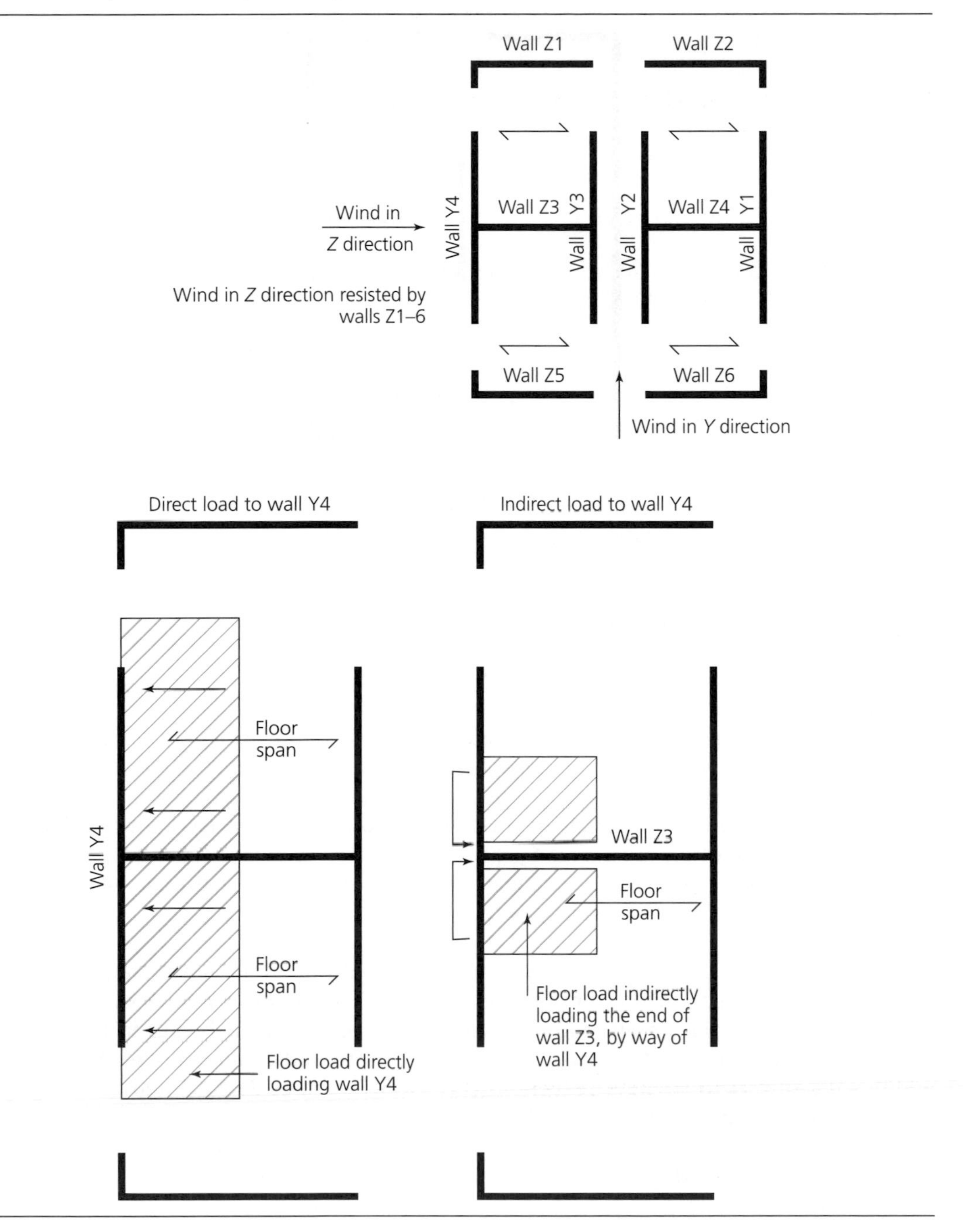

are spanning onto wall Y4, delivering a direct load. If wall Y4 was required to give resistance to wind Y, it would be preloaded by a uniformly distributed load of half the floor load. Wall Z3, in contrast, receives very little direct load. If it were required to resist wind Z, however, it could nevertheless mobilise an indirect floor load, given an adequate corner connection between the two walls. Thus, load distribution must be considered for each wall, and for each direction of wind load. The aim is to take advantage of all direct and indirect loads, in order to maximise the stabilising load. All the connections in an indirect load path have to be verified.

13.3.4 Wind load

The wind load is received by the external cladding, which passes the load to the floors by spanning vertically between them. The floors act as horizontal diaphragms (and have to be verified for that condition), and pass the loads on to the load-bearing walls. They in turn require sufficient racking, overturning and sliding resistance to transfer the loads down to the foundations.

Figure 13.5. Eccentricity of wind resistance

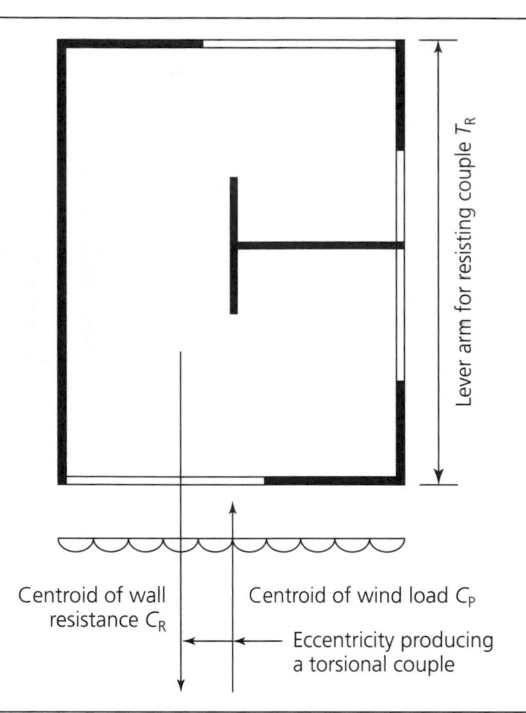

If the wall arrangement is symmetrical, and the walls are approximately equal in length (see Figure 13.4) then the wind load may simply be divided between them. However, the wall arrangement is more likely to be as shown in Figure 13.5. Here, preliminary calculations might show that the wall capacity in the Y direction is adequate, but the centroid of the resistance C_R does not line up with the centroid of the wind force C_P, producing an out-of-balance couple T_E. In this case, it is sometimes possible to invoke a resisting couple T_R from orthogonal walls, which under this wind direction are relatively lightly loaded, so that $T_R \geq T_E$.

The process of wind design for multi-storey blocks is usually iterative. A wall that seemed to have an adequate plan length might, for instance, prove to have a relatively small overturning resistance due to lack of permanent load, and, as noted in Section 13.1, breaks in floor continuity might restrict the options for load redistribution. Short walls, and walls with openings, are additionally subject to strength reductions (PD 6693-1, clauses 21.5.2.3 and 21.5.2.8, respectively) to allow for their relative lack of stiffness.

13.4. Frame loading

13.4.1 Vertical loads

The vertical loads on the frame are generally comprised of:

- The **self-weight** of the total fabric supported by the frame: EN 1991-1-1 gives mean densities (used as characteristic values) for many building materials. Partitions are either assessed on the basis of a layout or as a unit area allowance added to the floor weight. Note that a high allowance that is not fully used is unsafe, rather than safe, in racking and overturning calculations.
- **Superimposed loads:** from EN 1991-1-1. Note the reduction factor for multi-storey floor loading in the National Annex to EN 1991-1-1, clause NA 2.6.
- **Roof loads:** imposed loads are taken from EN 1991-1-1 and its National Annex, and snow loads are taken from EN 1991-1-3 and its National Annex.
- **Wind loads:** a possible vertical component of the wind load on the roof.

13.4.2 Horizontal loads

- **Wind loads:** generally on four orthogonal faces. Calculated in accordance with the rules in EN 1991-1-4.

13.5. Load cases and partial load factors for wall analysis

13.5.1 Load cases

The floor and wind loads are collected and transmitted to the ground by the structural walls within the plan layout. Each wall will have to be analysed and checked for racking capacity, overturning and sliding resistance, and panel strength. In general, it will be necessary to check the following load combinations in order to identify the critical loads for the analysis of each wall, as follows:

- **Load case 1: permanent load only plus wind.** The least vertical load combined with the characteristic lateral load gives the critical overturning, sliding and racking conditions. There is only one variable load, so there are no combination load cases. The duration of the load (since there is a wind component) will be instantaneous.
- **Load case 2: all loads.** The variable loads (principally wind and superimposed floor load) will be subject to the combination rule. One combination with the permanent load will give critical stud loads on the leeward side. Again, the duration of the load will be instantaneous, and the superimposed floor loads subject to a reduction for several storeys, as noted above.
- **Load case 3: all vertical loads.** The variable loads (principally floor and snow loads) will be subject to the combination rule, and multi-storey load reduction for the superimposed floor loads. The duration of load will be medium term (for floor superimposed loads), or short term (if snow is included). This load combination might generate critical load cases for vertical members less affected by wind, and is a relatively simple summation.

13.5.2 Partial load factors

The partial load factors γ_G and γ_Q depend upon the effect of an action. Their main value is used when the action has an unfavourable effect (e.g. creates stress in a member). Favourable effects arise from actions that tend to reduce the stress in a member, or stabilise a structure. Their values are given in the National Annex to EN 1990, Tables NA.A1.2(A) and NA.A1.2(B).

- **For variable actions in all load cases, γ_Q is either 1.5 (unfavourable) or 0 (favourable).**

For a permanent action used in a calculation of equilibrium, γ_G is either 1.1 (unfavourable) or 0.9 (favourable). However, frame calculations are not just about stability but go on to check strength. For this case, the National Annex gives values of 1.35/1.15 for a combined equilibrium/strength check, but with the proviso that setting γ_G to 1.0 for both the favourable and unfavourable parts of the load does not produce a less favourable effect. The permanent load of frames is almost always favourable, thus:

- **For permanent actions in an equilibrium/strength check, γ_G is 1.**
- **For permanent actions in a strength check, γ_G is 1.35.**

13.6. Method for the simplified analysis of wall diaphragms given in PD 6993-1, clause 21

13.6.1 Introduction

The definitions of the various components of a racking wall are shown in PD 6693-1, Figure 1. The total length of a racking wall may be interrupted by significant discontinuities, such as doors or large windows, which effectively divide the walls into diaphragms. Each diaphragm, in its complete height, is then analysed for:

- sliding resistance (PD 6693-1, clause 21.4.2) – this is the sum of the frictional resistance under design permanent load (adjusted if necessary by any wind uplift) and the lateral capacity of the fastener connections between the bottom rail and the structure below
- racking and overturning resistance (Section 13.6.2 in this guide)
- leeward end condition (Section 13.6.3 in this guide).

As noted in Section 9.2.4 in this guide, the analysis in PD 6693-1 applies to panels that are held down by fasteners through the bottom rail (Figure 13.6) and not by a windward stud strap, as used in Method A. Thus, any restraint provided by the rail fasteners must necessarily pass through the sheathing nails in the bottom rail to be effective.

Figure 13.6. Comparison of Method A and the PD 6693-1 diaphragm base fix

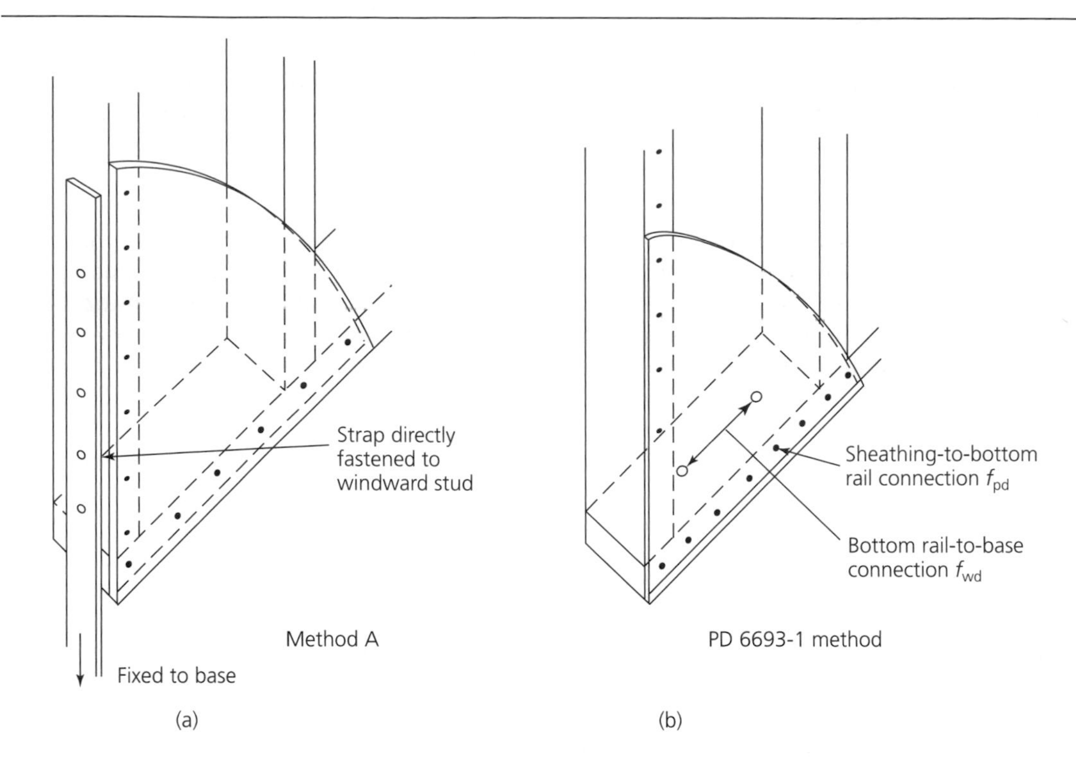

13.6.2 Racking and overturning analysis of wall diaphragms

The racking analysis of wall diaphragms given in PD 6693-1, clause 21.5, draws on the plastic lower bound method of analysis developed by Kallsner and Girhammar (2004), and uses an analytical model as shown in Figure 13.7. The bottom rail of the diaphragm is robustly fixed to the underlying supporting structure, and a destabilising horizontal force F is applied at the top at a lever arm of H.

The destabilising force will tend to cause rotation around the leeward corner R, which is resisted by the stabilising weight on the panel W, at a lever arm of X, and a group of sheathing nails giving

Figure 13.7. Racking analysis of wall diaphragms. (Based on PD 6693-1 (*Figure 4*). Reproduced with permission from PD 6693-1, © British Standards Institute, 2012)

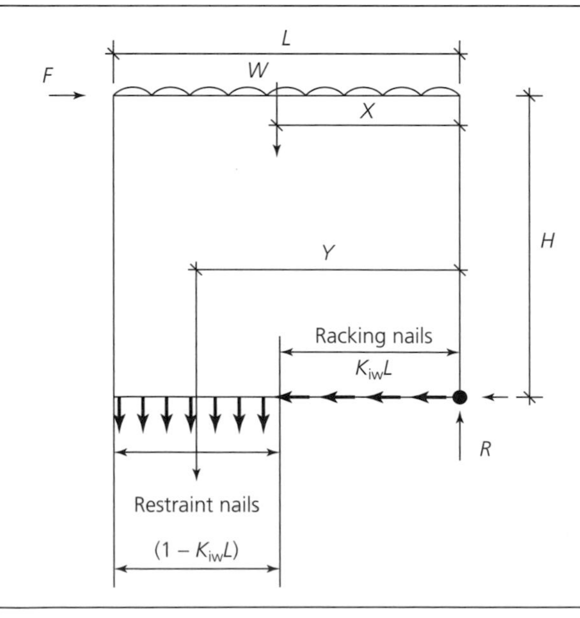

vertical restraint, at a lever arm of Y. The remaining sheathing nails can then act horizontally to resist the racking force F. It is now possible to determine F in terms of L, H, W and f_p, the capacity per unit length of the nails, and hence to determine K_{iw}, the proportion of nails which act in racking. The analysis of the PD 6693-1 model assumes that the sheathing is perfectly rigid, and that all the loads on the diaphragm are transferred to the leeward corner. It also makes the unconservative assumption that shear failure of the vertical connection between the sheathing panels and studs within the length of the diaphragm can be ignored.

For shorter diaphragms, or when there is a significant restraint requirement, the results of this analysis correlate reasonably well with tests. With longer diaphragms, or for relatively small uplift forces at the windward end, the method produces high vertical shear forces at the leeward end, and to allow for reduction in these due to the effect of yielding within the diaphragm length, an expression is given for the maximum leeward load to be designed for in PD 6693-1, clause 21.5.2.10.

The expression for K_{iw} is given in PD 6693-1, clause 21.5.2.5, and the various parameters in the expression are as follows:

- **Fastener strength (f_{pd}):** the design shear capacity per unit length of the perimeter fasteners to a sheathing sheet is given by the expression in clause 21.5.2.4 that enhances the individual fastener capacity by 20% (with nails at 50 mm c/c) up to 30% (with nails at 150 mm c/c). This is a correlation factor, derived from the results of tests on diaphragms. Where more than one sheathing sheet is used, the design shear capacity per unit length of the perimeter fasteners, (f_{pdt}), is derived from the expression in clause 21.5.2.2.
- **The influence of the bottom rail-to-base connection:** the sheathing-to-bottom rail connection (f_{pdt}), and the bottom rail-to-base tension connection (f_{wd}) are 'two links in a chain', and μ is the ratio of their strengths. The default fix through the bottom rail provided by framers is usually nails at 300 mm centres, which, in withdrawal, would give a low value of f_{wd}. If the racking calculation for the panel indicated that K_{iw} was less than 1, this would mean that some of the sheathing nails on the windward side would be required to resist uplift (see Figure 13.7), and to achieve the greatest tensile resistance per unit length, bottom rail fixings with f_{wd} equal to f_{pdt} will be required. For this condition μ will be 1. If uplift is not anticipated, for instance in diaphragms near the top of the racking structure where the default fix through the bottom rail is being used, f_{wd} will equal the design withdrawal capacity of the default fix per unit length and a value of μ less than 1 will apply.
- **Additional sheathing layers:** these may be added for extra racking resistance, either on the opposite side or the same side as the original, but their contribution is modified by the K_{comb} factor in clause 21.5.2.2.
- **Limitation of lateral deflection:** the relationship given in clause 21.5.2.3 effectively modifies the nail racking strength to restrict the lateral deflection of a diaphragm of height H to approximately $H/300$.
- **Openings:** as noted in Section 13.6.1 in this guide, significant openings, such as doors or large windows, which effectively divide a racking wall into individual diaphragms, are defined in clause 21.2.2. Openings that lie within these limits can be assessed by the expression in clause 21.5.2.8 for a value for $K_{opening}$, which reduces both the strength and the stiffness of the panel. Small openings, defined in clause 21.2.4, are deemed not to reduce strength or stiffness.

The calculated value of K_{iw} is then substituted in the expression in clause 21.5.2.1 to determine the racking strength of the wall diaphragm.

13.6.3 Leeward end conditions

As noted in Section 13.5.1, the critical racking and overturning condition for analysis is generally load case 1, which has the least vertical load combined with the full lateral load. However, load case 2, with additional vertical load, will generally produce the critical loading on the leeward end of the diaphragm. As for racking above, to allow for the reduction in the leeward stud compression forces due to the effect of yielding within the diaphragm length, the expression given under clause 21.5.2.10 in PD 6693-1 for $F_{c,d,leewd}$ is used. $F_{c,d,leewd}$ is allowed to be taken by all the studs within $0.1L$ of the end of the panel. In the same clause, permission is given to redistribute up to 50% of the load into wall returns, provided that the connection between them can be shown to be adequate.

The panel studs generally, but more particularly at the leeward end, should be checked as struts on their major axis, including any bending forces due to wind action. The minor axis may be assumed to be stabilised by the sheathing. The more usual critical condition, however, is on the sole plate, where the studs bear on the rails.

13.6.4 Transmission of forces through the floor zones

The forces at the base of a diaphragm at the first-floor level and above have to pass through the floor zone, and this condition has to be checked by the designer. In multi-storey frames, the floor zone is generally made up of solid timber (see Figure 13.1(d)), and so the transmission of the compressive forces is unlikely to be a problem. Tension forces could most simply be transmitted by through-bolts, but, as noted in Section 13.1, the floor zone is liable to shrinkage as the building dries out, and access after completion to retighten bolts may not be feasible. Possible solutions to tension force transmission at the first- and ground-floor levels are shown in the worked example below.

Example 13.1: analysis of multi-storey wall diaphragms

The wall arrangement is shown in Figure 13.8, and is made up of four diaphragms, of similar dimensions and in the same vertical plane. The wall is connected at its west end to the façade return wall B. The east end abuts a large window, which is a racking discontinuity. Wall C is also a racking wall.

Figure 13.8. Corner of a multi-storey block

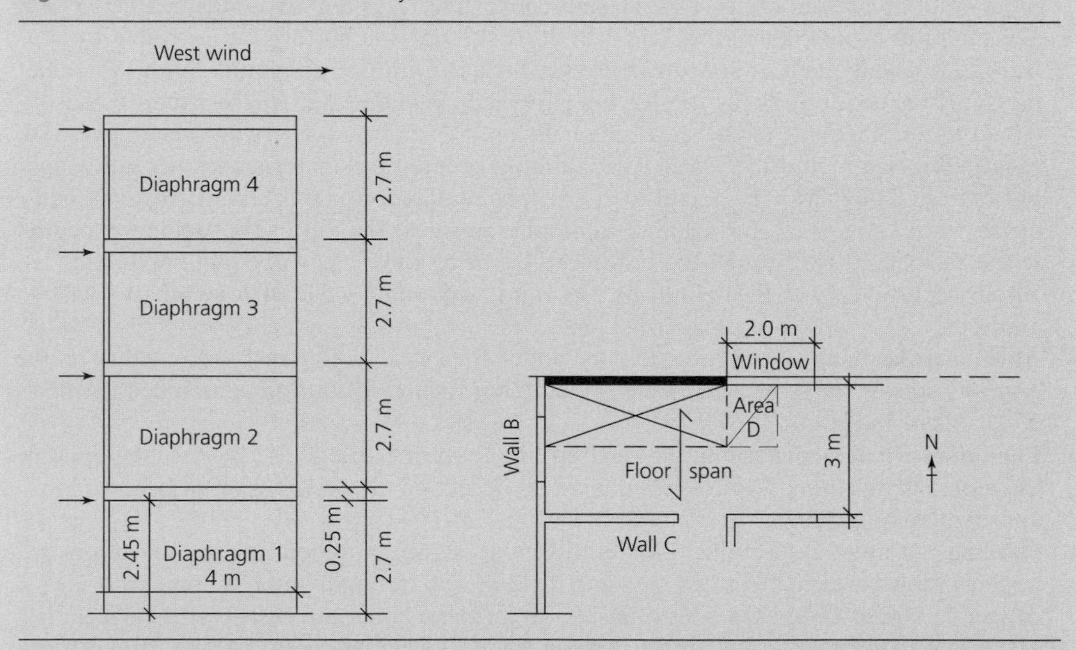

The floor spans between walls A and C, and area D loads the east end stud of wall A.

As wall A is an external wall, the structure functions in a service class 2 environment.

The building has a light-weight cladding, which contributes to the permanent load, but does not provide wind shielding to the structure. A perimeter upstand parapet protects the roof adjacent to wall A from significant uplift.

1. General

The analysis in this example is based on wind acting in a west-to-east direction, and for the purposes of covering the design procedures associated with these walls, the design effects from wind acting in the east-to-west direction are taken to be the same.

2. Characteristic and design loads

Table 13.1. Characteristic and design loads

Component	W: kN/m^2	Area, A: m^2	Characteristic load, WA: kN	Design load, γWA: kN		
				$\gamma_G = 1.0$	$\gamma_G = 1.35$	$\gamma_Q = 1.5$
Wall A (permanent)	0.7	$4 \times 2.7 = 10.8$	7.56	7.56	10.21	–
Wall B (permanent)	0.7	$2.7 \times 1.5 = 4.05$	2.84	2.84	3.83	–
Floor/roof (permanent)	0.8	$4 \times 1.5 = 6$	4.8	4.8	6.48	–
Floor (imposed)	1.5×0.8[a] 1.5×0.7[b]	6 6	7.2 6.3	–	–	10.8 9.45
Roof (imposed) – snow	0.6 0.6×0.5[b]	6 6	3.6 1.8	–	–	5.4 2.7

Wind: apportionment of design wind load on the building is given in Figure 13.9(a)
[a]Reduction factor for number of floors
[b]Psi factor for accompanying variables

3. Material specification and properties

Studs, rails: 140 × 38 mm, grade C16, maximum stud spacing 600 mm c/c.

Sheathing: oriented strand board (OSB/3), 9 mm thick.

Sheathing nails: 50 mm long, 2.85 mm-diameter smooth nails.

Framing nails: characteristic strength (based on the rules in Chapter 8 in this guide) = 0.57 kN (through a single sheathing)

k_{mod} value (based on rules in Chapter 2 in this guide) (strength class 2, duration of load instantaneous):

$k_{mod,1}$ (solid timber) = 1.1

$k_{mod,2}$ (OSB) = 0.9

$k_{mod} = \sqrt{k_{mod,1} \times k_{mod,2}} = 0.99$ (EN 1995-1-1, clause 2.3.2.1(2))

Design strength = 0.57 × 0.99/1.3 = 0.43 kN.

Design shear capacity per unit length of the perimeter nails (f_{pd}) based on PD 6693-1, clause 21.5.2.4, at:

150 mm spacing:	$0.43 \times (1.15 + 0.150)/0.150$	= 3.73 kN/m
75 mm spacing:	$0.43 \times (1.15 + 0.075)/0.075$	= 7.02 kN/m
50 mm spacing:	$0.43 \times (1.15 + 0.050)/0.050$	= 10.32 kN/m
2 sheets/50 mm spacing:	$(0.43 \times (1.15 + 0.050)/0.050) \times 1.5$	= 15.48 kN/m

4. Analysis of the wall under wind from the west

As noted in Section 13.6.1, it will be necessary to analyse three load cases:

- **Load case 1: permanent load only plus wind.** The critical case for racking, overturning and sliding. The loads are taken from Table 13.1, with $\gamma_G = 1.0$, γ_Q for wind = 1.5 and an instantaneous duration of load. The racking calculation is given below, in section 5.1 of this example. The equation also quantifies the need for any windward vertical restraint (i.e. if K_{iw} is less than 1), and therefore evaluates the overturning condition as well as the racking condition. Sliding calculations are given in this example in section 5.2 below.
- **Load case 2: all loads.** The critical case for compression forces on the leeward studs, with $\gamma_G = 1.35$, $\gamma_Q = 1.5$ and an instantaneous duration of load. Calculations are given in section 6 below.
- **Load case 3: all vertical loads.**This case gives the critical loading for studs in the central area of the diaphragms. Calculations are given in this example in section 7 below.

5. Load case 1

The racking wall diaphragms are shown in Figure 13.8 and the loads for each storey are marked on the wall elevation in Figure 13.9(a). At this stage, it is useful to carry out a simple analysis on the basis that the panels are completely rigid, and joined only at their ends with a compression or tension connection (Figure 13.9(b)). For this condition, since any uplift will be calculated on the basis of a single force at the end of the diaphragm, the value will be less than the line of nails that it represents. Nevertheless, it gives a general picture of the pattern of loading, and more particularly the degree of uplift that might occur at the windward end. Considering Figure 13.7, the sheathing nails in the bottom rail of diaphragm 1 have got to provide a racking resistance of approximately 34 kN, together with an uplift resistance of at least 14.2 kN. Their total capacity must therefore be at least (34 + 14.2)/4 kN/m (i.e. an f_{ptd} of 12 kN/m or so). This is a high value, but it can be achieved by using a second sheathing, and it gives a base value for entry into the K_{iw} formula.

5.1. Racking and overturning resistance

Diaphragm 1

$$H = 2.45\ \text{m}$$

$$L = 4\ \text{m}$$

μ will be given a maximum value of 1.0, as significant uplift resistance is predicted.

$$\begin{aligned} M_{d,stb} &= 4 \times 12.36 \times 2 = 98.88 \\ &\quad 2.84 \times 4 \times 4 = 45.44 \\ &\quad 144.32\ \text{kN m} \end{aligned}$$

The expression for K_{iw} is given in PD 6693-1, clause 21.5.2.5, and the function values to be used are:

$$\begin{aligned} M_{d,dst,top} &= 4.86 \times 3 \times 2.7 = 39.37 \\ &\quad 9.72 \times 2 \times 2.7 = 52.49 \\ &\quad 9.72 \times 2.7 = 26.24 \\ &\quad 118.1\ \text{kN m} \end{aligned}$$

Thus,

$$M_{d,stb,n} = M_{d,st} - M_{d,dst,top} = 144.32 - 118.1 = 26.22\ \text{kN m}$$

From the estimate above for f_{ptd}, try nails at 50 mm c/c, with a second sheathing layer overnailed at the same centres (see PD 6693-1, clause 21.5.2.2). This gives a strength of $f_{ptd} = 15.48$ kN/m.

Figure 13.9. (a) Load case 1; (b) simple rigid body analysis

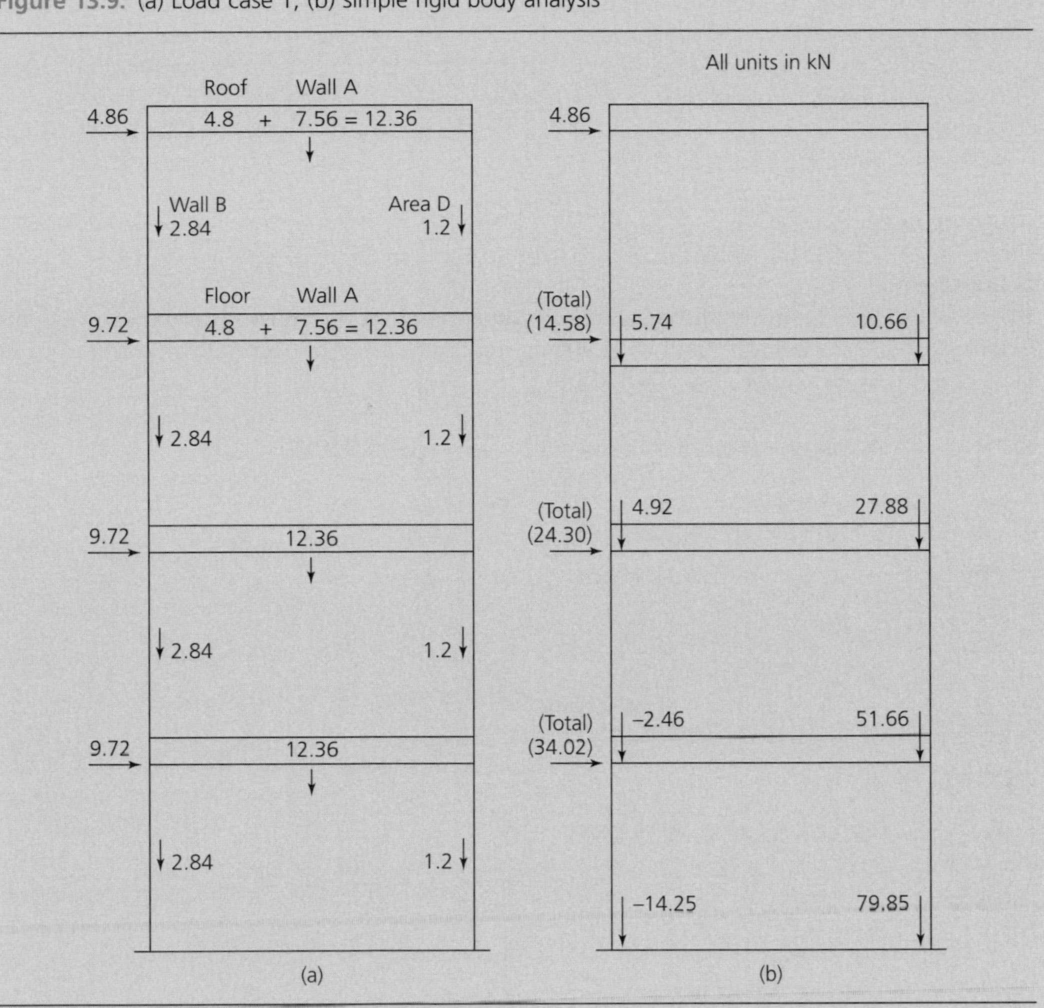

Inserting the above values into the relationship for K_{iw} in PD 6693-1, clause 21.5.2.5:

$$
\begin{aligned}
K_{1w} &= \min\{1,\ [1 + (H/\mu L)^2 + (2M_{d,stb,n}/\mu f_{d,f,t}L^2)]^{0.5} - (H/\mu L)\} \\
&= \quad 1,\ [1 + (2.45/4)^2 + (2 \times 26.22/(1 \times 15.48 \times 4^2))]^{0.5} - (2.45/4) \\
&= \quad 1,\ [1 + 0.375 + 52.44/247.68]^{0.5} - 0.613 \\
&= \quad 1,\ [1.587]^{0.5} - 0.613 = 1.26 - 0.613 = 0.647
\end{aligned}
$$

Since there are no openings in the diaphragm, $K_{opening} = 1$.

From PD 6693-1, clause 21.5.2.1, the design racking strength, $F_{1,V,Rd}$, is

$$F_{1,V,Rd} = K_{opening}K_{iw}f_{pdt}L = 1 \times 0.647 \times 15.48 \times L = 40.06 \text{ kN}$$

Required racking strength = 34.02 kN (Figure 13.9(b)), which is OK.

Required design uplift strength (see PD 6693-1, Figure 2):

$$\mu(1 - K_{iw})f_{pdt}L = 1 \times 0.353 \times 15.48 \times 4 = 21.86 \text{ kN}$$

The resistance to this uplift must be provided by anchoring the bottom rail to the foundation.

See section 8 in this example for diaphragm design.

Serviceability check. The serviceability requirement is given in PD 6693-1, clause 21.5.2.3, where

$$K_{1,w} f_{p,d,t} \leq 8(1 + K_{comb})(L/H)$$

$$0.647 \times 15.48 \leq 8 \times (1 + 0.5) \times (4/2.45)$$

which by inspection is OK.

Diaphragm 2

From Figure 13.9(b), the required racking resistance is 24.3 kN, with a small windward uplift in excess of 2 kN. Thus, the total nail capacity required in the bottom rail will need to be at least (24 + 2) = 26 kN, or 26/4 = 6.5 kN/m.

Try nails at 75 mm in a single sheathing layer: $f_{pdt} = 7.02$ kN/m.

$$M_{d,stb} = 3 \times 12.36 \times 2 = 74.16$$
$$3 \times 2.84 \times 4 = 34.08$$
$$108.24 \text{ kN m}$$

$$M_{d,dst,top} = 4.86 \times 2 \times 2.7 = 26.24$$
$$9.72 \times 2.7 = 26.24$$
$$52.48 \text{ kN m}$$

Thus,

$$M_{d,stb,n} = 108.24 - 52.48 = 55.76 \text{ kN m}$$

As before,

$$H = 2.45 \text{ m}, \ L = 4 \text{ m}, \ \mu = 1$$

$$K_{2w} = 1, \ [1 + (2.45/4)^2 + (2 \times 55.76/1 \times 7.02 \times 4^2)]^{0.5} - (2.45/4)$$

$$= 1.539 - 0.613 = 0.926$$

Thus, the design racking strength of this diaphragm is

$$F_{2,V,Rd} = 1 \times 0.926 \times 7.02 \times 4 = 26.0 \text{ kN}$$

which is OK.

Required design uplift strength (see PD 6693-1, Figure 2):

$$\mu(1 - K_{2w}) f_{pdt} L = 0.074 \times 7.02 \times 4 = 2.08 \text{ kN}$$

which is OK.

See section 8 below for diaphragm design.

Diaphragm 3

From Figure 13.9(b), the required racking resistance is 14.58 kN. No uplift is predicted, and so K_{3w} can be taken as 1, as all of the nails in the bottom rail can contribute to racking. The calculation may be done as follows.

Required racking resistance from Figure 13.9(a) = 14.58 kN.

Try nails at 150 mm centres ($f_{ptd} = 3.73$ kN/m from section 2 in this example).

Thus, the design racking strength of this diaphragm is

$$F_{3,V,Rd} = 1 \times 1 \times 3.73 \times 4 = 14.92 \text{ kN}$$

which is OK.

Diaphragm 4

This diaphragm has no uplift, and a smaller applied racking force than diaphragm 3 (nails at 150 mm centres). Use nails at 150 centres, as this is the maximum spacing allowed with the method (PD 6693-1, clause 21.1.3.2).

5.2. Sliding resistance (PD 6693-1, clause 21.4.2)

Racking force at base $\leq$ frictional resistance + lateral resistance of base rail fixings.

Design sliding force at base = 34.02 kN.

The total compressive force of the wall on the ground is the sum of the permanent loads (65.6 kN) plus the reaction to the uplift tie force (21.86 kN) = 87.46 kN.

Design frictional resistance = 0.4 × 87.46 kN = 34.98 kN.

The lateral resistance provided by those base rail fixings not providing tension resistance will not be required.

6. Load case 2

This case will give critical stud loads at the leeward end of the diaphragm. The loads for each storey are marked on the wall elevation in Figure 13.10(a), based on the component loads in Table 13.1, and a simple rigid body analysis is given in Figure 13.10(b).

Essentially, the horizontal load remains the same, but the vertical superimposed load has been added. In addition, the permanent design load has been calculated using $\gamma_G = 1.35$. There is now a small compressive load on the windward side, and the compressive load has significantly increased on the leeward side.

6.1. The capacity of the leeward studs

The compressive force at the leeward end of the diaphragm ($F_{c,d,leewd}$) is given by the expression in PD 6693-1, clause 21.5.2.10

$$F_{c,d,leewd} = 0.8W_{v,t,d}[(M_{d,dst,base}/M_{d,stb}) + (0.6/L)]$$

This force should be resisted by the compressive capacity of the studs within $0.1L$ of the leeward end.

Diaphragm 1

To enter the expression for $F_{c1,d,leewd}$:

- W_{vtd} is the total vertical design load acting on the wall
- $M_{d,stb}$ is the design stabilising moment about the leeward end of the wall from the design vertical load.
- $M_{d1,dst,base}$ (from Figure 13.9(a)):
  $4.86 \times 10.55 = 51.27$
  $3 \times 9.72 \times 5.15 = 150.17$
  201.44 kN m
- Thus, $F_{c1,leewd} = 0.8 \times 127.37 \times [201.44/256.9 + (0.6/4)]$
  $= 101.90 \times (0.78 + 0.15)$
  $= 94.77$ kN

See section 8 below for diaphragm design.

Figure 13.10. (a) Load case 2; (b) simple rigid body analysis

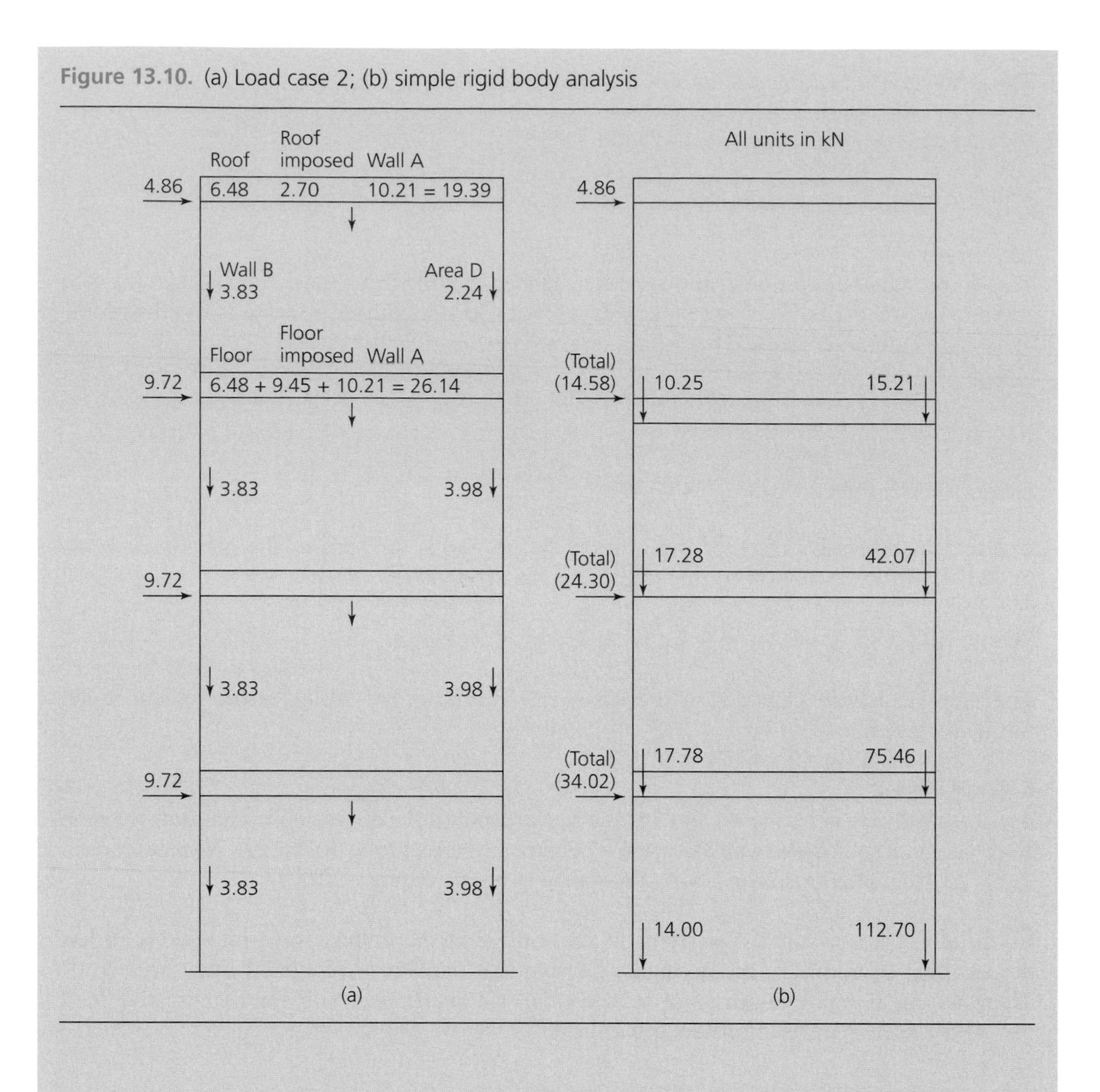

Table 13.2. Loading and overturning moments for diaphragm 1

Loads from Table 13.1	Loads from storey: kN	No. of storeys	Total load: kN	Lever arm: m	Moment: kN m
Permanent loads					
Wall A	10.21	4	40.84	2	81.68
Wall B	3.83	4	15.32	4	61.28
Floor	6.48	4	25.92	2	51.84
Area D	1.62	4	6.48	–	–
Imposed loads					
Roof	2.7	1	2.7	2	5.4
Floor	9.45	3	28.35	2	56.70
Roof area D	0.68	1	0.68	–	–
Floor area D	2.36	3	7.08		
		W_{vtd}	127.37	$M_{d,stb}$	256.9

Diaphragm 2

Enter expression for $F_{c2,d,leewd}$ as for diaphragm 1.

Table 13.3. Loading and overturning moment for diaphragm 2

Loads from Table 13.1	Loads from storey: kN	No. of storeys	Total load: kN	Lever arm: m	Moment: kN m
Permanent loads					
Wall A	10.21	3	30.63	2	61.26
Wall B	3.83	3	11.49	4	45.96
Floor	6.48	3	19.44	2	38.88
Area D	1.62	3	4.86	–	–
Imposed loads					
Roof	2.7	1	2.7	2	5.4
Floor	9.45	2	18.9	2	37.8
Roof area D	0.68	1	0.68	–	–
Floor area D	2.36	2	4.72		
		W_{vtd}	93.42	$M_{d,dst}$	189.3

- $M_{d2,dst,base}$ (from Figure 13.9(a)):

$$\begin{aligned} 4.86 \times 7.85 &= 38.15 \\ 9.72 \times 5.15 &= 50.06 \\ 9.72 \times 2.45 &= \underline{23.81} \\ & 112.02 \text{ kN m} \end{aligned}$$

- Thus, $F_{c2,leewd} = 0.8 \times 93.42 \times [112.02/189.3 + (0.6/4)] = 55.44$ kN

See section 8 below for diaphragm design.

Diaphragm 3

Enter expression for $F_{c3,d,leewd}$ as for diaphragm 1

Table 13.4. Loading and overturning moment for diaphragm 3

Loads from Table 13.1	Loads from storey: kN	No. of storeys	Total load: kN	Lever arm: m	Moment: kN m
Permanent loads					
Wall A	10.21	2	20.42	2	40.84
Wall B	3.83	2	7.66	4	30.64
Floor	6.48	2	12.96	2	25.92
Area D	1.62	2	3.24	–	–
Imposed loads					
Roof	2.7	1	2.7	2	5.4
Floor	9.45	1	9.45	2	18.9
Roof area D	0.68	1	0.68	–	–
Floor area D	2.36	1	2.36		
		W_{vtd}	59.47	$M_{d,dst}$	121.7

- $M_{d3,dst,base}$ (from Figure 13.9(a)):

$$\begin{aligned} 4.86 \times 5.15 &= 25.03 \\ 9.72 \times 2.45 &= \underline{23.81} \\ & 48.84 \text{ kN m} \end{aligned}$$

- Thus, $F_{c3,leewd} = 0.8 \times 59.47 \times [48.84/121.7 + (0.6/4)] = 26.23$ kN

See section 8 below for diaphragm design.

Diaphragm 4
As the leeward compression (Figure 13.10(b)) is only 15 kN or so, no special calculation is necessary.

PD 6693-1, clause 21.5.2.10 (Note 2), permits the check on leeward stud stresses to be waived for dwellings of two storeys or fewer, provided that there are a minimum of two studs within $0.1L$ of the leeward end. This permission could reasonably be interpreted as applying to the top two storeys of a four-storey residential block.

7. Load case 3

Vertical loads only are considered, providing the design case for studs near the centre of the diaphragm. Since there is no lateral load, the assumption is that the vertical loads simply pass down the studs. The leading variable will be the floor imposed load, with a medium-term duration of load.

The loading on wall A consists simply of its self-weight and half the span of the floors and roof.

7.1. The capacity of the central studs

The design load for a 600 mm length of wall at level 1 is:

	load	×	area	×	γ	×	storeys		
Wall A	0.7	×	(2.7 × 0.6)	×	1.35	×	4	=	6.12
Floor	0.8	×	(1.5 × 0.6)	×	1.35	×	4	=	3.89
Floor (imposed)	(1.5 × 0.8)	×	(1.5 × 0.6)	×	1.5	×	3	=	4.86
Roof (imposed)	(0.6 × 0.5)	×	(1.5 × 0.6)	×	1.5	×	1	=	0.41
									15.28 kN

The capacity of a single stud (140 mm × 30 mm) will be determined by the bearing capacity on the rail in compression perpendicular to the grain. Assuming that the studs are spaced at 600 mm centres, the length of the stud in bearing may be increased in accordance with EN 1995-1-1, clause 6.1.5, by 30 mm on each side, giving an effective length of 98 mm.

Thus, the effective bearing area of the stud is 140 × 98 mm = 13 720 mm^2.

The design compressive stress perpendicular to grain is

$$f_{c,90,d} = f_{c,90,k}k_{mod}/\gamma_m = 2.2 \times 0.8/1.3 = 1.35\ \text{N/mm}^2$$

Thus, the design effective bearing capacity is

$$13\,720 \times 1.35 = 18.57\ \text{kN}$$

which is OK.

(See the note below regarding additional lateral bending due to wind pressure.)

8. The design of the diaphragms

Each diaphragm has to be designed for the critical loading calculated in sections 5, 6 and 7 in this example to determine:

- The sheathing and nail spacing (see section 5.1 above).
- The stud arrangement. The design capacity of individual studs is determined either by their strength as struts (buckling on the major axis), or in end bearing on the rails. For 140 mm-deep storey height studs, end bearing is generally the limiting design condition (although for studs on external walls the effect of additional lateral bending due to wind pressures should also be considered).

- The required base anchorage. At level 1, a concrete substructure (or an equivalent construction) has been assumed. At the upper levels, it is assumed that the floor zone is infilled with solid timber.

Diaphragm 1

Sheathing and nail spacing. From section 5.1 in this example, the required design strength f_{ptd} is achieved with two layers of sheathing, each with nails at 50 mm centres.

Stud arrangement. From section 6.1 above, the diaphragm is to be designed for a compression load of 94.77 kN within each end zone of $(0.1L =)$ 400 mm. The design compression load applies at the base of the studs, because the stud loads are reduced over the height by the downward forces from the sheathing nails, transmitting the complementary vertical shear load to the horizontal wind racking load on the diaphragm. A possible arrangement is shown in Figure 13.11, using two triple studs.

Figure 13.11. Diaphragm and stud arrangements

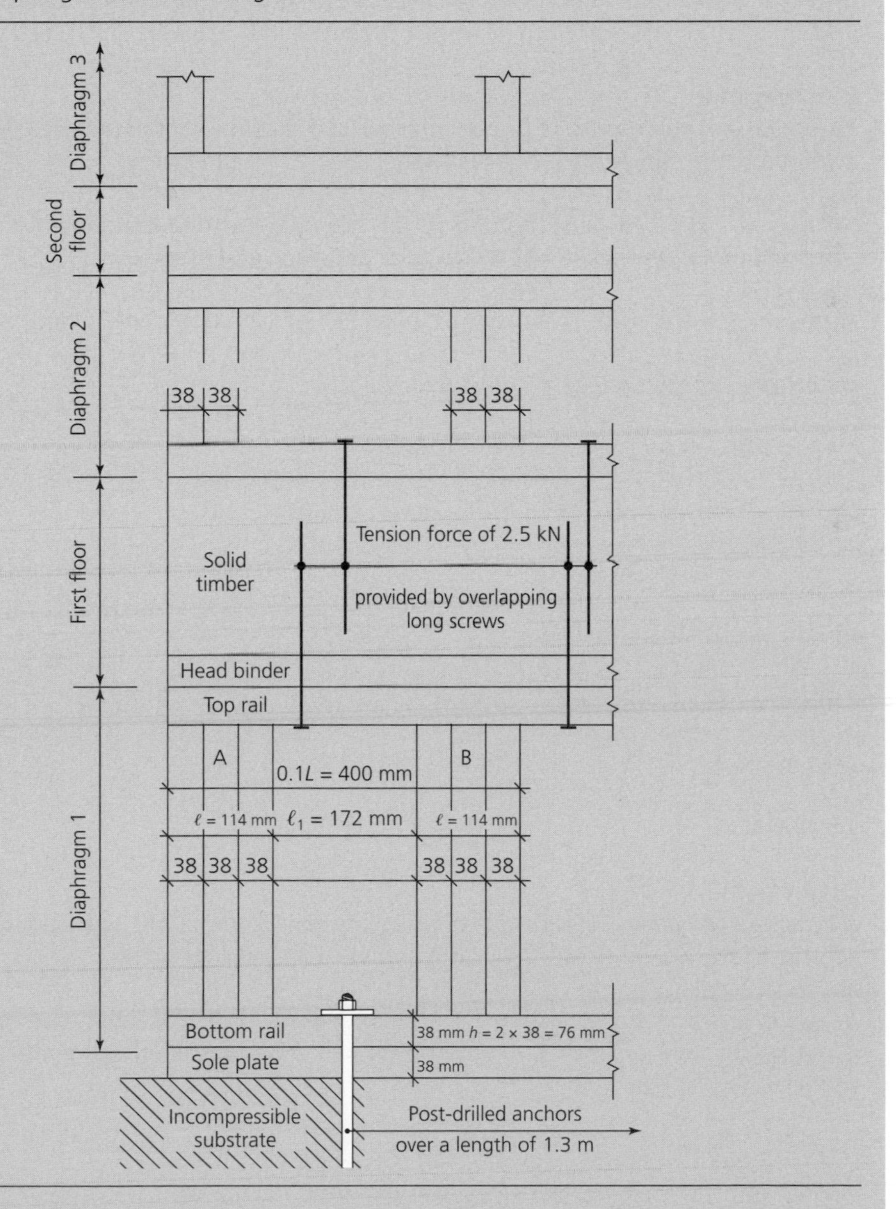

The rule for the effective contact length of the studs is given in EN 1995-1-1, clause 6.1.5(1) and Figure 6.2. In this case, the total effective length is

$$(3 \times 38) + 30 + (3 \times 38) + (2 \times 30) = 318 \text{ mm}$$

The effective contact area of the studs is therefore

$318 \times 140 = 44\ 520\ \text{mm}^2$

Under certain stud configurations, an increase in the design compressive stress $f_{c,90,d}$ is allowed by the factor given in EN 1995-1-1, clauses 6.5.1(3) and 6.5.1(4). In the arrangement shown in Figure 13.11 the factor may be taken as 1.25.

Thus, the design compressive strength of the stud group is

$A_{ef}f_{c,90,k}k_{c,90}k_{mod}\gamma_m = 44\ 250 \times 2.2 \times 1.25 \times 1.1/1.3/1000 = 103.59\ \text{kN}$

which is OK.

Bottom rail anchorage. As shown in Figure 13.11, drilled concrete anchors are installed through the bottom rail and into the concrete substructure over a length of 1.3 m. The anchors should be designed to have an aggregate withdrawal capacity in excess of 21.86 kN.

Diaphragm 2

Sheathing and nail spacing. From section 5.1 in this example, the required design strength is achieved with a single layer of sheathing, nailed at 75 mm centres.

Stud arrangement. From section 6.1 in this example, the diaphragm is to be designed for a compression load of 55.44 kN within each end zone of 400 mm.

A possible arrangement is shown in Figure 13.11, using two double studs.

As before, the total effective stud length is

$(2 \times 38) + 30 + (2 \times 38) + 60 = 242\ \text{mm}$

and the effective contact area of the studs is therefore

$242 \times 140 = 33\ 880\ \text{mm}^2$

In this arrangement, the value of $k_{c,90}$ will be 1.0.

Thus, the design compressive strength of the stud group is, as before,

$33\ 880 \times 2.2 \times 1.0 \times 1.1/1.3/1000 = 62.92\ \text{kN}$

which is OK.

Diaphragm 3

Sheathing and nail spacing. From section 5.1 in this example, the required design strength is achieved with a single layer of sheathing, nailed at 150 mm centres.

Stud arrangement. From section 6.1 in this example, the diaphragm is to be designed for a compression load of 26.23 kN within each end zone of 400 mm. Figure 13.11 shows two studs within the zone.

As before, the total effective stud length is

$38 + 30 + 30 + 38 + 30 = 166\ \text{mm}$

and the effective contact area is

$166 \times 140 = 23\ 240\ \text{mm}^2$

The design compressive strength of the studs is

$$23\,240 \times 2.2 \times 1.1/1.3/1000 = 43.26 \text{ kN}$$

which is OK.

(As noted in section 6.1 above, this check could have been waived.)

Diaphragm 4

From section 5.1 in this example, the required racking design strength is achieved with a single layer of sheathing, nailed at 150 mm centres. No end uplift occurs, and the vertical loading is nominal.

Studs at a maximum spacing of 600 mm throughout may be used, and the panel located by standard nailing through the bottom rail.

REFERENCES

Gulvanessian H, Formichi P and Calgaro J-A (2008) *Designers' Guide to Eurocode 1: Actions on Buildings: EN 1991-1-1 and -1-3 to -1-7*. ICE Publishing, London.

Kallsner B and Girhammar UA (2004) Influence of the framing joints on plastic capacity of partially anchored wood-framed shear walls. *Working Commission W18 – Timber Structures, Meeting 37*, Edinburgh.

Lancashire R and Taylor L (2011) *Timber Frame Construction*, 5th edn. TRADA Technology, High Wycombe.

Designers' Guide to Eurocode 5: Design of Timber Buildings
ISBN 978-0-7277-3162-3

http://dx.doi.org/10.1680/dtb.31623.199

Appendix A
Possible changes to EN 1995-1-1

General

The next full revision of EN 1995-1-1 is scheduled for publication after 2015. However, in 2010 CEN/TC 250, the Technical Committee of the European Committee for Standardisation (CEN) responsible for technical matters associated with this code, drawing on input from national code committees, identified some errors in the code as well as matters where it was felt that clarification of interpretation was required. The points identified have still to be fully discussed and may change in detail and content. When agreed the changes are expected to be issued in a corrigendum (amendment) statement or statements prior to the next full revision of the code.

Those matters that are considered of significance to design and are likely to be incorporated in amendment statements are briefly referred to in the following sections. The information is given to assist with the reader's understanding and interpretation of the code, but it is important to understand they have no status until issued under a formal amendment to EN 1995-1-1 or are referred to in PD 6693-1.

Items that may change

Each item is referred to using the relevant clause number given in EN 1995-1-1: 2004 + A2: 2012.

Clause 1.2

EN 10346 will replace EN 10147.

Clauses 2.2.3(3), 2.2.3(4) and 2.3.2.2(1)

There is conflict between the statement given in *clause 2.2.3(3)* and the content of *clause 2.2.3(4)*, and to clarify the requirement these clauses as well as *clause 2.3.2.2(1)* are to be amended as follows.

Clause 2.2.3(3):

> '(3) The final deformation u_{fin}, see figure 7.1, should be calculated by superimposing the creep deformation u_{creep} calculated using the quasi-permanent combination of actions, see EN 1990, 6.5.3(2)(c), on the instantaneous deformation u_{inst} calculated from 2.2.3(2). The creep deformation should be calculated using mean values of the appropriate moduli of elasticity, shear moduli and slip moduli and the relevant values of k_{def} give in Table 3.2.'

Clause 2.2.3(4):

> '(4) If the structure consists of members or components having different creep behaviour, the long-term deformation due to the quasi-permanent combination of actions should be calculated using the final mean values of the appropriate moduli of elasticity, shear moduli and slip moduli according to 2.3.2.2(1). The final deformation u_{fin} is then calculated by superimposing the instantaneous deformation due to the difference between the characteristic and the quasi-permanent combination of actions on the long term deformation.'

Clause 2.3.2.2(1):

> '(1) For the serviceability limit states, if the structure consists of members or components having different time-dependent properties, the final mean value of modulus of elasticity

$E_{mean,fin}$, shear modulus $G_{mean,fin}$ and slip modulus $K_{ser,fin}$ which are used to calculate the long-term deformation due to the quasi-permanent combination of actions, see EN 1990, 6.5.3(2)(c), should be taken from expression (2.7), (2.8) and (2.9).'

Clause 4.2, Table 4.1

The statement under note a is to be revised to read:

'If hot dip zinc coating is used on steel plates, Fe/Zn 12C should be replaced by Z275 and Fe/Zn 25C by Z350 in accordance with EN 10346. If hot dip zinc coating is used on dowel type fasteners, Fe/Zn 12C should be replaced by a layer of zinc of minimum 39 μm and Fe/Zn 25C by a layer of zinc of minimum 49 μm in accordance with EN ISO 1461.'

Clause 6.1.5(4)

The current description is meant to include for the effect of distributed loading on members in addition to local loads, and is to be revised. The proposed change to the text is likely to state:

'(4) For members on discrete supports loaded by distributed loads and/or by concentrated loads further away from the support than $\ell_1 = 2h$, see Figure 6.2b, the value of $k_{c,90}$ should be taken as:

- $k_{c,90} = 1.5$ for solid softwood timber
- $k_{c,90} = 1.75$ for glued laminated softwood timber provided that $\ell \leq 400$ mm

where h is the depth of the member and ℓ is the contact length.
(Note: A series of point loads acting at close centres (e.g. joists or rafters at centres < 600 mm) may be regarded as a distributed load.)'

Clause 6.1.8

The values given for the shape factor, k_{shape}, are to be replaced by:

'$k_{shape} = 1.2$ for a circular cross-section;

$$k_{shape} = \min\left\{\begin{matrix} 1 + \dfrac{0.05h}{b} \\ 1.3 \end{matrix}\right\} \text{ for a rectangular cross-section.'}$$

Clause 6.2.3(2)

The clause as written does not cover for cases where the member can fail by lateral torsional instability under major axis bending. As a conservative approximation when torsional instability applies, it is proposed that the effect of the tensile force is ignored and that the member is checked under bending using the design rules in *clause 6.3*. The following note will be added after *clause 6.2.3(2)*:

'Note: To check the instability condition, the method given in *6.3* can be used with $\sigma_{t,0,d} = 0$.'

Clause 6.5.2, Equation 6.60

The function b should read b_{eff}, as defined in *clause 6.1.7(2)*.

Clause 8.3.2(4)

The definition of t_{pen} is to be clarified as follows:

't_{pen} is the pointside penetration length or the length of the threaded part, excluding the point length, in the pointside member.'

Clause 8.3.2(6)

It will be clarified that the units of the characteristic withdrawal strength and characteristic pull-through strength given in *Equations (8.25)* and *(8.26)*, respectively, are in N/mm^2.

Clause 8.4(6)

The characteristic yield moment per staple leg is to be revised, and should read '$M_{y,Rk} = 150d^3$'.

Clause 8.4(7)
In the clause delete '—— n_{ef} according to 8.3.1.1(8)' and replace by '—— $n_{ef} = n$'.

Clause 8.6(3)
The minimum unloaded end distance, $a_{3,c}$, in *Table 8.5* is to be changed to read:

'$90° \leq \alpha \leq 150°$	$a_{3t}\ \lvert\sin\alpha\rvert$
$150° \leq \alpha \leq 210°$	max(3.5d; 40 mm)
$210° \leq \alpha \leq 270°$	$a_{3t}\ \lvert\sin\alpha\rvert$'

Clause 8.7.1
To clarify requirements when screws are used, *clauses (1)P, (4)* and *(5)* are to be replaced and *clause (6)* is to be changed as follows:

'(1)P The effect of the threaded part of the screw shall be taken into account in determining the load carrying capacity by using an effective diameter d_{ef} when determining the yield capacity and the embedment strength of the threaded part. The outer thread diameter d shall be used to determine spacing and end distances and the effective number of screws.
(4) For screws with an effective diameter $d_{ef} > 6$ mm, the rules in 8.5.1 apply.
(5) For screws with an effective diameter $d_{ef} \leq 6$ mm, the rules in 8.3.1 apply.
(6) For screws $6 < d_{ef} \leq 8$ mm minimum values for spacing, edge and end distances shall be determined by linear interpolation between Tables 8.2 and 8.4.'

Clause 8.7.2(4)
The first line should read:

'(4) For connections in softwood timber with screws in accordance with EN 14592 with'

Clause 8.8.5.1(1)
The design force, $F_{A,Ed}$, and the value of r are to be redefined as follows:

'$F_{A,Ed}$ is the design force, positive when tension, acting on a single plate at the centroid of the effective area (i.e. half of the total force in the timber member).
r is the distance from the centre of gravity of the effective plate area to the segmental plate area dA.'

Clause 8.8.5.1(2)
A clarification note is to be added as *clause 8.8.5.1(2)*, stating that only the component of F_{Ed} perpendicular to the timber surface should be reduced.

Clause 8.8.5.1(4)
The first three lines are to change plus *Equation (8.50)*, as follows:

'(4) Contact pressure between the timber members in chord splices in compression may, when $F_{Ed} \leq 0$, be taken into account by designing the single plate for a design force, $F_{A,Ed}$, and a design moment, $M_{A,Ed}$, according to the following expressions:

$$F_{A,Ed} = \frac{F_x}{|F_x|}\sqrt{F_x^2 + (F_{Ed}\sin\beta)^2} \qquad (8.50)$$

where

$$F_x = \frac{F_{Ed}\cos\beta}{2} + \frac{3|M_{Ed}|}{2h}\text{'}$$

Clause 8.8.5.2(1)
For clarification, the following note is to be added:

'NOTE: F_{Ed} can be reduced by the contact pressure determined in 8.8.5.1(3).'

Clause 8.9, Table 8.7

In the row for $a_{3,t}$, in *Table 8.7* delete '1.5d_c' in the last column, and replace with '2.0d_c'.

Clause 8.10, Table 8.8

In the row for $a_{3,t}$, in *Table 8.8* delete '2.0d_c' in the last column, and replace with '1.5d_c'.

Annex A (informative): block shear and plug shear failure at multiple dowel-type steel-to-timber connections

Replace *Equation A.7*, associated with modes (d)/(g), with:

$$`t_{ef} = t_1\left(\sqrt{2+\frac{4M_{y,Rk}}{f_{h,k}dt_1^2}}-1\right)'$$

Annex B.4 (informative): mechanical jointed beams

In *Equation B.9* in *clause B.4*, replace $(h_2)^2$ with $(h)^2$.

Designers' Guide to Eurocode 5: Design of Timber Buildings
ISBN 978-0-7277-3162-3

ICE Publishing: All rights reserved
http://dx.doi.org/10.1680/dtb.31623.203

Index